TECHNOLOGIES FOR VALUE ADDITION IN FOOD PRODUCTS AND PROCESSES

TECHNOLOGIES FOR VALUE ADDITION IN FOOD PRODUCTS AND PROCESSES

Edited by

Sankar Chandra Deka, PhD

Dibyakanta Seth, PhD

Nishant Rachayya Swami Hulle, PhD

Apple Academic Press Inc.
3333 Mistwell Crescent
Oakville, ON L6L 0A2, Canada

Apple Academic Press Inc.
1265 Goldenrod Circle NE
Palm Bay, Florida 32905, USA

First issued in paperback 2021

Exclusive worldwide distribution by CRC Press, a member of Taylor & Francis Group

ISBN 13: 978-1-77463-474-5 (pbk)
ISBN 13: 978-1-77188-798-4 (hbk)

Library and Archives Canada Cataloguing in Publication

Title: Technologies for value addition in food products and processes / edited by Sankar Chandra Deka, Dibyakanta Seth, Nishant Rachayya Swami Hulle.

Names: Deka, Sankar Chandra, 1965- editor. | Seth, Dibyakanta, 1980- editor. | Hulle, Nishant Rachayya Swami, 1985- editor.

Description: Includes bibliographical references and index.

Identifiers: Canadiana (print) 20190146826 | Canadiana (ebook) 20190146893 | ISBN 9781771887984 (hardcover) | ISBN 9780429242847 (ebook)

Subjects: LCSH: Food industry and trade—Technological innovations.

Classification: LCC TP370 .T43 2020 | DDC 664/.02—dc23

Library of Congress Cataloging-in-Publication Data

Names: Deka, Sankar Chandra, 1965- editor. | Seth, Dibyakanta, 1980- editor. | Hulle, Nishant Rachayya Swami, 1985- editor.

Title: Technologies for value addition in food products and processes / edited by Sankar Chandra Deka, Dibyakanta Seth, Nishant Rachayya Swami Hulle.

Description: Oakville, ON, Canada ; Palm Bay, Florida, USA : Apple Academic Press, 2020. | Includes bibliographical references and index. | Summary: "The new volume looks at some important emerging food processing technologies in light of the demand for functional food products and high-value and nutritionally rich products. Technologies for Value Addition in Food Products and Processes covers a selection of important recent developments in food processing that work to enrich or maintain nutritional value of food products, including such applications as non-thermal plasma, refractance window drying, extrusion, enzyme immobilization, and dry fractionation. Dry fractionation, in particular, has emerged as a sustainable alternative to wet processes in the last three decades for producing protein concentrates from legumes. The chapters on fish processing cover both traditional knowledge and advances in fish processing technologies. A chapter on bioethanol production discusses the past and present status of the industry, focusing on economic feasibility and environmental viability. The chapter on traditional foods discusses fermentation process and nutritional aspects of ethnic foods followed by the Rabha-Hasong, Mishing and Karbi communities of Assam, India. With the contribution from experts in their respective fields, this book provides new information on novel food processing technologies"-- Provided by publisher.

Identifiers: LCCN 2019028837 (print) | LCCN 2019028838 (ebook) | ISBN 9781771887984 (hardcover) | ISBN 9780429242847 (ebook)

Subjects: LCSH: Food industry and trade--Technological innovations.

Classification: LCC TP370.5 .T427 2020 (print) | LCC TP370.5 (ebook) | DDC 664--dc23

LC record available at https://lccn.loc.gov/2019028837

LC ebook record available at https://lccn.loc.gov/2019028838

Apple Academic Press also publishes its books in a variety of electronic formats. Some content that appears in print may not be available in electronic format. For information about Apple Academic Press products, visit our website at **www.appleacademicpress.com** and the CRC Press website at **www.crcpress.com**

About the Editors

Sankar Chandra Deka, PhD

Sankar Chandra Deka, PhD, is presently working as a Senior Professor in the Department of Food Engineering and Technology, Tezpur University Assam, India. He has more than 29 years of teaching and research experience. He has guided the theses work of more than 35 students for their BTech, MSc, and MTech degrees, and has worked with seven PhD students to date. Dr. Deka has successfully handled more than 15 research projects funded by various government funding agencies. He has published more than 96 research papers in journals of national and international repute and about 20 book chapters. His area of interest is food quality, food chemistry, and fermented foods and processing.

Dibyakanta Seth, PhD

Dibyakanta Seth, PhD, is presently an Assistant Professor in the Department of Food Engineering and Technology, Tezpur University, Tezpur, Assam, India. He has published several research papers in national and international journals. He has participated in national/international conferences and attended training programs in the area of food processing and has been bestowed with Young Scientist and Young Educator Awards at the International Conference on Food Properties held at Bangkok, Thailand, in 2016 and Sharjah, United Arab Emirates, in 2017, respectively. He has been associated with an e-course development program of the University Grants Commission (UGC), ePG-PATHSHALA in the capacity of Paper Coordinator as well as Content Writer on the subject of Unit Operations in Food Processing. He has handled two funded research projects and is an active member of scientific bodies such as Association of Food Scientists and Technologists (India) Mysore, Indian Society of Agricultural Engineers New Delhi, and Swedish South Asian Studies Network. His research interests include dairy engineering and technology, spray drying, optimization, and transfer processes. He teaches subjects such as fluid mechanics, mechanical operations in food processing, and dairy products technology at Tezpur University.

Nishant Rachayya Swami Hulle, PhD

Nishant R. Swami Hulle, PhD, is an Assistant Professor in the Department of Food Engineering and Technology, Tezpur University, Assam, India. He has published several research articles in international peer-reviewed journals, co-authored book chapters published by CRC Press, and has presented at various conferences. His areas of research interest are nonthermal processing, fruit, and vegetable processing, and extraction of bioactives from plant sources. He has received a Young Scientist Award at the Third International Conference on Food Properties (iCFP, 2018), Sharjah, United Arab Emirates. He teaches subjects such as fruit and vegetable process technology; processing technology of meat, poultry, and fish; and thermal operations in food processing at Tezpur University.

Contents

Contributors

C. Anandharamakrishnan
Indian Institute of Food Processing Technology (IIFPT), Ministry of Food Processing Industries, Government of India, Thanjavur, Tamil Nadu, India, E-mail: anandharamakrishnan@iifpt.edu.in

Aradhita Barmanray
Department of Food Technology, Guru Jambheshwar University of Science and Technology, Hisar, Haryana–125001, India, E-mail: dhitaray@gmail.com

Mamta Bhardwaj
Department of Food Engineering and Technology, Sant Longowal Institute of Engineering and Technology, Longowal–148106, Punjab, India, E-mail: mamta.bhardwaj20@gmail.com

Indu Bharti
Department of Food Technology, Guru Jambheshwar University of Science and Technology, Hisar, Haryana–125001, India

Suvendu Bhattacharya
Department of Food Engineering and Technology, Tezpur University, Tezpur, Assam, India; Formerly Chief Scientist, CSIR-Central Food Technological Research Institute, Mysore, India

Snehasis Chakraborty
Department of Food Engineering and Technology, Institute of Chemical Technology, Mumbai–400019 India, E-mail: sc.chakraborty@ictmumbai.edu.in

Binita Dev
Food Microbiology and Bioprocess Laboratory, Department of Life Science, National Institute of Technology, Rourkela–769008, Odisha

P. Divyang
Department of Food Quality Assurance, College of Food Processing Technology and Bio-Energy, Anand Agricultural University Anand–388110, Gujarat, India

Neeraj Ghanghas
Food Science and Technology Lab, Department of Food Science and Technology, National Institute of Food Technology Entrepreneurship and Management, Sonepat, HR, India

Namrata A. Giri
University Department of Chemical Technology (UDCT), Dr. Babasaheb Ambedkar Marathwada University, Aurangabad–431004, Maharashtra, India, E-mail: namrata_cft@yahoo.in

R. Jayabalan
Food Microbiology and Bioprocess Laboratory, Department of Life Science, National Institute of Technology, Rourkela–769008, Odisha, E-mail: jayabalanr@nitrkl.ac.in

B. S. Khatkar
Department of Food Technology, Guru Jambheshwar University of Science and Technology, Hisar, Haryana, India, E-mail: bskhatkar@yahoo.co.in

Nitin Kumar
Department of Food Engineering, National Institute of Food Technology Entrepreneurship and Management, Kundli, Sonepat, HR, India

Parveen Kumari
Department of Food Technology, Guru Jambheshwar University of Science and Technology, Hisar, Haryana, India

Mitali Madhumita
Department of Agriculture and Food Engineering, Indian Institute of Technology, Kharagpur, West Bengal, India

R. K. Majumdar
Professor, College of Fisheries, Central Agricultural University (I), Lembucherra, Tripura–799210, India

J. A. Moses
Indian Institute of Food Processing Technology (IIFPT), Ministry of Food Processing Industries, Government of India, Thanjavur, Tamil Nadu, India, E-mail: moses.ja@iifpt.edu.in

M. T. Mukilan
Food Science and Technology Lab, Department of Food Science and Technology, National Institute of Food Technology Entrepreneurship and Management, Sonepat, HR, India

Deep Prakash Parasar
Microbial Communication & Fungal Biology Group, Department of Biotechnology, Gauhati University, Guwahati–781014, Assam, India

Mamta Patel
Department of Food Quality Assurance, College of Food Processing Technology and Bio-Energy, Anand Agricultural University Anand–388110, Gujarat, India

Pramod K. Prabhakar
Department of Food Science and Technology, Food Science and Technology Lab, National Institute of Food Technology Entrepreneurship and Management, Sonipat, Haryana, India, E-mail: pramodkp@niftem.ac.in/, pkprabhakariitkgp@gmail.com

R. V. Prasad
Department of Food Quality Assurance, College of Food Processing Technology and Bio-Energy, Anand Agricultural University Anand–388110, Gujarat, India

S. R. Priyadarshini
Indian Institute of Food Processing Technology (IIFPT), Ministry of Food Processing Industries, Government of India, Thanjavur, Tamil Nadu, India

Nukasani Sagarika
Department of Food Quality Assurance, College of Food Processing Technology and Bio-Energy, Anand Agricultural University Anand–388110, Gujarat, India

B. K. Sakhale
University Department of Chemical Technology (UDCT), Dr. Babasaheb Ambedkar Marathwada University, Aurangabad– 431004, Maharashtra, India, E-mail: bksakhale@gmail.com

Hridip Kumar Sarma
Microbial Communication & Fungal Biology Group, Department of Biotechnology, Gauhati University, Guwahati–781014, Assam, India

D. C. Saxena
Department of Food Engineering and Technology, Sant Longowal Institute of Engineering and Technology, Longowal, Sangrur, Punjab, India, E-mail: dcsaxena@yahoo.com

Chandrima Shrivastava
Department of Food Engineering and Technology, Institute of Chemical Technology, Mumbai–400019 India, E-mail: chandrima95@gmail.com

Renuka Singh
Department of Food Engineering and Technology, Sant Longowal Institute of Engineering and Technology, Longowal, Sangrur, Punjab, India, E-mail: renukasliet@gmail.com

R. F. Sutar
Department of Food Quality Assurance, College of Food Processing Technology and Bioenergy, Anand Agricultural University Anand–388110, Gujarat, India

K. S. Yoha
Indian Institute of Food Processing Technology (IIFPT), Ministry of Food Processing Industries, Government of India, Thanjavur, Tamil Nadu, India

Abdelrahman Saleh Zaky
Department of Microbiology, Faculty of Agriculture, Cairo University, Giza, 12613, Egypt

Abbreviations

AA	ascorbic acid
AFEX	ammonia fiber explosion
AFM	atomic force microscope
AHA	American Heart Association
ALK	alkaline extraction
AM	additive manufacturing
ANF	anti-nutritional factors
ARP	ammonia recycling percolation
BSM	blackstrap molasses
BV	biological value
CA	citric acid
CAP	cold atmospheric plasma
CBP	consolidated bioprocessing
CD	celiac disease
CDC	Center for Disease Control and Prevention
CE	catechin equivalents
CFV	cut fruits and vegetables
CLEAs	cross-linked enzyme aggregates
CLEC	cross-linked enzyme crystals
CMC	carboxymethylcellulose
CMPA	cow's protein milk allergy
CO_2	carbon dioxide
COX	cyclooxygenase
CPCB	Central Pollution Control Board
DBD	dielectric barrier discharges
DCP	2,4-dichlorophenol
DDGS	distillers dried grain with solubles
DHA	docosahexaenoic acid
DMC	direct microbial conversion
DSC	differential scanning calorimetry
DSF	defatted soy flour
DTT	dithiothreitol
EAI	emulsion activity index
EB	elongation at break

EBP	ethanol blending program
EI	enzyme immobilization
EMR	electromagnetic radiation
EPA	eicosapentaenoic acid
eRHFs	extensively hydrolyzed rice protein formulas
FDA	Food and Drug Administration
FFV	flexible-fuel vehicles
FP	fish powder
FPD	flame photometric detector
FST	frequency sweep test
GAB	Guggenheim-Anderson-de Boer
GAE	gallic acid equivalent
GC	gas chromatograph
GC-MS	gas chromatography-mass spectroscopy
GHG	greenhouse gas
GI	glycemic index
GO	graphine oxide
GRAS	generally recognized as safe
HFCS	high-fructose corn syrup
HMF	hydroxyl-methyl furfural
HP	high pressure
HPAF	high-pressure assisted freezing
HPC	hydroxypropyl cellulose
HPIF	high-pressure induced freezing
HPP	high hydrostatic pressure
HPP	high-pressure processing
HPSF	high-pressure shift freezing
HQCF	high-quality cassava flour
HTST	high-temperature short time process
IBD	inflammatory bowel disease
IL	ionic liquid
iNOS	inducible nitric oxide synthase
IPPC	isoelectric precipitated protein concentrated
IR	infra-red
LAB	lactic acid bacteria
LAOS	large amplitude oscillatory shear
LDL	low-density lipoprotein
LIF	leukemia inhibitory factor
LPCP	low-pressure cold plasma
LTLT	low-temperature long time

LTP	low-temperature plasma
LVE	linear viscoelastic range
MAE	microwave-assisted extraction
MAP	modified atmosphere packaging
MC	methylcellulose
MGC	minimum gelation concentration
MMP-8	matrix metalloproteinase-8
MP	minimally processed
MTBE	methyl tertiary butyl ether
MW	microwave processing
NaCMC	sodium salt of carboxymethyl cellulose
NASA	National Aeronautics and Space Administration
NDF	neutral detergent fiber
NEI	northeastern region of India
NO	nitrogen oxide
NPN	non-protein nitrogen
NPR	net protein ratio
NPU	net protein utilization
NSI	nitrogen solubility index
NTP	non-thermal plasma
NWP	National Water Policy
OH	ohmic heating
PB	protein bodies
PDCAAS	protein digestibility-corrected amino acid score
PEF	pulsed electric field
PER	proteins efficiency ratio
PGA	propylene glycol alginate
PL	pulsed light
PLA	polylactic acid
PMJ	plasma micro-jet
PSU	practical salinity units
PTFE	polytetrafluoroethylene
PUFA	poly-unsaturated fatty acids
PVC	polyvinyl chloride
RF	radio frequency
RF	rice flour
RFP	radio-frequency plasmas
RS	resistant starch
RTS	ready to serve
RW	refractance window

SAA	soaking aqueous ammonia
SAOS	small amplitude oscillatory shear
SEM	scanning electron microscope
SEM	standard error of the mean
SF_6	sulfur hexafluoride
SHF	separate hydrolysis and fermentation
SME	specific mechanical energy
SO_2	sulfur dioxide
SPI	soy protein isolate
SPR	surface plasmon resonance
SSCF	simultaneous saccharification and co-fermentation
SSF	simultaneous saccharification and fermentation
TD	true digestibility
TPP/TVP	texturized plant/vegetable protein
TS	tensile strength
TSP	texturized soy protein
TST	temperature sweep test
TTA	total titratable acidity
TVBN	total volatile basic nitrogen
TVP	texturized vegetable protein
UHC	unburned hydrocarbon
USDA	United States Department of Agriculture
UV	ultraviolet rays
VLDL	very low-density lipoprotein
WHO	World Health Organization
WPC	whey protein concentrate
WVP	water vapor permeability
WVTR	water vapor transmission rate

Preface

The studies related to the food processing technologies are essential components of undergraduate and graduate studies. Food science and technology is a multidisciplinary research area that covers a wide number of subjects.

The present book presents the recent developments in food processing, including non-thermal plasma (NTP), refractance window (RW) drying, and enzyme immobilization (EI). The chapters on the applications of different processes like EI, extrusion, and dry fractionation, cover the recent developments in relevant domains. Dry fractionation has emerged as a sustainable alternative to wet processes in the last three decades, and it provides a more sustainable alternative for producing protein concentrates from legumes. The chapters on fish processing cover traditional knowledge and advances in fish processing technologies. The chapter on processing of plant products includes aonla, tuber crops, rice, and legumes. The chapter on bioethanol production discusses the past and present status of the industry with focus on the economic feasibility and environmental viability. An additional chapter discusses the traditional fermentation process and nutritional aspects of ethnic foods followed by the Rabha-Hasong, Mishing, and Karbi communities of Assam, India. The present book is intended to act as a reference source for students, researchers, and teachers.

We hope the contributions made by the learned authors will help students, teachers, and researchers working in the area of food processing technologies.

*—**Sankar Chandra Deka, PhD***
Dibyakanta Seth, PhD
Nishant R. Swami Hulle, PhD

CHAPTER 1

Rheology: A Tool to Predict Quality of Foods

MAMTA BHARDWAJ, RENUKA SINGH, and D. C. SAXENA

Department of Food Engineering and Technology, Sant Longowal Institute of Engineering and Technology, Longowal–148106, Punjab, India,
E-mail: mamta.bhardwaj20@gmail.com (Mamta Bhardwaj), renukasliet@gmail.com (Renuka Singh), dcsaxena@yahoo.com (D. C. Saxena)

ABSTRACT

The word "food" in itself is a complex system comprising of a wide range of biological components with various rheological characteristics. The diversity in these biological components in different food systems impart various compositional and structural variability to the food, thus, exhibiting different types of rheological behaviors viz. low viscosity fluids (e.g., milk), high viscosity fluids (e.g., ketchup) and hard solids (e.g., candies, and gel). The rheological behavior of food decides the stability and appearance of foods such as in the form of emulsions, pastes, and spreads, etc. Moreover, food quality, apart from its nutritional value, is a function of its rheological properties viz. structure and texture. The rheological characterization of food and food forming components is vital for predicting the food quality. Depending upon the form of a specific product (e.g., suspension, emulsion, gel, paste, liquid, solid, etc.) to be analyzed, a range of rheological techniques, tests, and equipments are available. Processing the rheological data in the form of models is vital to infer its physical significance in relation to the flow behavior. Therefore, the present chapter gives an insight into the application of rheological techniques, tests, and theoretical models to predict the quality of foods.

1.1 INTRODUCTION

Rheology is the study of flow and deformation of matter in response to the applied force wherein the relationship between applied stress or strain, deformation, and time is described. Viscosity and elasticity are two principle rheological parameters indispensable for describing the consistency of any material/product. Viscosity is "the resistance to flow," whereas elasticity is "the ability of a material to resist distortion and come back to its original shape." Apart from viscous and elastic materials, there is a third category viz. viscoelastic material exhibiting both viscous and elastic character. Food is a complex system, both compositionally and structurally. The difference in the proportion of major food constituents such as carbohydrates, proteins, fibers, fats, and others impart diverse rheological flow and deformation patterns. Therefore, food systems can range from low viscosity fluids (e.g., fruit juice) to hard solids (e.g., candy); the intermediate ones such as bread dough. Rheological tools may be detrimental for deciding the quality of the final food product as its quality is influenced by the majority of factors such as composition, processing parameters, engineering processes, etc. Moreover, the sensory attributes of any food product as perceived by the consumer are influenced by rheological properties such as creaminess, tenderness, hardness, juiciness, smoothness, etc. Thus, rheological characterization is exclusively important starting from the raw materials to the finished food product, i.e., prior to processing (raw materials), during processing (intermediates) and after processing (finished food product) (Tabilo-Munizaga, 2005). Apart from food quality, the rheological data are necessary for plant designs, heat, and mass transfer calculations, designing of mixers, extruders, pumps, etc. (Joshi and Ranade, 2003). Another area where rheology is of utmost concern is the new product development, in the last decade. An emphasis had been laid on the development of functional foods, edible films and coatings, low fat and diet foods, weaning foods and nutraceuticals foods, etc. To achieve this we are playing with the ingredient composition, addition or subtraction of some components, altering the concentrations such as use of fat replacer inulin to develop low fat ice cream (Akalin et al., 2008), addition of mango peel powder to develop high fiber biscuits (Ajila et al., 2008) and many more. These additions or subtractions have a direct influence on the textural properties, mouthfeel, and other characteristics detrimental for the acceptance of the product by the consumers. Hence, the rheological evaluation of food products can be applied for the selection of raw materials and process type, and most importantly, the quality control of the final product.

Putting in a nutshell, the study of rheology and its techniques are necessary for:

- Establishing a relationship between rheological properties and sensory perception;
- Selection of food ingredients for end uses;
- Effect on food quality due to compositional changes;
- Elucidating the textural quality;
- New product development;
- Effect on quality of final food product due to processing.

For rheological characterization, various tools and techniques are available. Also, a range of theoretical and empirical models are available in order to predict the material's performance during the course of processing and experimental conditions. Foods can be liquids, semi-solid, solid solids, and hard solids (Van Vliet et al., 2009; Foegeding et al., 2011). So as to effectively study the role of rheological tools in food quality prediction, it is imperative to have knowledge of basics of rheology, their classification, different models, and rheological tests. The following chapter, apart from these topics, covers the application of rheological tests in the prediction of food quality and its processes.

1.2 BASICS OF RHEOLOGY

Basic stress (τ) and strain (γ) relationships are keys to all rheological determinations and classifications. Stress is the force per unit area of a material, given by,

$$\tau = \frac{F}{A}$$

where 'τ' is the stress and 'F' is the force, and 'A' is an area on which force is applied. The unit of stress is N/m^2 or Pa.

The strain is the deformation induced in the material in response to the applied stress, i.e., the relative change in dimensions of the material due to the externally applied force; and it is dimensionless. Rheological materials on the basis of stress-strain behavior can be classified broadly as ideal elastic, ideal viscous, and viscoelastic materials. Most of the food materials are complex and posses both elastic and viscous properties, thus exhibiting viscoelastic characteristics. The nature of ideal elastic, ideal viscous, and viscoelastic materials is represented and formulated using spring, dashpot, and combination of the two (Table 1.1). Stress and strain relationship is governed by Hooke's law in the case of solids and Newton's law in case of fluids.

TABLE 1.1 Rheological Classification of Materials

Ideal viscous	Viscoelastic		Ideal elastic
Newton's law (perfect liquid)	Maxwell model (more like solid)	Kelvin-Voigt model (more like fluid)	Hooke's law (perfect solid)

In order to explain viscous, elastic, and viscoelastic materials, dashpot, spring, and the combination of the two is used. Nature of spring imitates to that of elastic materials, they store energy on the application of stress and then without deforming retain their origin shape. As per Hooke's law, applied stress is proportional to strain induced in the material, and the constant of proportionality is termed as elasticity, N/m^2 or Pa. Dashpot resembles to that of viscous fluids, where energy dissipation takes place when stress is applied. Newton's law governs the behavior of viscous fluids. Viscosity is the resistance to flow, and it occurs due to internal friction between the layers of fluid. For the fluids which deviate from Newtonian behavior, the apparent viscosity is the more significant term. Apparent viscosity depends on shear rate whereas it is constant for Newtonian fluids, therefore, referred to as Newtonian viscosity or simply viscosity. Table 1.2 gives the basic equations and mechanical models of viscous, elastic, and viscoelastic materials.

TABLE 1.2 Mechanical Models for Elastic, Viscous, and Viscoelastic Materials

Characteristic	Model	Representation element		Model equation
Elastic	Hooke's Law	Spring	E	$\gamma = \frac{1}{E}\tau$
Viscous	Newton's Law	Dashpot	η	$\dot{\gamma} = \frac{1}{\eta}\tau$
Viscoelastic	Kelvin-Voigt model (more elastic than viscous)	Spring and Dashpot in parallel	E η	$\tau = E\gamma + \eta\dot{\gamma}$
	Maxwell model (more viscous than elastic)	Spring and dashpot in series	η E	$\eta\dot{\gamma} = \tau + \frac{\eta}{E}\dot{\tau}$

τ – shear stress, γ – strain, $\dot{\gamma}$ – shear rate, η – viscosity, E – elasticity.

1.3 CLASSIFICATION OF FLUID FOODS

Newton's law of viscosity governs the classification of viscous and Visco-elastic fluid foods. Accordingly, there are two types of fluid foods: (a) Newtonian, and (b) Non-Newtonian (Figure 1.1).

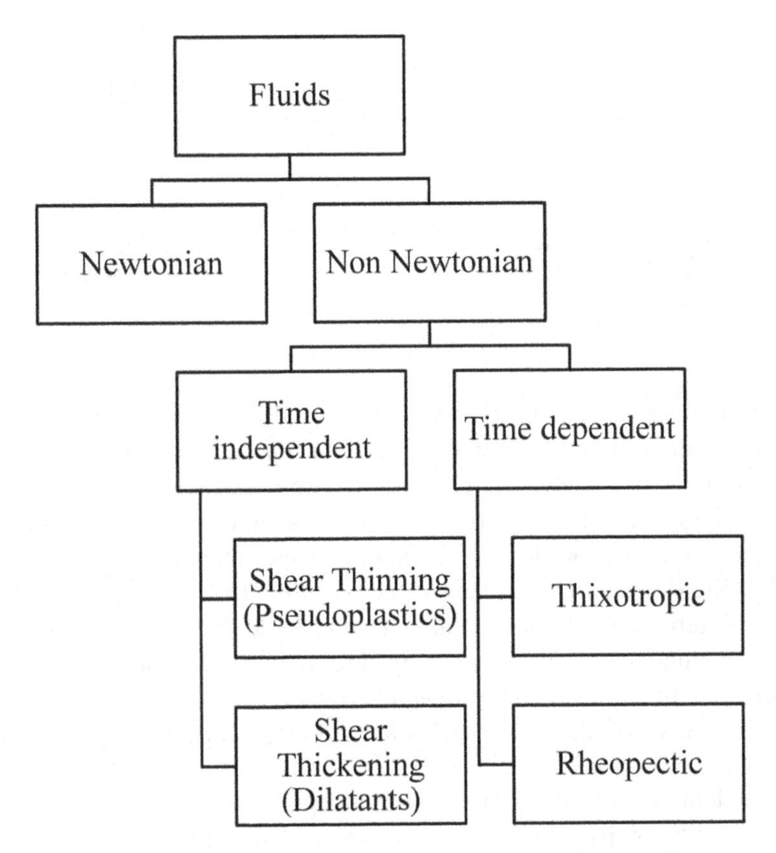

FIGURE 1.1 Classification of fluids.

Newtonian fluids follow Newton's law; stress and strain follow a linear relationship and viscosity. The behavior of Newtonian fluids can be explained exclusively by temperature and pressure, but Non-Newtonian fluids show a nonlinear relationship between shear stress and shear strain, viscosity, in this case, is referred to as apparent viscosity (Munson et al., 2014). In Table 1.1, fundamental stress-strain equations and viscosity relationship are represented. As far as Non-Newtonian fluid foods are concerned, the shear strain rate ($\dot{\gamma}$) is more significant to consider for calculations than the shear strain ($\dot{\gamma}$). Viscosity, in this case, is referred to as an apparent viscosity (η_a). There are a lot many examples of fluid foods which fall in the category of non-Newtonian fluids, such as fruit juice concentrates, mayonnaise, butter, cheese, starch paste, etc. (Table 1.3).

TABLE 1.3 Newtonian and Non-Newtonian Fluids

Fluid type	Viscosity	Fundamental Equation	Examples
Newtonian	Newtonian viscosity (Constant)	$\eta = \frac{\tau}{\gamma}$	Water, Mineral oil etc.
Non-Newtonian	Apparent viscosity (Variable)	$\eta = \frac{\tau}{\gamma}$	Cheese, mayonnaise, starch gel, etc.

η – viscosity; γ – shear strain; $\dot{\gamma}$ – shear strain rate; τ – shear stress.

Further, non-Newtonian fluids are classified into two subcategories: (a) time-dependent fluids, (b) time-independent fluids. Rheopectic and Thixotropic fluids come under the category of time-dependent non-Newtonian fluids (Figure 1.2). Rheopectic fluids show a time-dependent increase in viscosity, i.e., with an increase in time of stress, their viscosity increases. Practically, rheopectic fluids are rarely seen (Jethra, 1994). On the other hand, thixotropic fluids show behavior opposite to that of rheopectic fluids. They get thinner, i.e., their viscosity decreases as they are stressed over a period of time. A lot many examples of thixotropic materials exist as food products and their raw ingredients. Fruit pulp and fruit gels show thixotropic behavior.

Pseudoplastic and dilatant are time-independent non-Newtonian fluids. The viscosity of pseudoplastic decreases and that of dilatants increases with stress (Schowalter, 1978). The phenomenon of dilatancy is relatively rare. Starch suspensions in water and other solvents such as glycerol had shown shear thickening behavior as investigated by Williamson and Hecker (1931), and Hoffman (1982). Shear thickening is supposed to be arising from complicated interactions which may arise from hydrogen bonding, chain entanglement, or particle crowding, particularly in case of particulate suspensions and dispersions (Bagley and Dintzis, 1999). Pseudoplastics are very common in food systems; for example, from everyday life is ketchup, dairy cream, fruit purees, etc.

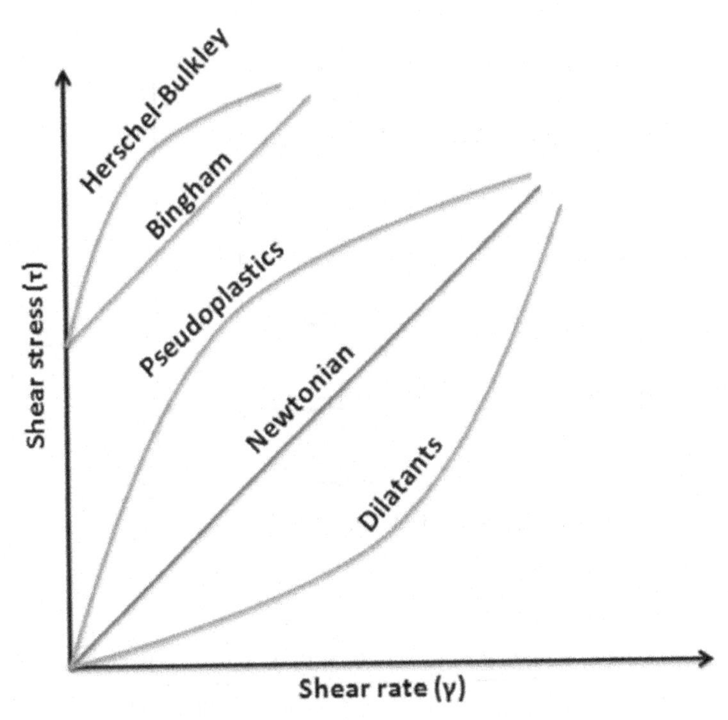

FIGURE 1.2 Rheograms for Newtonian and non-Newtonian fluids.

1.4 RHEOLOGICAL MEASUREMENTS

Rheological measurements, crucial for studying the food materials, can be broadly classified into the steady and dynamic rheological analysis. Figure 1.3 represents the classification of rheological measurements based upon the nature of experimentation and principle involved.

1.4.1 STEADY SHEAR PROPERTIES

Steady shear properties are based on rotational rheometric measurements. Steady shear rheology reflects the flow characteristics of the fluids, whether Newtonian or non-Newtonian. The sample is provided with a range of shear rate, the resulting stress is measured from the data of shear rate, and shear stress behavior of fluid is predicted. The data obtained is fitted on different rheological models (Table 1.4), thus deducing the nature of fluid, yield stress, and other parameters (Table 1.5).

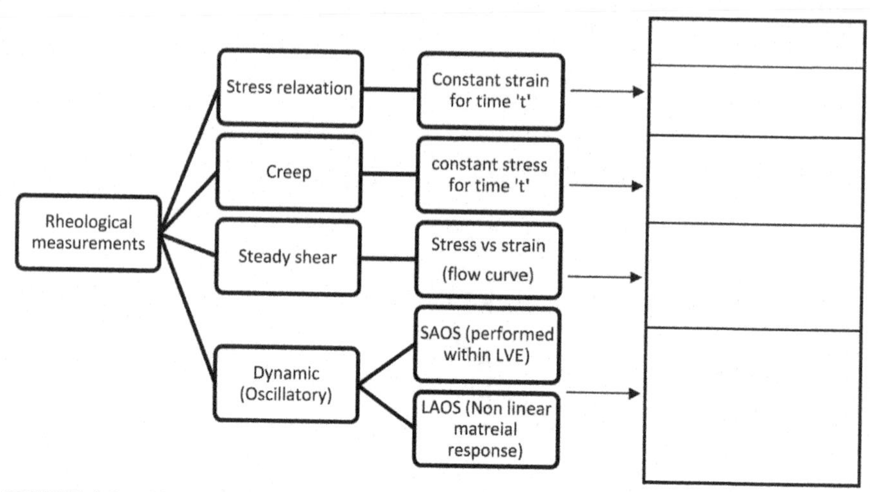

FIGURE 1.3 Classification of rheological measurements.

TABLE 1.4 Models for Steady Shear Flow Properties

Model	Equation	Applicable on	References
Power law model (Ostwald-de-waele)	$\tau = k\dot{\gamma}^n$	Oil, syrup, ketchup	Ghanaei et al., 2010; Coussat et al., 2005
Casson model	$\tau^{0.5} = \tau_o^{0.5} + \eta(\dot{\gamma})^{0.5}$	Jelly, chocolate, honey	Goncalves et al., 2010 Ofoli et al., 1987
Herschel Bulkley model	$\tau - \tau_O = k\dot{\gamma}^n$	Tomato puree, Xanthum Gum	Marcotte et al., 2011
Bingham plastic	$\tau - \tau_o = \eta\dot{\gamma}$	Mayonnaise, ketchup, mustard	Engin et al., 2004 Ofoli et al., 1987

τ – shear stress, $\dot{\gamma}$ – strain rate, η – viscosity, n – flow behavior index, k – consistency coefficient.

TABLE 1.5 Flow Properties of Fluids Defined by Flow Behavior Index (n), Consistency Coefficient (k) and Yield Stress (τ_o)

Newtonian fluids	Power-law fluids	Bingham plastic fluids	Herschel Bulkley fluids
$\tau_o = 0, n = 1, k = \mu$	$\tau_o = 0$	$\tau_o > 1, n = 1, k = \mu_p$	$\tau_o > 1, \infty > n > 0, k > 0$

In the case of Power law model, it uncovers the nature of non-Newtonian fluid, i.e., whether pseudoplastic or dilatant. If the obtained flow behavior index, n, is greater than 1, then the food material is dilatant in nature, and if less than 1, then it is pseudoplastic (Figure 1.4).

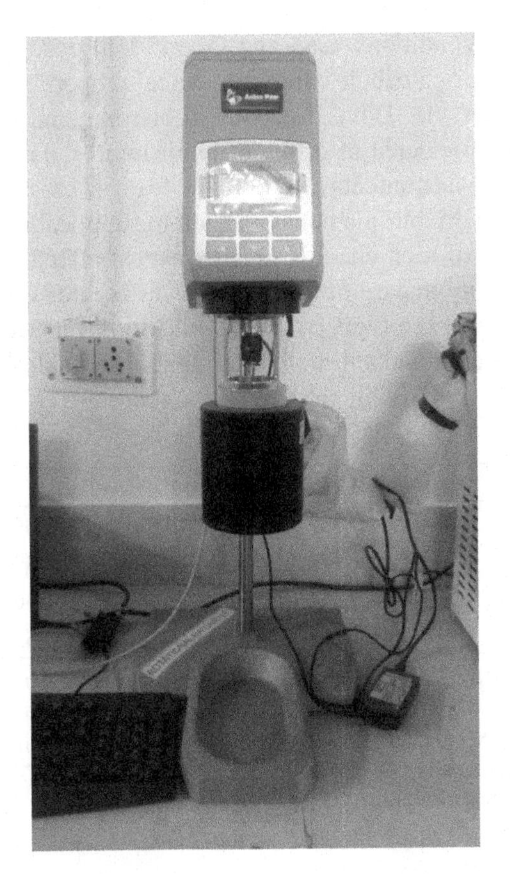

FIGURE 1.4 Rotational rheometer.

1.4.1.1 COMPOSITIONAL CHANGES

In order to predict the aftermath of altering the composition of any food material, shear steady testing can reveal about the effect on the quality parameters. The same has been quoted over here with the example of chocolate. Chocolate is a multiphase product consisting of a dispersed and a continuous phase (Afoakwa et al., 2009). Cocoa butter occurs as a continuous phase, and nonfat components form the dispersed phase. The nonfat components are sugar, cocoa solids, and milk powder (Beckett, 2015). Chocolate is known for its unique eating characteristics as it melts in the mouth, and the flavor is released. So the sensory attributes and selection of processing conditions are affected by variation in the composition, i.e., type, and concentration of above-mentioned ingredients and the processing or manufacturing conditions. Gao and Guo (2015) evaluated the relationship between sensory attributes and rheological properties of four different types of chocolates viz. Dove, Leconte, Toblerone, and Cote dor. Steady shear tests were conducted at different temperatures. The four chocolates possessed different fat contents (Table 1.4). Steady shear rheology showed a shear thinning and thixotropic behavior with the applicability of the Casson model to the flow of melt chocolates. Fat content affected the viscosity of chocolates. Viscosity in case of Toblerone and Cote dor was lower than the other two because of the higher fat content. Moreover, Dove and Leconte had higher yield stress and anti-deformation ability (Table 1.6).

TABLE 1.6 Composition of Chocolate Samples

Samples	Moisture (%)	Ash (%)	Fat (%)
Dove	0.60 ± 0.06^{a}	2.00 ± 0.01^{b}	36.54 ± 1.10^{a}
Leconte	0.35 ± 0.05^{b}	1.62 ± 0.03^{c}	39.75 ± 0.20^{a}
Toblerone	0.55 ± 0.01^{a}	3.22 ± 0.16^{a}	50.70 ± 0.50^{a}
Cote dor	0.57 ± 0.00^{a}	1.75 ± 0.05^{c}	53.26 ± 0.96^{a}

Note: Values in the same row, followed by different superscript letters, are significantly different ($P < 0.05$).

Adapted from: Gao, X., Guo, T., Han, F., Tian, Y., & Zhang, Z., (2015). Rheological and sensory properties of four kinds of dark chocolates. *American Journal of Analytical Chemistry, 6*(13), 1010. https://creativecommons.org/licenses/by/4.0/

1.4.1.2 SENSORY PERCEPTION

Sensory perception is affected by the rheological properties of the product in question. In a study by Szczesniak and Farkas (1962), the relationship between mouthfeel and flow characteristics of gum solutions was established. The gum solutions showed shear thinning characteristics, and depending on that they were grouped into three groups. Group A, B, C comprised gums with very shear thinning, intermediate shear dependence and slight shear thinning, respectively. The sensory panelist rated group A as non-slimy and group C as extremely or very slimy.

1.4.2 DYNAMIC MECHANICAL ANALYSIS (OSCILLATORY)

Dynamic tests operate either at controlled strain or controlled stress conditions. The sample is subjected to strain or stress, which varies harmonically with time. The deformation or stress applied is mostly sinusoidal in nature. In these tests, one part of the rheometer is fixed, and the other one oscillates at a predefined strain, in response to this the time-dependent stress is monitored. This class of methods is ideal for characterization of viscoelastic fluids because information regarding relative viscous and elastic response of materials is deduced. Cone-plate geometry and parallel plate geometry is used for the measurements (Figures 1.5 and 1.6). When any material is subjected to strain, then the response in the form of stress is generated immediately or after a time lag (Table 1.6 and Figure 1.4). For instance, ideal elastic solids respond to applied strain immediately without any time lag, so the phase angle between strain and stress is zero degree. In other words, stress, and strain are in phase in case of elastic materials and out of phase for Newtonian (Ideal fluid) and Viscoelastic materials (Table 1.7).

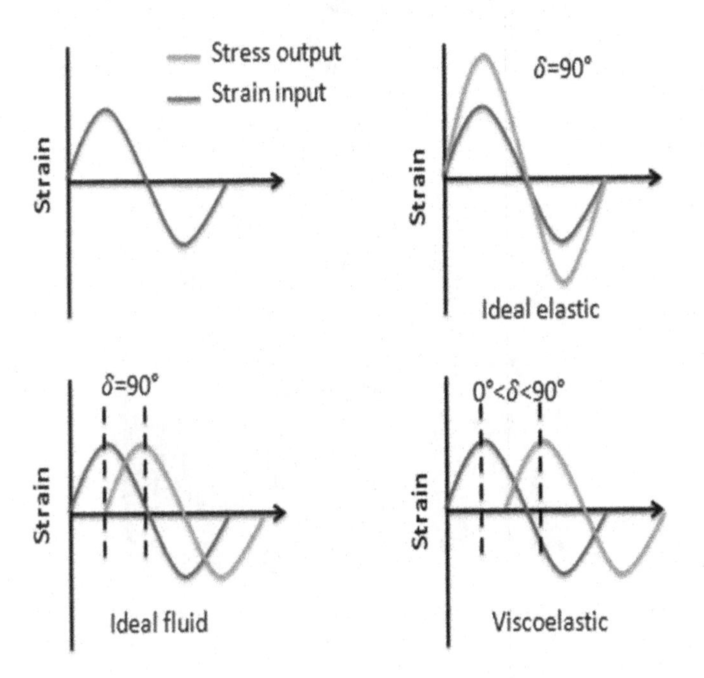

FIGURE 1.5 Stress response for different materials for a sinusoidal strain input.

Mathematically, strain input is given as,

$$\gamma(t) = \gamma_o \sin(\omega t)$$

As this strain input is a function of time, so represented in the form of strain rate (unit $^{-}$s) as,

$$\dot{\gamma} = \gamma_o \omega \cos \omega t$$

where, γ is the strain at time t, γ_o is the magnitude of strain, $\dot{\gamma}$ is the strain rate, and ω is the frequency of oscillation.

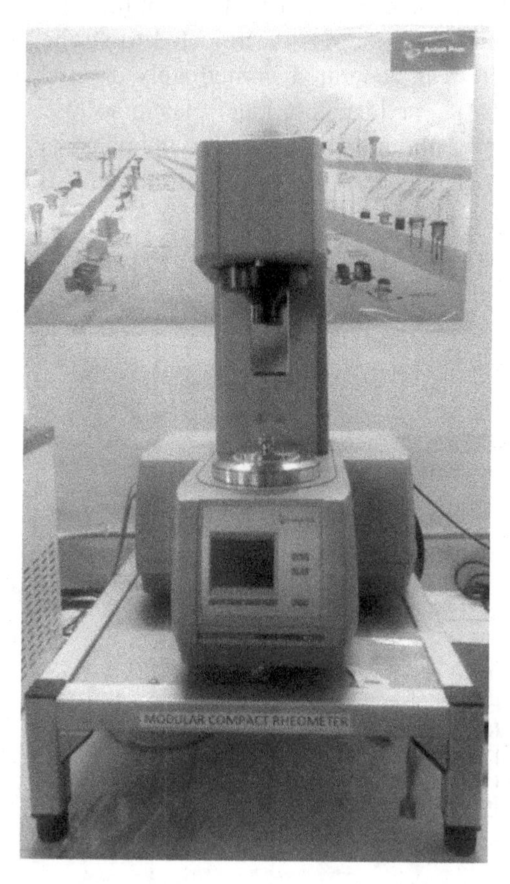

FIGURE 1.6 Dynamic testing rheometer.

TABLE 1.7 Phase Lag for Different Materials

Material	Phase lag between applied strain rate and stress generated
Ideal elastic	0°
Ideal fluid	90°
Viscoelastic	$0° < \delta < 90°$

In dynamic oscillatory tests, the output is obtained in the form of storage modulus (G'), loss modulus (G") and tan δ (Table 1.6). Storage modulus reflects the elastic property of the Viscoelastic fluid and relates to the energy stored by the fluid, whereas loss modulus represents the viscous behavior and the energy dissipated by the Viscoelastic fluid to the environment. tan δ relates the two moduli, and it is the ratio of energy dissipated to energy stored per cycle (Table 1.8).

$$\tan\delta = \frac{G''}{G'}$$

TABLE 1.8 Parameters Obtained from Dynamic Oscillatory Measurements

Parameter	Mathematical expression	Physical significance
Storage modulus (G')	$G' = \frac{\tau_o}{\gamma_o}\cos\delta$	Elastic component
Loss modulus (G")	$G'' = \frac{\tau_o}{\gamma_o}\sin\delta$	Viscous component
tan δ	$\tan\delta = \frac{G''}{G'}$	Energy lost to energy stored

tanδ values indicate the predominating behavior of the food materials, whether viscous or elastic. For instance, tan $\delta < 1$ indicates predominately elastic behavior and tan $\delta > 1$ indicates predominately the viscous behavior (Dong-Won Choi, 2012).

Dynamic oscillatory tests can be divided into two categories: a) small amplitude oscillatory shear (SAOS) and b) large amplitude oscillatory shear (LAOS). Out of these two, SAOS is a much-established method for elucidating the rheological properties as it performs within the linear viscoelastic range (LVE) for the sample in question (Hyun and William, 2011). Within the LVE range, the amplitude of the strain is so small that the rheological properties of Viscoelastic material are independent of strain amplitude (Olabide, 2015) and only depends on the sinusoidal frequency. If the strain is increased beyond the LVE region at a fixed frequency, the transition from linear to the nonlinear regime (i.e., LAOS) is seen and the structure of material changes irreversibly (Hyun and Wilhelm, 2011). But as far as soft solids and food materials are concerned, SAOS is used extensively for rheological characterization. In order to know the strength and stability of the food materials and to reveal the behavior of samples in terms of dominance either by the viscous or elastic component, oscillatory tests are crucial.

Dynamic oscillatory tests can be conducted as follows:

- Strain sweep test (to get LVE range);
- Frequency sweep test (FST);
- Temperature sweep test (TST).

In order to conduct the FST and TST, Strain sweep test is of utmost importance so as to conduct the testing within the LVE range of strain (Table 1.9).

TABLE 1.9 Dynamic Mechanical Tests and Operation Parameters

Test	Variable parameter	Constant	Remarks	Information deduced
Strain sweep test (AST)	Amplitude of strain	Frequency, temperature		LVE (Linear viscoelastic range)
Frequency sweep (FST)	Frequency ramp	Amplitude of strain, temperature	Strain-within the LVE region	Nature of fluid-Dilute solution, concentrate solution, week gel, strong gel (Steffe, 1996)
Temperature sweep (TST)	Temperature ramp	Amplitude of strain, frequency		Stability/transition of viscous or elastic components to temperature

Dynamic rheological tests are important in quality analysis of food products such as jams, jellies, cheese, mayonnaise, thickeners, and additives, starch dispersions and gel, candies, etc. Starch is well known for its gelling and thickening properties and is used extensively in food products. It affects the rheological properties and in turn, the quality of the products formed. Research on dynamic mechanical analysis on starch and rice flour (RF) blends (Jun-Hwan, 2010), wheat starch gel (Yousefi, 2015), Starch-galactomannan mixture (Dong-Won Choi, 2012), OSA substituted potato starch (Chui Won, 2016) has been done so as to aid in analysis of food quality as affected by rheological properties of these starches. FST of starch gel reveals that it possesses a weak gel-like structure (Yousefi, 2015). TST studies of starch dispersion give information on the temperature of transition from liquid-like to solid-like property. So, this temperature can be related to the gelatinization temperature of starch. In other food products such as ice-cream (Dogan, 2013), the addition of stabilizers affect the structure of ice cream, the addition of xanthan gum to ice cream showed that it further supported the strong gel-like structure at the applied frequency range, and an independent relationship between the storage and loss modulus.

1.4.2.1 STRESS RELAXATION

Stress relaxation and creep test are time-dependent rheological tests. An instantaneous strain is applied to the sample, and the change in stress is measured as a function of time. For instance, if a food material is applied and deformed to a fixed value of strain held for a predefined time, the stress required for maintaining the applied strain decreases with time (Cruz et al., 2014). For solid food and agricultural products, stress relaxation and creep test can be done using Texture analyzer, while for fluid foods, dynamic rheometer can be used (Figure 1.7).

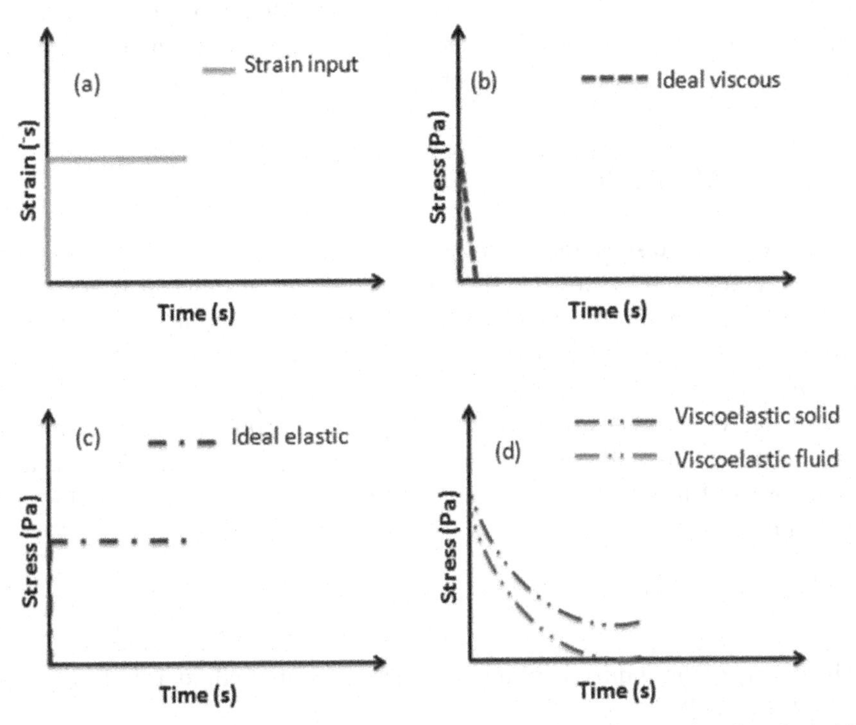

FIGURE 1.7 Strain input (a) response of ideal viscous (b), ideal elastic (c), and viscoelastic (d).

The ideal fluids respond to the applied strain instantaneously to infinite stress and decay rapidly, whereas ideal solids maintain a constant stress value. Viscoelastic fluids develop some stress which decays with time and finally becomes zero, but in the case of viscoelastic solids, residual stress is left in the material (Darby, 1976). In the stress relaxation test, the value of relaxation modulus (G) is deduced, given as:

$$G(t) = \frac{\tau}{\gamma_o}$$

where, τ is the stress measured, and γ_0 is the applied strain; and unit of relaxation modulus is Pascal (Pa). For comparative analysis usually, the double logarithmic plot of relaxation modulus versus time is useful.

Stress relaxation data is important for deciding the end use of flour types (soft and hard wheat flour) (Barajas, 2014), to check feasibility of substitution of different materials such as, sago starch blended in wheat flour for wheat based product development (Zaidul et al., 2003), to study dough characteristics (Li. et al., 2003), staling of corn tortilla (Limanond et al., 2002), to differentiate low and high protein cultivars from dough study (Safari-Ardi, 1998), to study the effect of time, temperature, and moisture on grains, such as rice kernels (Wang et al., 2017).

1.4.2.2 CREEP-RECOVERY MEASUREMENTS

Creep recovery test is also a time-dependent test which consists of two phases, creeps, and recovery. In this test, the strain induced is recorded during the creep and recovery phases by applying and removing constant stress for a predefined time (Figure 1.8). Data obtained from creep recovery is applied to different mechanical models such as the Maxwell model, Kelvin model, Linear model, and Burger model (a combination of two or more Maxwell and Kelvin components). The results of creep test are expressed in the form of creep compliance, J, which is a function of time, given as,

$$J(t) = \frac{\gamma(t)}{\tau_o}$$

where, γ(t) is the induced strain, τ_0 is the stress applied; and unit of creep compliance is Pa^{-1}.

Usually, the agricultural produce in the form of grains is stored in sacks one above the other in cover and plinth type of storage, or the lower grains in bin storage. So, a definite amount of load always acts on the grains/kernels. Creep recovery test is of great value to get data on the deformation induced and the quality disruption, i.e., breakdown of the kernels in such cases. Figueroa et al., (2013) studied the stress relaxation and creep recovery tests on wheat kernel versus dough so as to ease the selection of wheat kernels by millers or bakers.

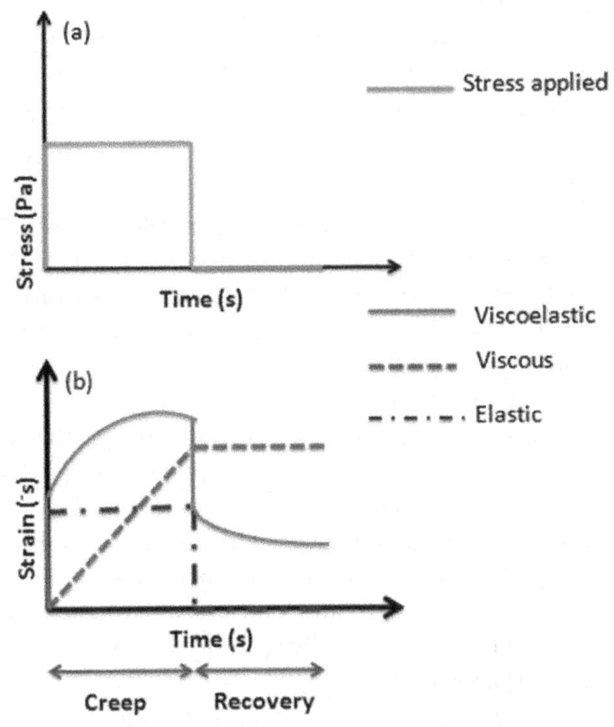

FIGURE 1.8 Creep recovery response for ideal elastic, ideal viscous, and viscoelastic food materials.

1.5 GENERALIZED APPLICATIONS OF RHEOLOGY

1.5.1 APPLICATION OF RHEOLOGY IN FRUIT JAMS

The jam comes under the category of intermediate moisture food, and according to norms, it should contain at least 45% of fruit and 60–70° Brix (CODEX, 2009). The increasing awareness of the harmful effect of sugars generated the demand to switch from sugars to the artificial sweeteners. The effect of addition of the artificial sweetener (sorbitol) was studied by Basu et al., (2010) in which they observed change in rheological property of jam with the help of dynamic rheology viscometer under the controlled stress and strain condition and found that value of G' is greater than G" which indicates the viscoelasticity of jam. A similar analysis was also observed for starch gel (Rosalina and Bhattacharya, 2001). The ($\tan \delta<1$) tells about the gel-like nature. For attaining the good quality of gel-like nature of jam 50–70% of sucrose is required. In the study done by Basu and Shivhare (2010), it was found that if the amount of sucrose exceeded the range of 60–70%, the jam was found to be softer with an excess amount of water.

In order to develop an efficient processing method/parameters for the manufacturing of jams and concentrates, it is necessary to study the rheological properties of raw ingredients and final products. Rheological measurements and rheological modeling of jams were studied by Basu et al., (2007) in order to reveal their significance in quality control. Rheological properties of raw ingredients under different processing conditions were found to affect the color, flavor, taste, texture, and nutritional values. Apart from quality parameters, rheological properties of fruit jams and concentrates are necessary to calculate pumping costs. Same has been studied by Deen et al., (2007). Apparent viscosity of the fruit jam was higher than that of fresh fruit puree; this was elucidated from the plots of apparent viscosity versus shear rate. This change was observed due to the change in sugar concentration. Temperature dependence on rheological properties was also found, causing a decrease in viscosity with an increase in temperature. Designing of equipments for specific operations in manufacturing and processing of jams and jellies are designed on the basis of affect on the viscosity as influenced by different shear rates (Javanmard et al., 2010).

Jams and jellies developed from different fruits bear different rheological properties, and the same has been investigated by Gao et al., (2011) for four different kinds of jams (Kewpie, strawberry, marmalade, and blueberry Dalfour). Jams possess non-Newtonian behavior with shear thinning properties. Apart from the compositional differences, the measuring parameters and conditions are also important. For instance, the non-Newtonian behavior of jams drives the use of models that defines an unstable non-laminar flow behavior. Due to variable/non-uniform particle size, it is hard to achieve stability, i.e., laminar flow. So testing conditions must be wisely chosen in order to reveal the exact rheological flow behavior of jams (Javanmard et al., 2010). Accordingly, Herschel Bulkley model was found to be one of the best defining the flow behavior of jams.

Jams posses a mosaic structure because its ingredients possess variation in their flow behavior. For instance, pectin one of the important component of jam exhibits a non-Newtonian behavior, whereas its serum possesses a Newtonian behavior. Therefore, in order to explain this mosaic structure, jam's weak-law model was used by Bhattacharya (1999) where fruit jam was considered as a viscoelastic gel, with the three-dimensional network having weak interactions; also the interaction between micro-structure of the fruit gels and its rheological properties was expressed.

1.5.2 APPLICATION OF RHEOLOGY IN FRUIT JUICES

The rheological properties of juices majorly get affected by the duration of storage and packaging. Hence, the rheological study of juices needs to be done as a function of storage time, temperature, and preservative used to increase shelf-life in packaging form. To study the changes in the rheological behavior of litchi juice during the storage period and at the time of production, a comparative matrix is used. In which the shear rate and flow behavior index of both stored/packed juice and fresh juice are compared. The shear rate study is done according to the power-law equation, whereas flow behavior index is measured by Rheostat 2 viscometer with coaxial cylinder attachment (Sharma et al., 2014). The rheological behavior of freshly prepared canned tomato product was studied at different break temperature, and it was found that break temperature has a direct effect on viscosity (Xu Shi et al., 1986). Justification of observation can be explained with the nature of pectin present in the product, a higher degree of inactivation of pectin enzyme polygalacturonase and pectin esterase at a higher temperature (Luh and Daoud, 1971).

Fruit juices mainly consist of pulp, which is an insoluble phase, and dispersed phase solution is known as serum. This dispersed and insoluble phase is present in the form of fruit tissue, cell wall, and insoluble form of polymer cluster. Hence, fruit juice's rheological property can be interpreted on the basis of the interaction between each phase. For determining the effect of high pressure (HP) homogenization on the tomato serum, a controlled stress rheometer with cone-plate geometry was used (Augosto et al., 2012). Homogenization was found to affect the yield stress and consistency coefficient of juice suspensions, such as that of tomato suspensions. It was found that smaller the particle size, larger is the interaction forces which ultimately cause an increase in the consistency coefficient and yield stress (Bayod et al., 2007; Yoo and Rao, 1994).

Apart from shear stress and flow behavior index, viscosity, and activation energy are determining factors for process consistency and quality of juice. It was observed that the increase in viscosity is higher if stored at low temperature as compared to the samples stored at ambient temperature, concluding that viscosity decreased with the increase in temperature for juices (Goksel et al., 2013). In another study, it was elucidated that viscosity is directly dependent on inter-molecular forces between water solutes and sugar molecules, which consists of hydrogen bonds and intermolecular forces, both forces depend upon temperature and concentration (Manjunatha and Raju, 2013).

1.5.3 APPLICATION OF RHEOLOGY IN BAKERY

There are different types of processes involved in producing different bakery products like baking, fermentation, and mixing in which rheology plays an important role; because of this, the study of rheology becomes mandatory for maintaining the structure, repeatability of the process, and quality of final products. During processing structural characteristic of the product should be controlled in order to maintain the rheological property of the baked product. From the various study, it has been observed that most preferred method for testing rheological behavior of dough is shear oscillation dynamic method. The oscillatory dynamic method is used under deformation conditions and shows the relation between performances of end-user (Angioloni, and Collar, 2008). The plateau region is the limited range of frequency, which is not sensitive to HMW gluten, which is considered to be responsible for variations in baked product quality. The branching of polymer and their interaction with the polymer chain determines the rheology of HMW (High Molecular Weight) Polymers (Shuey, 1975).

TST study of the dough gives clarity about the effect of water on rheological characteristics of dough. One of the important ingredients for dough formation is water. Water acts as a plasticizer in dough formation and has a negative effect on G' and G." When strain is lower than 0.1%, G" is lesser than G' but if the greater strain is used, the ratio becomes reversed because of solid viscoelastic conversion to elastoviscous liquid (Rasper, 1993). Angioloni and Rosa (2005) showed that during temperature variation of dough between the ranges of 25–90°C, G' values decrease down slowly but at temperature 75°C, it started to increase rapidly. Hence starch gelatinization, gluten cross-linking, or both, can be considered as an advanced explanation for thermally induced rheological changes.

1.5.4 APPLICATION OF RHEOLOGY IN DAIRY

Cheese is one of the important dairy products used globally. Sensory quality, such as mouthfeel of cheese is directly affected by the amount of fat content in it. But nowadays, health consciousness is driving to develop low-fat cheese. The effect of this alteration on the sensory attributes can be established by using rheological tools. Rogers et al., (2010) investigated the effect of fat content on textural properties of cheese by calculating the storage modulus as affected by fat content and temperature, At lower fat concentrations and lower temperature, the rigidity (storage modulus) of cheese was more than at ambient temperature. The hardness, complex viscosity, and elastic modulus decrease in low fat, high moisture mozzarella cheese due to high levels of moisture in nonfat substances (Tunick et al., 1993).

1.6 CONCLUSION

Rheological tools act as a milestone for ensuring the texture and quality of food products. Rheological characterization aids in improving the existing products, by analyzing the effect of additives, thickeners, functional components, substitutes, etc. added to the food products. Steady shear properties reveal about the flow behavior of the food materials. Dynamic mechanical analysis is an advanced form of rheology which allows deducing the effect of temperature, strain or strain rate, stress, frequency, etc. on the quality attributes of the product by providing with the data of storage modulus and loss modulus. The information on the relative proportion of viscous and elastic component within a material, and the transition of material from solid-like to liquid-like or vice-a-versa is deduced from the magnitudes of storage and loss moduli during the testing. Stress relaxation and creep recovery tests give information on the rigidity or flexibility of the food materials such as dough or grains/kernels thus revealing the maximum limit of stress which they can withstand without breakdown.

KEYWORDS

- **fluids**
- **rheological models**
- **rheology**
- **viscosity**

REFERENCES

Afoakwa, E. O., Paterson, A., Fowler, M., & Vieira, J., (2009). Comparison of rheological models for determining dark chocolate viscosity. *International Journal of Food Science & Technology*, *44*(1), 162–167.

Agirre-Olabide, I., Elejabarrieta, M. J., & Bou-Ali, M. M., (2015). Matrix dependence of the linear viscoelastic region in magnetorheological elastomers. *Journal of Intelligent Material Systems and Structures*, *26*(14), 1880–1886.

Ajila, C. M., Leelavathi, K. U. J. S., & Rao, U. P., (2008). Improvement of dietary fiber content and antioxidant properties in soft dough biscuits with the incorporation of mango peel powder. *Journal of Cereal Science*, *48*(2), 319–326.

Akalın, A. S., & Erişir, D., (2008). Effects of inulin and oligofructose on the rheological characteristics and probiotic culture survival in low-fat probiotic ice cream. *Journal of Food Science*, *73*(4), M184–M188.

Angioloni, A., & Rosa, M. D., (2005). Dough thermo-mechanical properties: Influence of sodium chloride, mixing time and equipment. *J. Cereal Sci.*, *41*, 327–331.

Angioloni, A., & Collar, C., (2008). Functional response of diluted dough matrixes in high-fiber systems: A viscometric and rheological approach. *Food Research International*, *41*(8), 803–812.

Augusto, P. E., Ibarz, A., & Cristianini, M., (2012). Effect of high-pressure homogenization (HPH) on the rheological properties of a fruit juice serum model. *Journal of Food Engineering*, *111*(2), 474–477.

Bagley, E. B., & Dintzis, F. R., (1999). Shear thickening and flow-induced structures in foods and biopolymer systems. In: *Rheology Series* (Vol. 8, pp. 63–86). Elsevier.

Basu, S., & Shivhare, U. S., (2010). Rheological, textural, microstructural and sensory properties of mango jam. *Journal of Food Engineering*, *100*(2), 357–365.

Basu, S., Shivhare, U. S., & Raghavan, G. S. V., (2007). Time-dependent rheological characteristics of pineapple jam. *International Journal of Food Engineering*, *3*(3), 1–10.

Bayod, E., Månsson, P., Innings, F., Bergenståhl, B., & Tornberg, E., (2007). Low shear rheology of concentrated tomato products. Effect of particle size and time. *Food Biophysics*, *2*(4), 146–157.

Beckett, S. T., (2015). *The Science of Chocolate*. Royal Society of Chemistry.

Bhattacharya, S., (1999). Yield stress and time-dependent rheological properties of mango pulp. *Journal of Food Science*, *64*(6), 1029–1033.

Codex, (2009). *Codex Standard 296*. Standard for Jams, Jellies and Marmalades.

Cruz, R. M., Khmelinskii, I., & Vieira, M., (2014). *Methods in Food Analysis*. CRC Press.

Darby, R., (1976). *Viscoelastic Fluids*. M. Dekker.

Deen, N. G., Annaland, M. V. S., Van Der Hoef, M. A., & Kuipers, J. A. M., (2007). Review of discrete particle modeling of fluidized beds. *Chemical Engineering Science*, *62*(1/2), 28–44.

Figueroa, J. D. C., Hernández, Z. J. E., Rayas-Duarte, P., & Peña, R. J., (2013). Stress relaxation and creep recovery tests performed on wheat kernels versus doughs: Influence of glutenins on rheological and quality properties. *Cereal Foods World*, *58*(3), 139–144.

Foegeding, E. A., Daubert, C. R., Drake, M. A., Essick, G., Trulsson, M., Vinyard, C. J., & Van De Velde, F., (2011). A comprehensive approach to understanding textural properties of semi-and soft-solid foods. *Journal of Texture Studies*, *42*(2), 103–129.

Gao, X., Guo, T., Han, F., Tian, Y., & Zhang, Z., (2015). Rheological and sensory properties of four kinds of dark chocolates. *American Journal of Analytical Chemistry*, *6*(13), 1010.

Goksel, M., Dogan, M., Toker, O. S., Ozgen, S., Sarioglu, K., & Oral, R. A., (2013). The effect of starch concentration and temperature on grape molasses: Rheological and textural properties. *Food and Bioprocess Technology*, *6*(1), 259–271.

Hoffman, R. L., (1982). Discontinuous and dilatant viscosity behavior in concentrated suspensions III. Necessary conditions for their occurrence in viscometric flows. *Advances in Colloid and Interface Science*, *17*(1), 161–184.

Hyun, K., Wilhelm, M., Klein, C. O., Cho, K. S., Nam, J. G., Ahn, K. H., & McKinley, G. H., (2011). A review of nonlinear oscillatory shear tests: Analysis and application of large amplitude oscillatory shear (LAOS). *Progress in Polymer Science, 36*(12), 1697–1753.

Javanmard, M., & Endan, J., (2010). A survey on rheological properties of fruit jams. *International Journal of Chemical Engineering and Applications, 1*(1), 31–37.

Jethra, R., (1994). Viscosity measurement. *ISA Transactions, 33*(3), 307–312.

Joshi, J. B., & Ranade, V. V., (2003). Computational fluid dynamics for designing process equipment: Expectations, current status, and path forward. *Industrial & Engineering Chemistry Research, 42*(6), 1115–1128.

Li, W., Dobraszczyk, B. J., & Schofield, J. D., (2003). Stress relaxation behavior of wheat dough, gluten, and gluten protein fractions. *Cereal Chemistry, 80*(3), 333–338.

Limanond, B., Castell-Perez, M. E., & Moreira, R. G., (2002). Modeling the kinetics of corn tortilla staling using stress relaxation data. *Journal of Food Engineering, 53*(3), 237–247.

Luh, B. S., & Daoud, H. N., (1971). Effect of break temperature and holding time on pectin and pectic enzymes in tomato pulp. *Journal of Food Science, 36*(7), 1039–1043.

Luyten, H., (1988). *The Rheological and Fracture Properties of Gouda Cheese (Doctoral Dissertation, Luyten).*

Magaña-Barajas, E., Ramírez-Wong, B., Torres-Chávez, P. I., & Morales-Rosas, I., (2012). Use of the stress-relaxation and dynamic tests to evaluate the viscoelastic properties of dough from soft wheat cultivars. In: *Viscoelasticity-From Theory to Biological Applications*. In Tech.

Manjunatha, S. S., & Raju, P. S., (2013). Modeling the Rheological behavior of tender coconut (*Cocos nucifera* L) water and its concentrates. *International Food Research Journal, 20*(2).

Menjivar, J. A., (1989). Fundamental aspects of dough rheology. In: Faridi, H. A., & Faubion, J. M., (eds.), *Dough Rheology and Baked Product Texture* (pp. 1–28). New York: Van Nostrand Reinhold.

Munson, B. R., Okiishi, T. H., Rothmayer, A. P., & Huebsch, W. W., (2014). *Fundamentals of Fluid Mechanics*. John Wiley & Sons.

Nindo, C. I., Tang, J., Powers, J. R., & Takhar, P. S., (2007). Rheological properties of blueberry puree for processing applications. *LWT-Food Science and Technology, 40*(2), 292–299.

Prentice, J. H., (1987). Cheese rheology. In: Fox, P. F., (ed.), *Cheese: Chemistry, Physics and Microbiology, I General Aspects* (pp. 299–344). Elsevier Applied Science, London.

Rani, U., & Bains, G. S., (1987). Flow behavior of tomato ketchups. *Journal of Texture Studies, 18*(2), 125–135.

Rasper, V. F., (1993). Dough rheology and physical testing of dough. In: *Advances in Baking Technology* (pp. 107–133). Springer, Boston, MA.

Rogers, N. R., McMahon, D. J., Daubert, C. R., Berry, T. K., & Foegeding, E. A., (2010). Rheological properties and microstructure of Cheddar cheese made with different fat contents. *Journal of Dairy Science, 93*(10), 4565–4576.

Rosalina, I., & Bhattacharya, M., (2001). Flow curves, stress relaxation and creep measurements of starch gels. *Journal of Texture Studies, 32*(4), 247–269.

Safari-Ardi, M., & Phan-Thien, N., (1998). Stress relaxation and oscillatory tests to distinguish between doughs prepared from wheat flours of different varietal origin. *Cereal Chemistry, 75*(1), 80–84.

Schowalter, W. R., (1978). *Mechanics of NonNewtonian fluidd.* Pergamon.

Sharma, R., Manikantan, M. R., Ranote, P. S., & Singh, T., (2014). Rheological behavior of litchi juice concentrates during storage. *International Food Research Journal, 21*(3).

Shuey, W. C., (1975). Practical instruments for rheological measurements on wheat products. *Cereal Chemistry, 52*(3), 42–81.

Szczesniak, A. S., & Farkas, E., (1962). Objective characterization of themouthfeell of gum solutions. *Journal of Food Science*, *27*(4), 381–385.

Tabilo-Munizaga, G., & Barbosa-Cánovas, G. V., (2005). Rheology for the food industry. *Journal of Food Engineering*, *67*(1/2), 147–156.

Taneya, S., Izutsu, T., & Sone, T., (1979). Dynamic viscoelasticity of natural cheese and processed cheese. In: Sherman, P., (ed.), *Food Texture and Rheology* (pp. 369–383). Academic Press, London.

Tkachuk, R., & Hlynka, I., (1968). Some properties of dough and gluten in D2O. *Cereal Chem.*, *45*(1).

Tunick, M. H., Mackey, K. L., Shieh, J. J., Smith, P. W., Cooke, P., & Malin, E. L., (1993). Rheology and microstructure of low-fat Mozzarella cheese. *International Dairy Journal*, *3*(7), 649–662.

Van Vliet, T., Van Aken, G. A., De Jongh, H. H., & Hamer, R. J., (2009). Colloidal aspects of texture perception. *Advances in Colloid and Interface Science*, *150*(1), 27–40.

Vitalli, A. A., & Rao, M. A., (1982). Flow behavior of guava puree as a function of temperature and concentration. *Journal of Texture Studies*, *13*(3), 275–289.

Wang, P., Wang, L. J., Li, D., Huang, Z. G., Adhikari, B., & Chen, X. D., (2017). The stress-relaxation behavior of rice as a function of time, moisture and temperature. *International Journal of Food Engineering*, *13*(2).

Watanabe, A., Larsson, H., & Eliasson, A. C., (2002). Effect of physical state of nonpolar lipids on rheology and microstructure of gluten-starch and wheat flour doughs. *Cereal Chemistry*, *79*(2), 203–209.

Williamson, R. V., & Hecker, W. W., (1931). Some properties of dispersions of the quicksand type. *Industrial & Engineering Chemistry*, *23*(6), 667–670.

Xu, S. Y., Shoemaker, C. F., & Luh, B. S., (1986). Effect ofbreakk temperature on rheological properties and microstructure of tomato juices and pastes. *Journal of Food Science*, *51*(2), 399–402.

Yoo, B., & Rao, M. A., (1994). Effect of unimodal particle size and pulp content on rheological properties of tomato puree. *Journal of Texture Studies*, *25*(4), 421–436.

Zaidul, I. M., Karim, A. A., Manan, D. M. A., Azlan, A., Nik, N. N. A., & Omar, A. M., (2003). Stress relaxation test for sago–wheat mixtures gel. *International Journal of Food Properties*, *6*(3), 431–442

CHAPTER 2

Food Extrusion Technology and Products

SUVENDU BHATTACHARYA

Department of Food Engineering and Technology, Tezpur University, Tezpur, Assam, India; Formerly Chief Scientist, CSIR-Central Food Technological Research Institute, Mysore, India

ABSTRACT

The technology of extrusion of foods has grown rapidly in the last four decades mainly because this technology can pragmatically produce a variety of interesting products with varying size, unique shape and attractive texture in a cost-effective manner. The common industrial products include breakfast cereals, snacks, texturized plant protein (meat analog), flatbread, confectionery, and pasta products.

Extrusion technology possesses several worth-mentioning advantages over similar processing systems. An extruder can mix-cook-shape at a single operation, and therefore reduces the number of the equipment for processing. Hence, the requirement of space, labor, and energy is markedly lower to similar cooking-forming processing systems of similar throughput. Besides, the creation of attractive shape, and the development of multi-layered and filled products is possible for commercial success.

The present book chapter deals with the different types of extruder, extruder components and their functions, and commercially important extruded products. The chapter also mentions the new areas of extrusion technology such as 3D printing and turbo extrusion.

2.1 INTRODUCTION

The technology of extrusion and extrusion cooking is an important area of food processing wherein several products are manufactured that differs widely in their functional behavior. Extrusion of food combines several unit operations like conveying, thermo-mechanical change and degradation, mixing, and shaping into a single step. Though the extruded products started production about four decades ago, the advancement of the understanding on this interesting subject, precise, and product-specific design of hardware, use of different raw materials and their combinations, improved process controls, and a large number of research publications have permitted the use of a wide range of raw materials to make various products in a cost-effective manner. The common industrial products at present include breakfast cereals, snacks, texturized plant protein (meat analog), flatbread, confectionery, and pasta products. Many raw materials are subjected to extrusion technology to develop extruded products. Thus, it is appropriate to discuss the raw materials linking specific products. The present book chapter thus deals with different raw materials used, types of the extruder, extruder components and their functions, and commercially important extruded products.

2.2 EXTRUDER

An extruder is a screw pump in which an Archimedes screw rotates in a tightly fitting stationary sleeve or barrel. The rotation of the screw moves the particulate food forward and experiences shearing, compression, and mixing. The dissipation of mechanical energy due to the rotation of the screw or externally added thermal energy makes the feed viscoelastic. Finally, the feed exists through a restriction called die.

Various designs are possible for an extruder. The single screw extruders have only one screw while the twin-screw extruder has two intermeshing screws; these screws may rotate in the same direction (corotating) or opposite direction (counter rotating). The food industries prefer the fully intermeshing corotating twin-screw extruders because of technical advantages. The cost-effective single-screw extruders are generally employed for developing expanded snacks and texturized plant/vegetable protein (TPP/TVP) while complex products like filled snacks and multilayered products require the costlier twin-screw extruders.

2.2.1 EXTRUDER COMPONENTS

An extruder contains several components, such as:

1. **Screw:** The flighted screw is the heart of an extruder wherein variations in the screw design are possible concerning flight height, pitch,

and thickness, helix angle, the diameter of the screw, length-to-diameter (L/D) ratio, the presence of reverse pitch screw elements, mixing disks, kneading block, etc.

2. **Barrel:** It houses the extruder screw keeping a minimum gap between the tip of the screw and inner diameter of the barrel. Occasionally, the barrel has grooves to impart additional shearing action on the feed during extrusion.
3. **Die:** The opening at the downstream of an extruder is called die nozzle which usually houses the die. A large extruder may have several die openings. A rotary cutter containing several knives is placed at the dead end to cut the emerging continuous strands to an appropriate size.
4. **Feeding System:** The powdery feed is usually fed to the extruder employing a feeding screw or vibratory feeding device employing the principle of constant gravimetric or volumetric feeding.
5. **Drive:** A powerful motor with a facility to reduce and control the speed of the extruders is used.
6. **Heating System:** Direct injection of steam and external heating through barrels are usually followed when extrusion cooking is desired. Extra heat input is absent for the extrusion system producing pasta products.
7. **Safety and Control Systems:** An extruder is expected to have safety devices for out of range temperature, pressure, and torque. The temperature control is mandatory to maintain uniform product quality.
8. **Take Away Device:** Once the extrudates come out of the extruder and cut to the desired size by the rotary knives, a takeaway system in the form of the belt conveyor or pneumatic system is employed.

2.2.2 EXTRUSION AND EXTRUSION COOKING

The process of extrusion is the act of an extruder to convey the powdery material through a die of desired size and shape. The preparation of pasta products is an example of extrusion. The addition of thermal energy during extrusion creates a cooked food product, and the process is called extrusion cooking, and the examples include expanded snacks like a cornball, corn curl, etc.

2.3 EXTRUSION PROCESSING

2.3.1 RAW MATERIALS USED

Many raw materials from different sources are used as the feed for extrusion to manufacture a wide range of product that varies in size, shape, the extent of cooking, cellular structure, texture, density, and consumer acceptability (Figure 2.1). Usually, the nearly free-flowing dry/partially moist powder is fed to the extruder, which becomes a viscoelastic mass due to water/steam addition and under shear, pressure, and thermal treatments. The physical and chemical properties of the raw material such as the average particle size, moisture, protein, and fat contents, hardness, and cohesiveness of particles affect the extrusion process and the extruded product. Apart from the main raw materials (Table 2.1), the role of additives (plasticizer, lubricant, binder, nutrient/fortification, and expansion agents, etc.) markedly alters the characteristics of the extrudates (Bhattacharya, 2012). The additives/ingredients are usually added at three phases, such as before, during, and post-extrusion stages depending on the requirement of the finished product. Further, the ingredients may also be classified according to their functional roles (Guy, 2001); these are (a) structure-forming materials, (b) dispersed-phase filling materials, (c) plasticizers and lubricants, (d) soluble solids, (e) nucleating agents, (f) coloring materials, and (g) flavoring substances.

2.3.2 PRE-PROCESSING AND POST-EXTRUSION TREATMENTS

Though not mandatory, the use of pre-extrusion treatment is common wherein the dry powdery feed is mixed with water and/or steam, so that hydration and partial cooking/pre-gelatinization are achieved in addition to obtaining a uniformly mixed feed. Screw conveyors and mixing devices with paddles are often employed, and the resultant partially wet cooked/heated mix is directly transferred to the extruder cooker.

Post-extrusion treatments such as drying, baking, frying, roasting, toasting, flavoring, coating, etc. are employed to achieve a safe moisture content, converting the ready-to-cook/serve extrudates into the ready-to-eat product, additional expansion to make the finished product crisp and soft, and development of roasted/toasted flavor. Dip coating, spraying, and enrobing are normally employed where continuous drum and inclined coating pan are employed. Typical flavoring and coating ingredients include oil/fat, salt, spice mix, liquid chocolate, sugar solutions, honey, starch dispersions, gum

Arabic dispersions, etc. The common ingredients and processes employed are shown in Table 2.2.

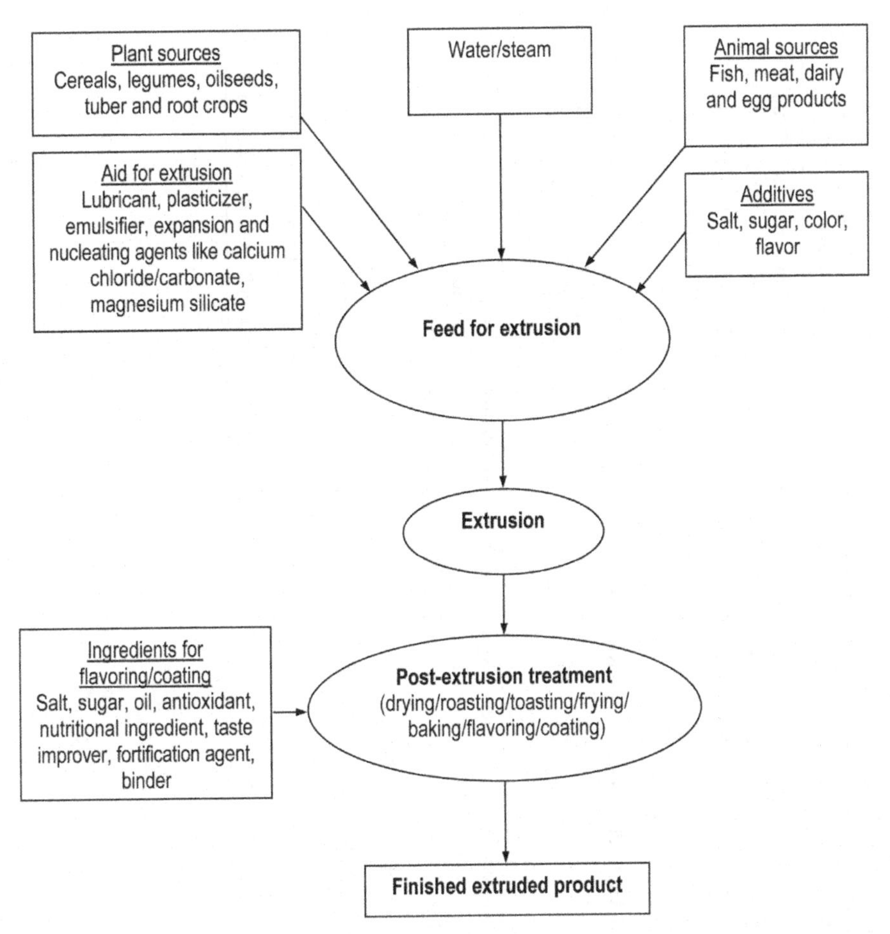

FIGURE 2.1 Schematic diagram of the extrusion process and the raw materials used. (*Source*: Reprinted with permission from Bhattacharya, 2012. © 2012 Taylor & Francis.)

2.4 EXTRUDED PRODUCTS

The extruded products can be classified based on the type of raw material employed. The categories may be cereal, pulse, oilseed, fish/meat, etc. However, a pragmatic categorization may be based on the convenience of use; these are (a) ready-to-eat, (b) half-product/intermediates

TABLE 2.1 Extruded Products and the Raw Materials Used

Product group	Product	Raw materials used	References
Modified starch/grain flour	Raw materials for specialty foods/convenience foods, baby foods, weaning foods	Corn, potato, wheat, tapioca, rice starch, cereal grit/flour	Hagenimana et al., (2006); Altan et al., (2009), Fletcher et al., (1985); Cai and Diosady (1993); Della Valle et al., (1995)
Texturized vegetable/soy protein (TVP/TSP)	Soy chunk/nugget	Defatted soy flour	Harper (1981)
Expanded snack/product	Blended extruded product	Rice, corn, soy, whey, potato	Molina et al., (1983); Ascheri et al., (1998); Konstance et al., (1998); Ding et al., (2005) and (2006); Stojceska et al., (2009); Chang et al., (1998)
Muscle food	Restructured fish/meat mince	Mince from fish/meat/poultry, and wheat/cassava flour/soy protein isolate/wheat gluten	Surasani (2016); Alvarez et al., (1990); Noguchi (1990)
Soy flour	Trypsin inhibited soy flour	Defatted soy flour	Van den Hout et al., (1998)
Legumes, nut	Expanded extrudates	Pinto bean, pigeon pea, almond, pre-gelatinized cereal flour, chickpea, green gram	Balandrán-Quintana et al., (1998); De Pilli et al., (2005); Rampersad et al., (2003); Bhattacharya (1997)
Pasta products	Legume and cereal pasta	Wheat, pea flour, egg	Zardetto and Rosa (2009); Wang et al., (1999)

TABLE 2.2 Common Post-Extrusion Treatments and Raw Materials Used for Selected Products

Name of product	Type of treatment	Machinery employed	Raw materials used during post-extrusion treatment
Texturized vegetable protein	Drying	Continuous belt dryer	Nil
Low-density cereal/legume/tuber and root crop-based snack	Drying	Continuous belt/rotary inclined dryers	Nil
	Roasting	Air/contact inclined surface roaster	Nil
	Flavoring	Batch inclined drum/enrober/continuous rotary drum with spray/powder dispensing system	Salt, oil, sugar, flavors, binders, antioxidants, color, minerals, and vitamins
Breakfast cereals	Roasting/toasting	Air/contact surface roasting/toasting	Nil
	Coating	Batch or continuous drum/enrober	Sugar, malt powder, flavor, liquid glucose, oil
Soy granules	Size reduction and drying	Granulators/disintegrator/rotary slicer and continuous dryer	Flavor, salt

Source: Bhattacharya, S., (2012). Raw materials for extrusion cooking. In: Maskan, M., & Altan, A., (eds.), *Advances in Food Extrusion Technology* (pp. 69–85). CRC Press, Boca Raton, Florida, USA, with Permission.

(ready-to-process) products requiring further processing like baking/toasting/frying before consumption, and (c) processed raw material to be used for product development.

2.4.1 PASTA PRODUCTS

The pasta products are extrusion-shaped products which need further cooking at the consumer end. The simple design of the pasta press and in the absence of any heating facility, the machine only does continuous kneading and shaping of the dough. The shaped dough strands are subjected to slow drying to achieve safe moisture content for storage until used. The easy to process, low price of the product, and long shelf-life make the pasta products a popular food. A few of these products are fried in oil to produce attractive snacks. Pasta products which include macaroni, spaghetti, and vermicelli are made from a mixture of flour/semolina and water; other ingredients/additives such as egg yolk, salt, minerals, and vitamins, protein-rich ingredients, etc., are often added. The use of appropriate dies allows preparing attractive shapes like tubes, shells, elbows, stars, wheels, rings, etc. The prepared pasta products are either immediately cooked and consumed, or dried for consumption at a later time (Figure 2.2).

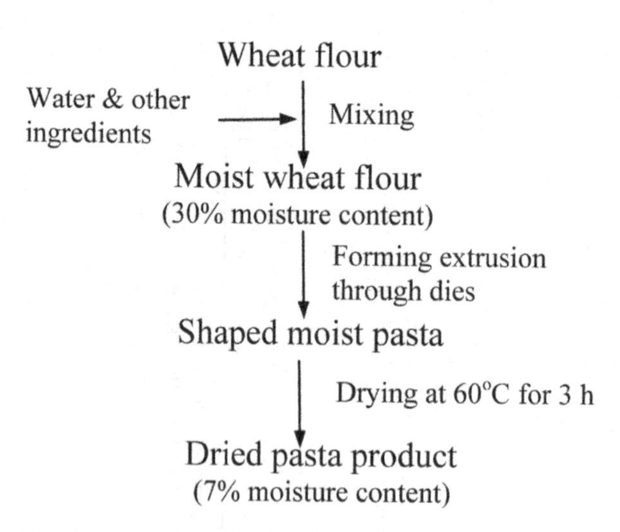

FIGURE 2.2 Flowchart for the preparation of pasta products.

The conventional raw material for pasta products is wheat while the use of other cereals such as rice and corn as well as legumes, and their blends are also employed. Understanding of the rheological behavior of doughs is thus necessary related to product development with particular reference to inter-relating dough behavior and their handling characteristics.

2.4.2 SNACKS

Ready-to-eat extrusion-puffed snack foods are a popular commodity in many countries. The bulk density of the snacks sometimes goes below 100 kg/m^3 and possesses a crisp texture and attractive taste. After extrusion cooking, the collets are dried/baked/toasted to less than 4 or 5% moisture content followed by coating with flavors, salt, and oil (Figure 2.3). Higher moisture containing collets may be fried directly after extrusion cooking. A variety of shapes such as collets, balls, curls, tubes, wheels, rings, sticks, and scoops are available. Different flavors like chilly, cheese, barbecue, onion, garlic, etc. are applied during the post-extrusion step to offer an appealing taste. Though corn grit/flour is commonly used, snacks are also produced from potato, wheat, and rice flour along with mixtures of modified and unmodified starches. In another approach, the feed is first cooked in an extruder and then formed in a second extruder followed by drying. The shelf-stable pellets are called semi- or half-products; the product may be puffed separately or even be puffed/cooked prior to consumption by consumers employing a baking/microwave oven. The dual or co-extrusion involves the simultaneous extrusion of two materials forming a single product; the meat paste filled pillow shaped snack is an example.

The blending of cereals and pulses/oilseeds are often practiced to improve the flavor and taste of the product or increase the protein content. Another advantage is the inactivation of the anti-nutritional factors (ANF) present in the pulses/oilseeds, and inactivating some of the enzymes such as urease in soybean and lipase in rice bran. Several pulses/oilseeds have been subjected to extrusion cooking like green gram and Bengal gram (chickpea) (Bhattacharya, 1997; Bhattacharya and Prakash, 1994), kidney bean, cowpea (Van der Poel et al., 1992), lentil, and faba bean (Lombardi et al., 1991), pinto bean, pigeon pea, defatted soybean concentrate/isolate, etc. (Rampersad et al., 2003). These products are suitable for nutritional intervention programs.

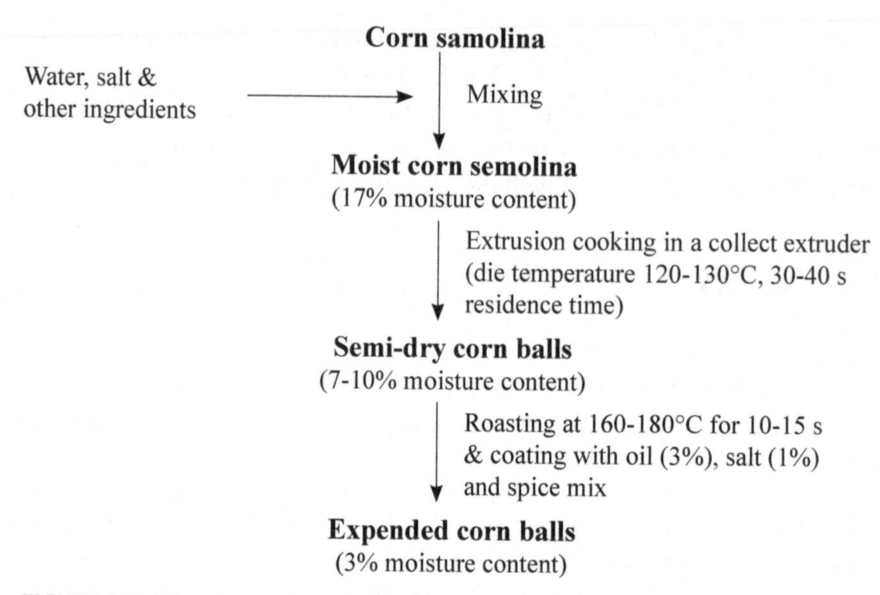

FIGURE 2.3 Flowchart for the preparation of spiced ready-to-eat cornballs.

2.4.3 TEXTURIZED VEGETABLE PROTEIN (TVP)

A single-screw extruder operating at a high temperature and pressure is employed for the production of TVP and meat extenders. The extrudates are dried to safe moisture of about 6% by using a multi-pass belt conveyor dryer. After hot water hydration, the product is used to extend/replace meat for use as pizza toppings, meat sausages and for developing fabricated food formulations. The other type of product is the meat analogs that can be used in place of meat. These products exhibit fiber formation due to extrusion cooking of defatted soybean flour and consequent alignment when passing through the restriction or die. The desirable proximate composition of defatted soy flour (DSF) to have a minimum protein content of 45 to 50% and less than 1% fat along with a high nitrogen solubility index (NSI) between 50 and 70. The addition of a small quantity of cereal flour in the feed is helpful to obtain a faster rehydrating product. The other ingredients that can be added are permitted colors, flavors, minerals, emulsifiers, and cross-linking enhancers such as elemental sulfur and calcium chloride.

2.4.4 BREAKFAST CEREALS

Corn and wheat are the main commodities for the manufacture of breakfast cereals though other grains like rice, oat, sorghum or a combination of grains like multi-grain cereals are also used (Eastman et al., 2001). The advantages of multigrain combinations include the possibility of diverse texture, micronutrient content, and lowering the cost of extrudate feed. Extruded corn flakes, sweet in taste and/or flavored with chocolates, are popular among the consumers. Breakfast cereals made out of whole grain possessing an appropriate level of fiber is now a popular brand.

2.4.5 CONFECTIONERY PRODUCTS

The use of cooker extruders in confectionery production has started in the last two decades. Several confectionery processing systems occupy long process lines, large floor space, and are time-consuming processes. Thus, the extruder cookers are employed to reduce the number of individual unit operations, of which some of them are batch processes. The common uncontrollable secondary reactions such as sugar inversion, Maillard reaction, sugar crystallization, enzymatic changes, and starch gelatinization can be controlled efficiently in an extruder cooker (Frame, 1994). Micronizing of cocoa beans before extrusion cooking helps shell removal due to steam generation from the cotyledon. The surface heating reduces the microbiological count without allowing excess fat migration into the shell. Powder feeding hoppers, in addition to air cannons, prevent bridging/funneling, which is a major confectionery ingredient problem.

Starches are frequently used in many confections such as licorice, toffees, gums, jellies, and pastes. Starches are also present in natural confectionery materials such as cocoa. During the preparation of starch confections, preconditioners can provide up to 80% of the thermal input to make marked reductions in the extruder frictional input demand. Pre-conditioners increase the extruder capacity and reduce its wear. However, the viscosity of the feedstock increases, and they reduce back leaking from the barrel when using steam injection in the case of gums and jellies. They also provide good control over the temperature and moisture content in the extruder barrel. Desirable flavor development, removal of bitter aftertastes, and textural benefits are also possible.

No single cooker extruder is suitable to produce all confectionery products. Single- and twin-screw extruders produce entirely different results and products even when using the same formulation and screw profile. The cooker extruder can perform many processing jobs such as gelatinization, texturization, heating, cooling, melting, blending, and deciding the size and shape. The output of the extruder cooker is higher than similar equipment of the same price for licorice production. A case study for the production of licorice is presented by Frame (1994).

2.5 QUALITY CHARACTERIZATION OF EXTRUDED PRODUCT

Different methods and measurements are employed to control the quality of extruded foods; the selection of the tests depends on the finished product. The common tests are related to sensory attributes and physical, physicochemical, nutritional, and structural characteristics. These quality characteristics are dependent on the composition of feed, type of product, and process variables. In specific, the indices like torque, specific mechanical energy (SME) input, and rheological status are the conventional parameters measured during extrusion. The physical, physicochemical, and sensory characteristics of the product include the extent of expansion, water absorption and water holding capacities, textural features, density, rheological status of product in water, moisture content, porosity, extent of gelatinization and retrogradation, and sensory attributes like appearance, texture, flavor, and overall acceptability. The nutritional characteristics are the protein and fat contents, energy content, mineral, and vitamin contents, ANF, residual enzyme activity, starch, and protein digestibility, amino acid balance, microbial status, etc. The morphological/microstructural features of the extruded products are size and shape of air cells, the thickness of cell wall, and fiber formation and alignment (Dubey and Bhattacharya, 2015).

2.6 FUTURE AREAS OF RESEARCH AND DEVELOPMENT

Scale-up and modeling are the subjects that provide an in-depth knowledge of the subject and the processing technologies. General predictive modeling on food extrusion is difficult because ingredients are diverse and can vary considerably. Thus, modeling tends to be product specific, and the new product development is towards the trial and error approach. An example may be cited concerning the production of extruded corn flakes wherein the equipment manufacturers recommend different methods. Further, their machines usually vary in terms of screw design particularly the production capacity, screw profile, power transmission, screw tip/barrel clearance, die geometry, etc. making the task of scale-up difficult.

The principle of additive manufacturing (AM) involving 3D printing has a good potential to design and fabricate complex food geometries, attractive texture, desired nutritional status and superior consumer acceptance of the product (Godoi et al., 2016). The 3D chocolate products can be developed using the principle of melting extrusion. The other example is 3D hydrogel with consecutively deposited layers that may vary in color and taste. The technology is expected to boost the innovation in the area of extruded product development.

The turbo extrusion technology operates with a very short residence time (less than 10 s) compared to conventionally used twin-screw extruders having a higher residence time (20–60 s). The claims of turbo-extrusion technology include the development of product having improved texture, and the ability to handle an extruder feed with higher fat and sugar contents (Heinz, 1996; Seker, 2011).

The other new areas of extrusion technology are (a) extruders to serve as an alternative processing method to manufacture different traditional foods that are appearing to be difficult to mechanize at present in an hygienic way, (b) extrusion with supercritical fluids, (c) high moisture extrusion of muscle protein, (d) use of extruder cooker as a continuous reactor, (e) replacing commonly used processing systems (such as baking and frying technology) by extrusion technology, (f) creating new foods with unique sensory features, and (g) automation in extrusion processing. It is expected that this proven technology will bring some more new food products having innovative features and nutritional superiority.

KEYWORDS

- **anti-nutritional factors**
- **defatted soy flour**
- **nitrogen solubility index**
- **rice flour**
- **specific mechanical energy**
- **texturized vegetable protein**

REFERENCES

Altan, A., McCarthy, K. L., & Maskan, M., (2009). Effect of extrusion cooking on functional properties and *in vitro* starch digestibility of barley-based extrudates from fruit and vegetable by-products. *Journal of Food Science, 74*(2), 77–86.

Alvarez, V. B., Smith, D. M., Morgan, R. G., & Booren, A. M., (1990). Restructuring of mechanically deboned chicken and nonmeat binders in a twin-screw extruder. *Journal of Food Science 55*(4), 942–946.

Ascheri, J. L. R., Bernal-Gomez, M. E., & Carvalho, C. W. P., (1998). Production of snacks from mixtures of rice and sweet potato flours by thermoplastic extrusion. I. Chemical characterization, expansion index and apparent density. *Alimentaria, 293*, 71–77.

Balandrán-Quintana, R. R., Barbosa-Cánovas, G. V., Zazueta-Morales, J. J., Anzaldúa-Morales, A., & Quintero-Ramos, A., (1998). Functional and nutritional properties of extruded whole pinto bean meal (*Phaseolus vulgaris* L.). *Journal of Food Science 63*(1), 113–116.

Bhattacharya, S., (1997). Twin-screw extrusion of rice-green gram blend: Extrusion and extrudate characteristics. *Journal of Food Engineering, 32*, 83–99.

Bhattacharya, S., (2012). Raw materials for extrusion cooking. In: Maskan, M., & Altan, A., (eds.), *"Advances in Food Extrusion Technology"* (pp. 69–85). CRC Press, Boca Raton, Florida, USA.

Bhattacharya, S., & Prakash, M., (1994). Extrusion of blends of rice and chickpea flours: A response surface analysis. *Journal of Food Engineering, 21*, 315–330.

Cai, W., & Diosady, L. L., (1993). Model for gelatinization of wheat starch in a twin-screw extruder. *Journal of Food Science, 58*(4), 872–875.

Chang, Y. K., Silva, M. R., Gutkoski, L. C., Sebio, L., & Da Silva, M. A. A. P., (1998). Development of extruded snacks using Jatoba (*Hymenaea stigonocarpa Mart*) flour and cassava starch blends. *Journal of the Science of Food and Agriculture, 78*, 59–66.

De Pilli, T., Severini, C., Baiano, A., Derossi, A., Arhaliass, A., & Legrand, J., (2005). Effects of operating conditions on oil loss and properties of products obtained by co-rotating twin-screw extrusion of fatty meal: Preliminary study. *Journal of Food Engineering, 70*, 109–116.

Della Valle, G., Boché, Y., Colonna, P., & Vergnes, B., (1995). The extrusion behavior of potato starch. *Carbohydrate Polymers, 28*(3), 255–264.

Ding, Q. B., Ainsworth, P., Plunkett, A., Tucker, G., & Marson, H., (2006). The effect of extrusion conditions on the functional and physical properties of wheat-based expanded snacks. *Journal of Food Engineering, 73*(2), 142–148.

Ding, Q. B., Ainsworth, P., Plunkett, A., Tucker, G., & Marson, H., (2005). The effect of extrusion conditions on the physiochemical properties and sensory characteristics of rice-based expanded snacks. *Journal of Food Engineering, 66*(3), 283–289.

Dubey, R. K., & Bhattacharya, S., (2015). Extrusion processing of foods. In: Bhattacharya, S., (ed.), *"Conventional and Advanced Food Processing Technologies"* (pp. 75–97). Wiley, Oxford, UK.

Eastman, J., Orthoefer, F., & Solorio, S., (2001). Using extrusion to create breakfast cereal products. *Cereal Foods World, 46*(10), 468, 470–471.

Fletcher, S. I., Richmond, P., & Smith, A. C., (1985). An experimental study of twin-screw extrusion-cooking of maize grits. *Journal of Food Engineering, 4*(4), 291–312.

Frame, N. D., (1994). In: Frame, N. D., (ed.), *The Technology of Extrusion Cooking* (pp. 52–72). Blackie, London.

Godoi, F. C., Prakash, S., & Bhandari, B. R., (2016). 3d printing technologies applied for food design: Status and prospects. *ournal of Food Engineering, 179*, 44–54.

Guy, R. C. E., (2001). *Raw Materials for Extrusion Cooking: Chapter 2* (pp. 5–28). Woodhead Publishing Ltd, Cambridge, UK.

Hagenimana, A., Ding, X., & Fang, T., (2006). Evaluation of rice flour modified by extrusion cooking. *Journal of Cereal Science, 43*(1), 38–46.

Harper, J. M., (1981). *Extrusion of Foods* (Vol. 2, pp. 89–112). CRC Press, Boca Raton, Florida, USA.

Heinz, S., (1996). *Cooker-Extruder Apparatus and Process for Cooking-Extrusion of Biopolymers*. US Patent 5567463.

Konstance, R. P., Onwulata, C. I., Smith, P. W., Lu, D., Tunick, M. H., Strange, E. D., & Holsinger, V. H., (1998). Nutrient-based corn and soy products by twin-screw extrusion. *Journal of Food Science, 63*(5), 864–868.

Lombardi, B. G., Di Lullo, G., & Carnovale, E., (1991). *In-vitro* iron dialysability from legumes: Influence of phytate and extrusion cooking. *Journal of the Science of Food and Agriculture, 55*(4), 599–605.

Molina, M. R., Braham, J. E., & Bressani, R., (1983). Some characteristics of whole corn: whole soybean (70:30) and rice: whole soybean (70:30) mixtures processed by simple extrusion cooking. *Journal of Food Science, 48*(2), 434–437.

Noguchi, A., (1990). Recent research and industrial achievement in extrusion cooking in Japan. High temperature/short time (HTST) processing, guarantee for high-quality food with long shelf-life. In: Zeuthen, P., Cheftel, J. C., Eriksson, C., Gormley, T. R., Linko, P., & Paulus, K., (eds.), *"Processing and Quality of Foods"* (pp. 1203–1214). Elsevier Applied Science, London.

Rampersad, R., Badrie, N., & Comissiong, E., (2003). Physico-chemical and sensory characteristics of flavored snacks from extruded cassava/pigeon pea flour. *Journal of Food Science, 68*(1), 363–367.

Seker, M., (2011). Extrusion of snacks, breakfast cereals and confectioneries. In: Maskan, M., & Altan, A., (eds.), *Advances in Extrusion Technology*, (pp. 168–208). CRC Press (Taylor and Francis), Boca Raton.

Stojceska, V., Ainsworth, P., Plunkett, A., & Ibanoglu, S., (2009). The effect of extrusion cooking using different water feed rates on the quality of ready-to-eat snacks made from food by-products. *Food Chemistry, 114*, 226–232.

Van den Hout, R., Jonkers, J., Van Vliet, T., & Van Zuilichem, D. J., (1998). Influence of extrusion shear forces on the inactivation of trypsin inhibitors in soy flour. *Food and Bioproducts Processing, 76*(3), 155–161.

Van der Poel, A. T. F. B., Stolp, W., & Van Zuilichem, D. J., (1992). Twin screw extrusion of two cowpea varieties: Effect of temperature and moisture level on antinutritional factors and protein dispersibility. *Journal of Science, Food and Agriculture, 58*, 83–87.

Wang, N., Bhirud, P. R., Sosulski, F. W., & Tyler, R. T., (1999). Pasta-like product from pea flour by twin-screw extrusion. *Journal of Food Science, 64*(4), 671–678.

Zardetto, S., & Rosa, M. D., (2009). Effect of extrusion process on properties of cooked, fresh egg pasta. *Journal of Food Engineering, 92*, 70–77.

CHAPTER 3

Non-Thermal Plasma (NTP) Applications for Food Decontamination Technology

R. V. PRASAD, R. F. SUTAR, NUKASANI SAGARIKA, P. DIVYANG, and MAMTA PATEL

Department of Food Quality Assurance, College of Food Processing Technology and Bio-Energy, Anand Agricultural University Anand–388110, Gujarat, India

ABSTRACT

The food industry is in a constant search for new technologies to improve the commercial sterilization process of agricultural commodities. Safety of food is one of the major concerns prevailing for the food industries, regulatory bodies, and customers. Microorganisms and pathogens that cause spoilage are a major problem concerning the food processing industries as they have an unfavorable impact on the health. Thermal processing of food is the most common and effective method for eliminating pathogens and pesticides residues. Though the thermal processing method is inexpensive, environmental friendly, and preservative free, it results in some undesirable changes related to the nutritional and organoleptic properties of foods. Plasma treatment is one of the novel and efficient technique for pesticide and microbial load removal from agricultural product surfaces. Pulsed electric field (PEF), High-pressure processing and advanced oxidation processes such as ultrasound, ozone, ionizing radiation, non-thermal plasma (NTP), etc. are some of the novel non-thermal technologies. A considerable number of studies have reported the efficacy of non-thermal cold plasma for the breakdown of pesticides and pathogens/microbial population in food and water; however, information regarding the toxicity of breakdown products is limited. In this chapter, we provide a critical analysis of the literature of cold plasma generation system, their types, and their effect on pesticide and microorganisms residues during non-thermal processing of fruits and vegetables. We also identify the opportunities for further development and provide guidelines for future research.

3.1 INTRODUCTION

Food spoilage and food poisoning are mainly caused by microorganisms, and usually, exist in processed as well as unprocessed foods. Because of these issues, controlling the presence of microbes have become an important factor to ensure safety and shelf life of foods. Food-borne illness, which is caused mainly by the consumption of food contaminated with pathogenic bacteria, is a worldwide concern. The presence and subsequent growth of microorganisms in food in addition to improper storage not only results in spoilage but also leads to a reduction of food quality. The microbial safety in ready to eat products is a cause of big concern not only for the consumers and food industries but also for the regulatory agencies. Despite the application of good hygienic practice and good manufacturing practice during the post-harvest period, occurrences of high initial loads of pathogenic bacteria increase the risk of outbreaks of foodborne diseases. The number of documented outbreaks of foodborne diseases has increased in recent years with *Salmonella* spp., *Listeria Monocytogenes*, and *Escherichia coli* being responsible for the largest number of outbreaks and deaths.

The fresh fruits and vegetables are often consumed without washing or with minimal processing. At present preference is being given to pesticide-free, organic fruits, and vegetables in both developed and developing countries. However, only a minor population is willing to pay for such produce. Guidelines for pesticide applications to standing crops and appropriate harvest times do exist. At the same time, the use of pesticides is poorly regulated and often dangerous. In developing countries, the easy availability of pesticides is becoming a popular method for self-harm. In addition to that, harvests are often rushed to the market, leaving large amounts of pesticide residues on fruits and vegetables (Chen et al., 2013).

Household food processing operations such as washing and peeling in general, are effective at eliminating pesticide residues to a great deal which also witnessed in the case with conventional industrial-scale food processing reduced a load of pesticides during handling and following common food processing operations at household as well as industrial scale. It is, therefore, necessary to monitor the pesticide residues on fruit and vegetables, because they can put human health at greater risk of various diseases. Probable risks for the environment and for humans due to the widespread use of pesticides is described in Figure 3.1.

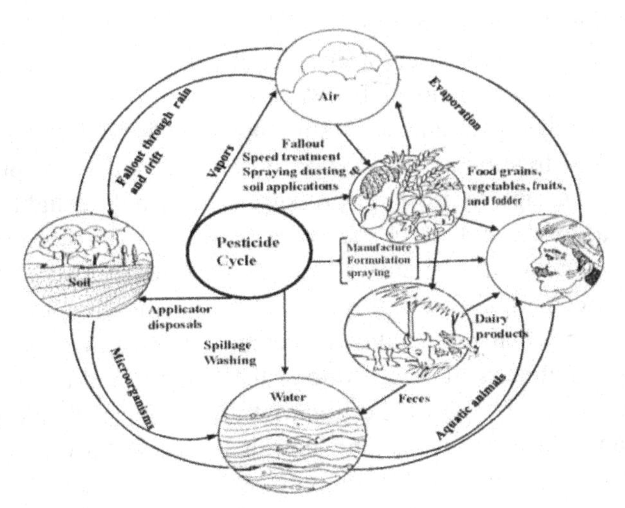

FIGURE 3.1 Risk caused to the environment and human due to the usage of pesticides. (*Source:* Reprinted with permission from Randhwaa, M. A., Ahmed, A., Javed, M.S. (2014). © 2014 Elsevier.)

Many researchers have reviewed the effects of many common unit operations such as baking, fermentation, malting, milling, drying, and thermal processing of agricultural produce. Plasma treatment may offer a novel and efficient method for pesticide removal from agricultural product surfaces (Bonnechere et al., 2012; Kaushik et al., 2009; Keikotlhaile et al., 2010; and Bajwa and Sandhu, 2014).

Commercially, fresh fruits and vegetables need to be decontaminated prior to packaging. The aim of controlling pesticide residues is to protect consumer health from their possible side effects. Several chemical and physical technologies have found to be efficient in reducing bacterial contamination in fresh produces (Parish et al., 2003). Chlorination (50–200 mg) of fresh produce is commonly used by various manufacturers. However, due to increase in safety concerns regarding the formation of potentially carcinogenic chlorinated compounds in water, and its limited efficiency in reducing foodborne pathogens on fresh produce, alternative methods have been sought out by the food industry that can ensure safety and environment friendly at the same time (Baur et al., 2004; Oliveira et al., 2012).

Thermal processing of food not only leads to undesirable changes in food materials but also diminishes the content or bioavailability of some bioactive compounds such as ascorbic acid (AA), phenolic compounds or carotenoids are severely diminished. Other chemical technologies, including washing with organic acids (e.g., citric, and ascorbic), hydrogen peroxide, and application of ozone are also available. However, limitations of these techniques, such as low direct antibacterial activity, pH dependence, and influence on sensory parameters been reported.

The food industry is currently studying non-thermal techniques such as ozone based treatments, irradiation, ultraviolet radiation, pulsed electric field (PEF), high hydrostatic pressure (HPP), pulsed light (PL), cold plasma, ultrasounds, and novel packaging practices. All these technologies have great potential in the field of minimally processed (MP) foods. Also, non-thermal technologies are preservation treatments that are effective at ambient or sub-lethal temperatures, thereby not affecting the nutritional and quality parameters of food. Among the non-thermal techniques, cold plasma technology is one of the minimal processing technologies which has got many benefits. At present, non-thermal plasma (NTP) has been considered as the novel non-thermal processes for decontamination of fresh produce (Lee et al., 2006; Critzer et al., 2007; Niemira and Sites, 2008). This is because plasma generates many different reactive species, such as O, O_2^+, O_3, OH^-, NO, NO_2, which have strong oxidative effects that can help remove contaminants, such as microbes and pesticides, from the surface of food products (Heo et al., 2014; Tiwari et al., 2009; Ramos et al., 2013; Niemira, 2008).

3.2 PLASMA SCIENCE AND TECHNOLOGY

The term 'plasma' usually refers to a partially or wholly ionized gas with a net neutral charge. It is generally composed of photons, ions, and free electrons as well as atoms in their fundamental or excited states. It is considered as the fourth state of matter in the world. It can be generated in a wide range of temperature and pressure by means of coupling energy to the gaseous medium. The generated plasma energy can be thermal, mechanical, nuclear, radian, or carried by an electric current. These energies dissociate the gaseous molecules into a collection of ions, electrons, charge–neutral gas molecules and other species. Density and temperature of the electrons are changed based on the type of energy supply and amount of energy transferred to the plasma (Misra et al., 2011).

Plasma is classified into two types, namely thermal and NTP based on the relative energetic levels of electrons and heavy species of the plasma.

Very less power is required to obtain NTP at atmospheric pressures and at ambient temperatures of 30–60°C. The typical explanation for plasma generation at atmospheric pressure include the corona discharge, dielectric barrier discharges (DBD), radio-frequency plasmas (RFP), and the gliding arc discharge. But for thermal plasma which is being generated at higher pressures, require high power, and also a thermal equilibrium should exist between the electrons and the heavy species. Plasma generation at atmospheric pressure is of particular interest in food processing industries both technically and industrially, as this does not require any extreme conditions. Cold plasma systems intended for use in food processing fall generally into three categories, and they are briefly described below.

3.2.1 REMOTE TREATMENT PLASMA

In remote treatment, cold plasma systems, plasma generated from a source is moved onto the surface of the objects to be treated. The plasma is mainly driven by a flow of the feed gas or manipulated by magnetic fields. The main advantage of this type of system is that the surface to be treated is placed at a physically separate point of generation (Chirokov et al., 2005). This simplifies the design as well as operation of the device, and increases the flexibility with respect to the shapes and sizes of objects to be treated.

3.2.2 DIRECT TREATMENT PLASMA

In the direct treatment plasma systems, the plasma generation equipment supplies active plasma directly on to the object to be treated and provides higher concentrations of active agents. Here the target is relatively close to the site of cold plasma generation and exposed to the plasma before active species recombine and are lost (Laroussi and Lu, 2005). Systems of this type can operate in pulsed mode, with plasma generated at pulse frequencies of hundreds or thousands of times per second. As there is little or no intervening normal atmosphere between the plasma generation apparatus and the target object, the level of UV radiation generated by recombination is relatively high at the treated surface. An important factor is that at sufficiently high voltages, commodities with high water activity and internal moisture content can be made to conduct electricity. Direct treatment cold plasma systems must be designed to exacting specifications to avoid concentrated electrical conduction through the product to be treated.

3.2.3 ELECTRODES CONTACT TREATMENT PLASMA

In the electrode contact systems, the surface to be sterilized is placed between the two electrodes. In these systems, the product is exposed to the broadest combination of the active antimicrobial agents, at the highest possible intensity of free electrons, radicals, ions, and UV radiation. Electrode contact cold plasma systems have a physical constraint based on the spacing between the electrodes. Although changes in the feed gas composition or the design of one or both electrodes can provide some flexibility, these systems are limited to commodities that physically fit between the electrodes (BenGadri et al., 2000). Systems of this type may be best suited for smaller or flatter commodities, such as nuts, berries, seeds, shell eggs, etc.

3.3 TYPES OF PLASMA GENERATION SYSTEM

3.3.1 GLIDING ARC COLD PLASMA

Plasma generated by ionizing air or gas mixture by virtue of applied voltage is known as gliding arc cold plasma system. The ionization voltage for any gas mixture is determined by the configuration of the distance between the electrodes (the gap width) and the gas pressure between them (Niemira and Gutsol, 2010). In this system, lowering the pressure of the gas also lowers the voltage required to ionize it. Cold plasma systems that operate under reduced pressure, therefore, require less power to generate antimicrobial active plasmas. This fundamental aspect of ionization potentials has led to a set of cold plasma generation technologies that use reduced pressure treatment chambers to generate plasma and further distribute it on to the food surface to be treated. The schematic diagram of gliding arc cold plasma system is shown in Figure 3.2.

3.3.2 MICROWAVE PUMPED COLD PLASMA

In microwave pumped cold plasma, the steps of particle formation in gas phases are supported by the ionization and dissociation of the compounds, as well as a particle charging in the plasma. The variation of particle size is correlated with the concentration of nuclei in the plasma. Cold plasma systems that use microwaves to ionize the treatment chamber gas have the advantage of using technology similar to that found in conventional microwave ovens. This allows for a more rapid technology development curve. In a study using microwave-pumped cold plasma within a one-meter long unit, glass slides treated with *Bacillus subtilis* or *Staphylococcus aureus* were treated for very short times. Reductions of 5.2 and 3.7 log obtained after six seconds of treatment in air at 1.6 mbar (160 Pa) (Tran et al., 2008). A study comparing air, argon, and ammonia as feed gases treated *Aspergillus niger, Bacillus stearothermophilus, B. subtilis*, and *Saccharomyces cerevisiae* in a microwave-pumped cold plasma chamber, evacuated to between 10 Pa and 50 Pa (Feichtinger et al., 2002). Although all three types of gases were ultimately effective in this cold plasma system, air yielded greater log reductions than ammonia or argon. The microorganisms were reduced by 6 logs within 30 seconds (air), 60 seconds (ammonia), and 180 seconds (argon). Thus UV production by oxygen species significantly contributed to the antimicrobial efficacy.

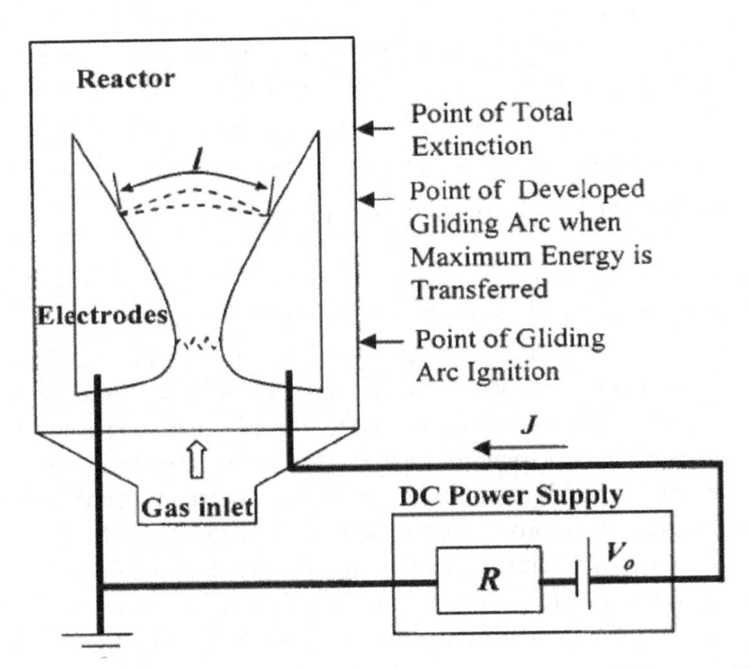

FIGURE 3.2 Schematic diagram of a gliding arc cold plasma system (Kalra et al., 2005).

3.3.3 RADIO FREQUENCY (RF)-DRIVEN COLD PLASMA

Radio frequency (RF) cold plasma systems ionize gases through the use of rapidly cycling electrical impulses, operating at various power and voltage settings. Along with microwave-based systems, the technologies associated with RF electronics have been in use for many years. Frequencies for these systems can range from several Hz to high MHz. Lassen et al., (2005) used an RF system operating at (13.56 MHz) to treat *B. stearothermophilus* spores. Various atmospheres were tested: O_2, Ar: H_2 (50%:50%, 15%:85%, 25%:75%, and 5%:95%), O_2:H_2 (50%:50% and 95%:5%) and O_2:CF_4 (88%:12%). RF-driven cold plasma systems can take different forms. For example, the system can be arranged to place the treated food between the electrodes, forming an electrode contact system. One example is a study where, Almonds inoculated with E. coli treated with cold plasma from a dielectric barrier discharge apparatus operating at frequencies of 1.0–2.5 kHz (Deng et al., 2007). The plasma feed gas was air, and the almonds were positioned in the space between the dielectric material and the ground electrode. A treatment for 30 s at 25 kV, 2.0 kHz yielded reductions of 4 logs, and treatment at a higher frequency, 2.5 kHz, resulted in a 5-log reduction. Alternatively, the RF-driven electrodes may hold above the food product, with the cold plasma blown onto the surface to be treated. An example of this type of cold plasma system is the plasma jet.

3.3.4 *ATMOSPHERIC PRESSURE COLD PLASMA*

The preceding section discussed cold plasma systems that depend on reduced partial pressure to facilitate ionization and plasma generation. It is important to note that, apart from the limitations on throughput imposed by batch processing, not all food commodities can tolerate vacuum conditions. Cold plasma systems that operate at one atmosphere (1 bar, 100 kPa) do not require air-tight vacuum chambers and are therefore simpler to build from that standpoint. Without the need for vacuum-tight chamber doors or gaskets, material to be processed can be moved through a treatment zone via conveyor. However, ambient pressure processing imposes new challenges for cold plasma systems because of the difficulty of ionization.

3.4 MICROBIAL INACTIVATION BY NTP

Cold plasma is one of the latest technologies used for destruction or inactivation of microorganisms (Sharma, 2009). The destruction of microorganisms during plasma treatment occurs as the microorganisms are exposed to radical bombardment on the surface of the cell with great intensity. The bombardment of the radicals causes damage on the surface of the cell which the microorganisms are unable to repair quickly. This further leads to the destruction of the living cell rapidly. This is termed as "etching" (Pelletier, 1992). The phenomenon of lesion formation is because of accumulation of electrostatic forces on the exterior surface of the living cell. The efficiency of NTP mainly depends upon two factors, i.e., substrate type and characteristics of microorganisms, which includes load, type, and physiological state of microorganisms (Stratakos and Koidis, 2015).

Different researchers studied the strength of cold plasma on microbial reduction in surface of fresh fruits and vegetables, legumes and pulses, potatoes as well as from plastic trays, paper cups and aluminum foil, and noted significant reduction in microbial population such as *Escherichia coli, Salmonella, Listeria monocytogens, Aspergillus parasiticus, Penicillium,* and *Erwinia carotovora,* respectively (Perni et al., 2008; Selcuk, 2008; Moreau, 2007; Yun, 2010). Studies have been conducted to disinfect *Salmonella* on fresh fruit and vegetable slices using direct current, atmospheric-pressure air cold plasma micro-jet (PMJ). The results showed that PMJ showed effective inactivation of *Salmonella* after 2 min of plasma treatment. The physiochemical properties of the slices before and after plasma treatment found to be within an acceptable range. Reactive oxygen species, the major bactericidal agents in plasma are detected by electron spin resonance spectroscopy and optical emission spectroscopy.

In another study of decontaminating and inactivating pathogens from the surface of fruits and vegetables, bacterial inactivation about 90% in carrot slices, 60% in cucumber slices, 40% in pear slices observed in half second treatment using a direct-current atmospheric-pressure cold plasma and they concluded that treatment of 4 seconds is sufficient to inactivate all pathogens from the surface of fruits and vegetables (Wang et al., 2012). In a similar study of apples, mangoes, and melon slices inactivation of all pathogens from the surface of treated slices achieved in a matter of minutes only using gliding arc cold plasma (Perni et al., 2008; Niemira and Sites, 2008).

Vleugels (2005) used the biofilm-forming bacterium *Pantoeaagglomerans* on bell peppers to show that atmospheric He-O_2 plasma is an effective inactivation agent without causing unacceptable levels of discoloration to the peppers, and that they are superior to low-pressure ultraviolet sources.

In a similar study, to eliminate pathogenic fungal contamination such as *Aspergillus parasiticus* and *Penicillium* spp. from the surface of hazelnut, peanut, pistachio, and seed surfaces using low-pressure cold plasma (LPCP) with sulfur hexafluoride (SF_6) and LPCP with air gas mixture. Also, the effectiveness of plasma against aflatoxins was also determined. The results showed that plasma treatment of 5 min with air gases resulted in 1-log reduction of *A. parasiticus* whereas in plasma treatment of 5 min with SF_6 resulted in 5-log reduction in fungal population (Barsan et al., 2008). A reduction of 3-log for both fungal species *A. Parasiticus and Penicillumspp* from seed surface was achieved with a 15-minute treatment of LPCP with SF_6 mixture (Selcuk, 2008). Also, LPCP–SF_6 plasma treatment worth 20minutes resulted in a 20% reduction in total aflatoxin content while 50% reduction in total aflatoxin content achieved using LPCP–air gas plasma treatment (Barsan et al., 2008).

Also, in the year 2009, Song et al., evaluated the efficacy of atmospheric pressure plasma on *Listeria monocytogenes* and *Listeria innocua*, sliced cheese, sliced ham and sliced ready to eat meat inoculated by different strains of Listeria spp. The experiment was performed with plasma parameters variations in input power of 50–150 W, and exposure time of 60–90 sec. More than eight log reductions achieved in 120 sec at 150 W. In contrast, reduction of strains after 120 sec ranged from 0.25 to 1.73 log CFU/g were observed in sliced ham. The exposure time required for inactivating 90% of *Listeria monocytogenes* population, from the survival curves of 75, 100, 125, 150W of APP treatments were 71.43, 62.50, 19.65, and 17.27 sec for *Listeria monocytogenes* in sliced cheese, and 476.19, 87.72, 70.92, and 63.69 sec for sliced ham. Similarly, Rod et al., (2012) observed reduction of *L. innocua* ranging from 0.8–0.4 to 1.6–0.5 log CFU/g with no significant effects of time and intensity when multiple treatments at 15.5 and 62 W for 60 sec were conducted.

Gurol et al., in the year 2012 studied atmospheric corona discharge of low-temperature plasma (LTP) system and evaluated its ability for killing *Escherichia coli* in milk with different fat contents. Plasma was applied at a regular time intervals of 0, 3, 6, 9, 12, 15 and 20 min and a significant 54% reduction in the population of *E. coli* cells after only 3 min was observed regardless of the fat content of the milk. The initial pre-plasma bacterial count of 7.78 logs CFU/ml in whole milk decreased to 3.63 log CFU/ml after 20 min of plasma application. No viable cells detected after one-week examination in whole milk samples and remained for over six-week storage period.

Furthermore, scientists investigated the effects of the growth phase, growth temperature, and chemical treatment regime on the inactivation of *S. typhimurium* by nitrogen cold atmospheric plasma (CAP). The efficacy of CAP treatment for decontaminating the surfaces of lettuce, strawberry, and potato inoculated with *S. typhimurium* was evaluated. They found that the rate of inactivation of *S. typhimurium* was independent of the growth phase, growth temperature, and chemical treatment regime. Under optimal conditions, 2 min treatment resulted in a 2.71 log-reduction of *S. typhimurium* viability on membrane filters whereas a 15 min treatment was necessary to achieve 2.72, 1.76 and 0.94 log-reductions of viability on lettuce, strawberry, and potato respectively (Fernández et al., 2013).

An experiment conducted using dielectric barrier discharge atmospheric cold plasma against *Escherichia coli, Salmonella enteric* and *Listeria monocytogenes* inoculated on cherry tomato and strawberry surfaces for evaluating the efficacy of the plasma system. ACP treatment for 10, 60 and 120 sec resulted in a reduction of *Salmonella, E. coli,* and *L. monocytogenes* populations of tomato to undetectable levels from initial populations of 3.1, 6.3, and 6.7 log10 CFU/sample, respectively. Treatment time for 300 sec resulted in a reduction of *E. coli, Salmonella, and L. monocytogenes* populations by 3.5, 3.8, and 4.2 log CFU/sample, respectively (Ziuzina et al., 2013). Pasquali in the year 2016 also observed a significant reduction of 1.35 log MPN/cm^2 for E. coli after 15 min of treatment and approximately 2-log CFU/cm^2 for *L. monocytogens* after 30 min of treatment in radicchio leaves (Chicory).

Zhang et al., in the year (2013) experimented with Low-pressure cold oxygen plasma for decontamination of *Salmonella enterica subsp. Enterica serovar typhimurium* str. LT2 (*Salmonella typhimurium* LT2) on spinach, lettuce, tomato, and potato surfaces have shown oxygen plasma was more effective than washing with 3% H_2O_2 on eliminating *S. typhimurium* LT2 on green leafy vegetables. Tomato wax cuticle layer found to be chemically changed, which they justified with oxidation reaction and decomposition of carbon chains, which readily and completely be removed by water.

Later in the year 2014, Kim et al., conducted an experiment for decontamination of *Aspergillus flavus* and *Bacillus cereus* spores using microwave powered cold plasma with varying treatment time on red pepper powder. The results showed that a reduction of 2.5 ± 0.3 log spores/g and 3.4 ± 0.7 log spores/g was observed by the cold plasma treatments with nitrogen at 900 W and 667 Pa for 20 min and 30 min, for *Aspergillus flavus and Bacillus cereus* spores, respectively. In another research, strawberries were treated using 60 kV Dielectric Barrier Discharge Atmospheric Cold Plasma, where strawberries were kept inside a sealed package containing ambient air (42% relative humidity) and treated for 5 min plasma treatment. Results revealed 2-log reduction of aerobic mesophilic bacteria, yeast, and mold within 24 h of post-ACP treatment. They also studied decontamination of strawberries inside a sealed package with two different gas mixtures of 65%O_2 + 16% N_2 + 19% CO_2 and 90% N_2 + 10% O_2 using DBD cold plasma and achieved reduction in microflora population by 3-log CFU/g from an initial count of 5 log CFU/g after 300 sec treatment, mentioning that gas mixtures did not play any role in the inhibition of microflora (Misra et al., 2014a).

Also in the year 2014, Baier studied on atmospheric pressure plasma-jet on corn salad, cucumber, apple, and tomato at fixed power of 8W and gas flow of Ar mixed with 0.1% O_2. The gas flow rate of 5 l/min for 60 sec was applied keeping the plasma jet at a distance of 17 mm from treatment surface. After treatment antibacterial tests were performed on corn salad, cucumber, apple, and tomato and reduction in *Escherichia coli* achieved by 4.1 ± 1.2, 4.7 ± 0.4, 4.7 ± 0, and 3.3 ± 0.9 log units, respectively, after 60 sec treatment time.

3.5 EFFECTS OF NTP ON PESTICIDES

Food is the basic necessity of life and food contaminated with toxic pesticides is associated with severe effects on the human health. Hence it is pertinent to explore strategies that address this situation of food safety especially for the developing countries where pesticide contamination is widespread due to indiscriminate usage, and a major part of the population lives below poverty line (Kaushik et al., 2009).

Pesticides, namely Azoxystrobin, Cyprodinil, Fludioxonil, and Pyriproxyfen on strawberries analyzed after 60, 70 and 80 kV treatments for 1 to 5 minutes of exposure time using DBD cold plasma, where the levels of azoxystrobin, cyprodinil, fludioxonil, and pyriproxyfen decreased by a maximum of 69%, 45%, 71 and 46% respectively after 5 min of treatment at 80 kV. In a similar study, degradation behavior of diazinon insecticide by air-dielectric barrier discharge (DBD) plasma was investigated where, three different plasma discharge powers (0.4, 0.15, and 0.75 W) were applied to the cucumber surfaces for 10 minutes wherein results they observed an increase in plasma discharge power increases pesticide reduction rates. Optimum plasma treatment time and power are found to be 10 min at 0.75 W, respectively (Mishra et al., 2014b; Dorraki et al., 2016).

In the year 2010, Bai et al., studied the degradation of dichlorvos and omethoate organophosphorus pesticides sprayed onto maize samples. Also, they demonstrated the successful degradation of dichlorvos pesticides coated on glass slides. Kim et al., (2007) reported the decomposition of paraoxon and parathion with an atmospheric pressure and radio-frequency plasma.

In similar research where degradation of organophosphorus pesticides such as diazinon, and chlorpyrifos in apples and cucumbers were studied using Dielectric Barrier Discharge (DBD) plasma treatments for different time intervals ranging from 2 minutes to 10 minutes with 10 and 13 kV voltage. Apple and cucumber slices of (3 × 3 cm square, 0.5 cm thick) were cut with a knife and were dipped into pesticides solution having different concentrations of 500 and 1000 ppm for 2 minutes and allowed to air dry. Every sample exposed to cold plasma introduced to the cold plasma Monopole DBD device with two voltages, 10 and 13 kV, for 2, 4, 6, 8, and 10 min of the time interval. After treatments, residual pesticides extracted from the treated samples using Liquid-Liquid extraction known as solvent extraction and partitioning were analyzed using GC-MS and capillary column (non-polar) Hp-5MS (Phenyl dimethylsiloxane) 30 mm in length, 250 mm in inner diameter and 0.25 μm in the thickness of immobile phase. The highest percentage of detoxification of diazinon achieved with cold plasma treatment of 10 min for 500 ppm at 13 kV and 10 kV. This reduced the residue of diazinon to 87.38% and 63.67%, respectively. These values for 1,000 ppm at 13 kV and 10 kV were 16.54% and 18.42%, respectively. Similarly, for chlorpyrifos highest amount of detoxification achieved with cold plasma treatment of 10 min for 500 ppm at 13 kV and 10 kV. This reduced the remaining of diazinon to 86.96% and 32.55%, respectively. These values for 1,000 ppm at 13 kV and 10 kV were 7.53% and 44.33%, respectively (Sayyedeh et al., 2017). Results are shown in graphs (Figure 3.3).

Li et al., (2013) studied, nitenpyram residues degradation using DBD cold plasma from wastewater. Nitenpyram is a neonicotinoid pesticide used in agriculture and for treating domestic animals. The removal of nitenpyram from wastewater is, therefore, a major environmental issue, not only due to its toxicity, but also since, it is difficult to degrade in wastewater. Different treatments with varying time and power supply applied to nitenpyram wastewater sample and reduction in the final pesticide contents were calculated. For a discharge time of 180 min and an input power of 80 W, 140 W or 200 W, the degradation of NTP was 66.7%, 75.1% or 82.7%, respectively.

Bai et al., (2009) studied the degradation of dichlorvos and omethoate using O_2 cold plasma. In this study, maize purchased from the market was fortified with dichlorvos and omethoate aqueous solution at normal form concentration by spray. Afterward, the maize samples were exposed to O_2 plasma for treatment times of 30, 60, 90, and 120 s at discharge power levels of 30, 60, 90, and 120 W with different O_2 flux. Each maize sample was withdrawn from the plasma reaction chamber immediately after plasma treatment. Dichlorvos and omethoate residues detected and quantified using gas chromatography (GC) equipped with a flame photometric detector (FPD). Results indicated that the fraction of dichlorvos and omethoate decreases sharply with an increasing plasma treatment time up to 120 s, regardless

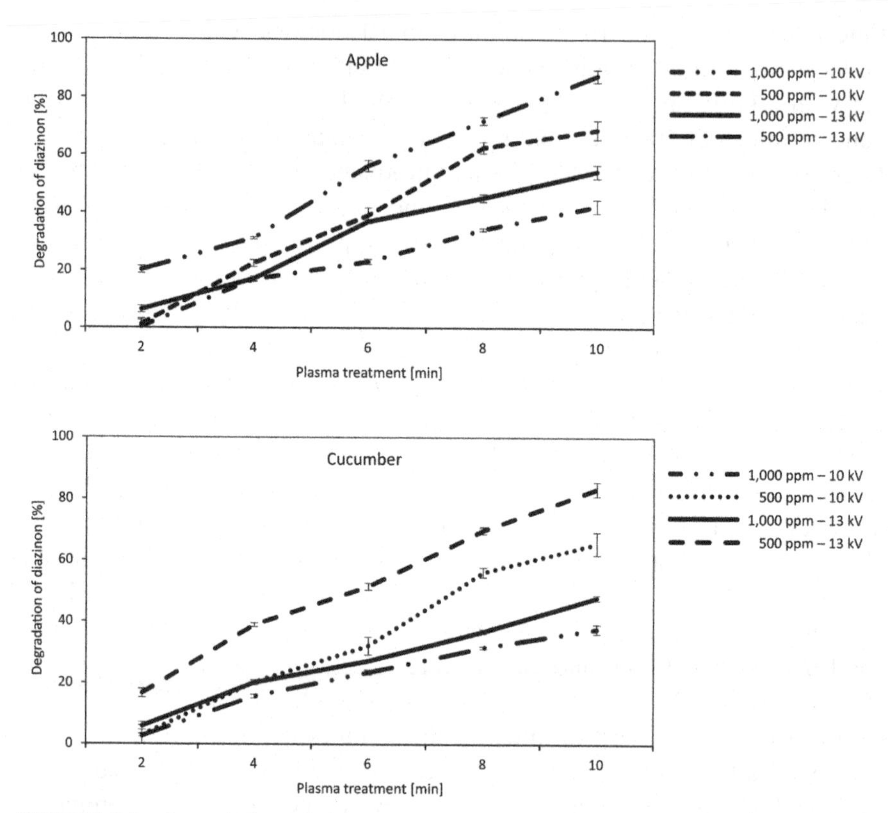

FIGURE 3.3 Degradation of diazinon with plasma treatment. (*Source*: Reprinted with permission from Mousavi et al., 2017).

of the sample position. This indicates that the treatment time is the most influential parameter, and the longer treatment time can speed up the reduction efficiency of pesticide. Almost 90% of the initial amounts of dichlorvos and omethoate removed after 120 s in the discharge zone.

Reddy et al., (2014) used dielectric barrier discharge–NTP with a varying voltage of 14 kV, 16 kV, and 22 kV oxidation process for the mineralization of a pesticide endosulfan from aqueous medium. The typical results showed that the synergy between plasma excitation of endosulfan followed by the catalytic action of cerium oxide, which not only improved the conversion, but also increased the mineralization efficiency. Catalytic plasma approach showed a threefold increase in mineralization. Degradation followed first-order kinetics, and the rate of degradation is proportional to power input and reciprocal to initial endosulfan concentrations.

3.6 LIMITATIONS OF COLD PLASMA TECHNOLOGY IN FOOD PROCESSING

- Key limitations of cold plasma are the relatively early stage of technology development, the variety, and complexity of the necessary equipment, and the largely unexplored impacts of cold plasma treatment on the sensory and nutritional qualities of treated foods.
- Also, the antimicrobial modes of action for various cold plasma systems vary depending on the type of cold plasma generated.

3.7 CONCLUSION

With the rise of the concept of "green consumerism," meeting the consumer demand for nutritious and fresh food other than providing food safety has increased interest in non- thermal preservation methods. Cold plasma is an emerging novel technology in the present era. Cold plasma technology is gaining fame for its unique characteristics like treatment in low or ambient temperature for a short period of time, which helps in retaining the integrity and quality of food products. Cold plasma has proved to be efficient in sanitizing equipment for inactivating the foodborne pathogens as well as pesticides from fresh produce and packaging materials. It also helps in improving certain manufacturing processes, acts as an active packaging technique, and retards browning reaction in fruits and vegetables. For optimization and scale up to commercial treatment levels, it requires a complete understanding of the chemical processes. Nevertheless, this area of technology shows promise and is the subject of active research to enhance efficacy. Being a cold treatment it is effective in retaining the texture, sensory, and functional properties of foods. Thus, cold plasma is a promoting technique for food processing in the near future. The emerging technology, NTP enhances the shelf life and safety of the food by inactivating the microorganisms and degradation of pesticides. Considering the reported results by direct plasma exposure and circulation, it concluded that the confined environment enhances the efficient irradiation of plasma by eliminating airflow. This system can be applied to the store the agricultural products in a fresh manner and also the exclusion of harmful materials on the products. Very limited numbers of studies have demonstrated the successful decontamination of pesticides and microorganisms using NTP. In India, no such work reported so far. There is a need to optimize the suitable non-thermal system which generates effective plasma production for potential cold plasma treatment for the degradation of pesticide residues, decontamination of microorganism on fruits, vegetables, and spices.

KEYWORDS

- **cold atmospheric plasma**
- **ionizing radiation**
- **non-thermal processing**
- **oxidation processes**
- **plasma treatment**

REFERENCES

Bai, Y., Chen, J., Mu, H., Zhang, C., & Li, B., (2009). Reduction of dichlorvos and omethoate residues by O_2 plasma treatment. *Journal of Agricultural and Food Chemistry, 57*(14), 6238–6245.

Bai, Y., Chen, J., Yang, Y., Guo, L., & Zhang, C., (2010). Degradation of organophosphorus pesticide-induced by oxygen plasma: Effects of operating parameters and reaction mechanisms. *Chemosphere, 81*(3), 408–414.

Baier, M., Görgen, M., Ehlbeck, J., Knorr, D., Herppich, W. B., & Schlüter, O., (2014). Non-thermal atmospheric pressure plasma: Screening for gentle process conditions and antibacterial efficiency on perishable fresh produce. *Innovative Food Science & Emerging Technologies, 22*, 147–157.

Bajwa, U., & Sandhu, K. S., (2014). Effect of handling and processing on pesticide residues in food- a review. *J. Food Sci. Technol., 51*, 201–220.

Barsan, P., Basaran-Akgul, N., & Oksuz, L., (2008). Elimination of *Aspergillus parasiticus* from nut surface with low-pressure cold plasma (LPCP) treatment. *Food Microbiology, 25*(4), 626–632.

Baur, S., Klaiber, R., Hammes, W. P., & Carle, R., (2004). Sensory and microbiological quality of shredded, packaged iceberg lettuce as affected by pre-washing procedures with chlorinated and ozonated water. *Innov. Food Sci. Emerg. Technol., 5*, 45–55.

Bonnechere, A., Hanot, V., Bragard, C., Bedoret, T., & Van Loco, J., (2012). Effect of household and industrial processing on the levels of pesticide residues and degradation products in melons. *Food Addit. Contam. Part A, Chem. Anal. Control Expo. Risk Assess., 29*, 1058–1066.

Chen, J. Y., Lin, Y. J., & Kuo, W. C., (2013). Pesticide residue removal from vegetables by ozonation. *Journal of Food Engineering, 114*(3), 404–411.

Chirokov, A., Gutsol, A., & Fridman, A., (2005). Atmospheric pressure plasma of dielectric barrier discharges. *Pure Appl. Chem., 77*(2), 487–495.

Critzer, F., Kelly-Wintenberg, K., South, S., & Golden, D., (2007). Atmospheric plasma inactivation of foodborne pathogens on fresh produce surfaces. *Journal of Food Protection, 70*(10), 2290.

Deng, S., Ruan, R., Mok, C. K., Huang, G., Lin, X., & Chen, P. (2007). Inactivation of *Escherichia coli* on almonds using nonthermal plasma. *Journal of Food Science. 72*(2), 62–66.

Dorraki, N., Mahdavi, V., Ghomi, H., & Ghasempour, A., (2016). Elimination of diazinon insecticide from cucumber surface by atmospheric pressure air-dielectric barrier discharge plasma. *Biointerphases, 11*(4), 1007–1116.

Feichtinger, J., Schulz, A., Walker, M., & Schumacher, U. (2003). Sterilisation with low-pressure microwave plasmas. *Surface and Coatings Technology. 174–175*, 564–569.

Fernandez, A., Noriega, E., & Thompson, A., (2013). Inactivation of *Salmonella enterica-serovar typhimurium* on fresh produce by cold atmospheric gas plasma technology. *Food Microbiology, 33*(1), 24–29.

Gadri, R. B., Roth, J. R., Montie, T. C., Kelly-Wintenberg, K., Tsai, P. P. Y., Helfritch, D. J., et al., (2000). Sterilization and plasma processing of room temperature surfaces with a one-atmosphere uniform glow discharge plasma (OAUGDP). *Surface and Coatings Technology, 131*, 528–542.

Gurol, C., Ekinci, F. Y., Aslan, N., & Korachi, M., (2012). Low-temperature plasma for decontamination of E. coli in milk. *International Journal of Food Microbiology, 157*(1), 1–5.

Heo, N. S., Lee, M. K., Kim, G. W., Lee, S. J., Park, J. Y., & Park, T. J., (2014). Microbial inactivation and pesticide removal by remote exposure of atmospheric air plasma in confined environments. *Journal of Bioscience and Bioengineering, 117*(1), 81–85.

Kalra, C. S., Gutsol, A. F., & Fridman, A. A., (2005). Gliding arc discharges as a source of intermediate plasma for methane partial oxidation. *IEEE Transactions on Plasma Science, 33*(1), pp. 32–41. doi: 10.1109/TPS.2004.842321.

Kaushik, G., Satya, S., & Naik, S. N., (2009). Food processing a tool to pesticide residue dissipation-a review. *Food Research International, 42*, 26–40.

Keikotlhaile, B. M., Spanoghe, P., & Steurbaut, W., (2010). Effects of food processing on pesticide residues in fruits and vegetables: A meta-analysis approach. *Food and Chemical Toxicology, 48*, 1–6.

Kim, S. H., Kim, J. H., & Kang, B. K., (2007). Decomposition reaction of organophosphorus nerve agents on solid surfaces with atmospheric radio frequency plasma generated gaseous species. *Langmuir (the ACS Journal of Surfaces and Colloids), 23*(15), 8074–8078.

Kim, J. E., Lee, D. U., & Min, S. C., (2014). Microbial decontamination of red pepper powder by cold plasma. *Food Microbiology, 38*, 128–136.

Laroussi, M., & Lu, X., (2005). Room temperature atmospheric pressure plasma for biomedical applications. *Applied Physics Letters*, 87.

Lassen, K. S., Nordby, B., & Grün, R., (2005). The dependence of the sporicidal effects on the power and pressure of RF-generated plasma processes. *J. Biomed. Mater. Res. Part B Appl. Biomater., 74*, 553–559.

Lee, K., Paek, K., Ju, W. T., & Lee, Y., (2006). Sterilization of bacteria, yeast, and bacterial endospores by atmospheric-pressure cold plasma using helium and oxygen. *Journal of Microbiology, 44*(3), 269–275.

Li, S. P., Jiang, Y. Y., Cao, X. H., Dong, Y. W., Dong, M., & Xu, J., (2013). Degradation of nitenpyram pesticide in aqueous solution by low-temperature plasma. *Environmental Technology, 34*(12), 1609–1616.

Misra, N., Tiwari, B., Raghavarao, K. S. M. S., & Cullen, P., (2011). Non-thermal plasma inactivation of foodborne pathogens. *Food Engineering Reviews, 3*, 1–12.

Misra, N. N., Patil, S., Moiseev, T., Bourke, P., Mosnier, J. P., Keener, K. M., & Cullen, P. J., (2014). In-package atmospheric pressure cold plasma treatment of strawberries. *Journal of Food Engineering, 125*, 131–138.

Misra, N. N., Pankaj, S. K., Walsh, T., O'Regan, F., Bourke, P., & Cullen, P. J., (2014). In-package nonthermal plasma degradation of pesticides on fresh produce. *Journal of Hazardous Materials, 271*, 33–40.

Moreau, M., Feuilloley, M., Veron, W., Meylheuc, T., Chevalier, S., Brisset, J. L., & Orange, N., (2007). Gliding arc discharge in the potato pathogen Erwiniacarotovora subsp. atroseptica: Mechanism of lethal action and effect on the membrane-associated molecules. *Applied Environmental Microbiology, 73*(18), 5904.

Niemira, B. A., & Sites, J., (2008). Cold plasma inactivates *Salmonella stanley* and *Escherichia coli* O157: H7 inoculated on golden delicious apples. *Journal of Food Protection, 71*(7), 1357–1365.

Niemira, B. A., & Gutsol, A., (2010). Nonthermal plasma as a novel food processing technology. In: Zhang, H. Q., Barbosa, C. G., Balasubramaniam, V. M., Dunne, P., Farkas, D., & Yuan, J., (eds.), *Non-Thermal Processing Technologies for Food* (pp. 271–288). Blackwell Publishing, Ames, IA.

Oliveira, M., Vinas, I., Anguera, M., & Abbeys, M., (2012). Fate of Listeria monocytogenes and Escherichia coli O157: H7 in the presence of background natural microbiota on conventional and organic lettuce. *Food Control, 25*, 678–683.

Parish, M. E., Beuchat, L. R., Suslow, T. V., Harris, L. J., Garret, E. H., Farber, J. N., & Busta, F. F., (2003). Methods to reduce/eliminate pathogens from fresh and fresh-cut produce. *Comprehensive Reviews in Food Science and Food Safety, 2*, 161–173.

Pasquali, F., Stratakos, A. C., Koidis, A., Berardinelli, A., Cevoli, C., Ragni, L., & Trevisani, M., (2016). Atmospheric cold plasma process for vegetable leaf decontamination: A feasibility study on radicchio (red chicory, *Cichorium intybus* L.). *Food Control, 60*, 552–559.

Pelletier, J., (1992) Sterilization by plasma processing. *Aggressologie, 33*, 457–477.

Perni, S., Liu, D. W., Shama, G., & Kong, M. G., (2008). Cold atmospheric plasma decontamination of the pericarps of fruit. *Journal of Food Protection, 71*(2), 302–308.

Ramos, B., Miller, F. A., Brandao, T. R. S., Teixeira, P., & Silva, C. L. M., (2013). Fresh fruits and vegetables—An overview on applied methodologies to improve its quality and safety. *Innovative Food Science and Emerging Technologies, 20*, 1–15.

Randhwaa, M. A., Ahmed, A., Javed, M.S. (2014). Wheat contaminants (pesticides) and their dissipation during processing. *Wheat and Rice in Disease Prevention and Health* (pp. 263–277). Academic Press.

Reddy, P. M. K., Mahammadunnisa, S., & Subrahmanyam, C., (2014). Catalytic non-thermal plasma reactor for mineralization of endosulfan in aqueous medium: A green approach for the treatment of pesticide-contaminated water. *Chemical Engineering Journal, 238*, 157–163.

Rod, S. K., Hansen, F., Leipold, F., & Knochel, S. (2012). Cold atmospheric pressure plasma treatment of ready-to-eat meat: inactivation of *Listeria innocua* and changes in product quality. *Food Microbiology, 30*(1), 233–238.

Mousavi, S. M., Imani, S., Dorranian, D., Larijani, K., Shojaee, M. (2017). Effect of cold plasma on degradation of organophosphorus pesticides used on some agricultural products. *Journal of Plant Protection Research, 57*(1), 25–35.

Selcuk, M., Oksuz, L., & Basaran, P., (2008). Decontamination of grains and legumes infected with *Aspergillus* spp. and *Penicillium* spp. by cold plasma treatment. *Bioresource Technology, 99*(11), 5104–5109.

Sharma, A., Collins, G., & Pruden, A., (2009). Differential gene expression in *Escherichia coli* following exposure to non-thermal atmospheric pressure plasma. *Journal of Applied Microbiology, 107*, 1440–1449.

Song, H. P., Kim, B., Choe, J. H., Jung, S., Moon, S. Y., Choe, W., & Jo, C., (2009). Evaluation of atmospheric pressure plasma to improve the safety of sliced cheese and ham inoculated by 3-strain cocktail *Listeria monocytogenes*. *Food Microbiology, 26*(4), 432–436.

Stratakos, A. C., & Koidis, A., (2015). Suitability, efficiency and microbiological safety of novel physical technologies for the processing of ready-to-eat meats and pumpable products. *International Journal of Food Science and Technology, 50*(6), 1283–1302.

Tiwari, B. K., O'Donnell, C. P., & Cullen, P. J., (2009). Effect of non-thermal processing technologies on the anthocyanin content of fruit juices. *Trends in Food Science & Technology, 20*(3/4),137–145.

Tran, N., Amidi, M., & Sanguansri, P., (2008). Cool plasma for large scale chemical-free microbial inactivation of surfaces. *Food Australia, 60*(8), 344–347.

Vleugels, M., (2005). Atmospheric plasma inactivation of biofilm-forming bacteria for food safety control. *Ieee Transactions on Plasma Science, 33*(2), 824–828.

Wang, R. X., Nian, W. F., Wu, H. Y., Feng, H. Q., Zhang, K., Zhang, J., & Fang, J., (2012). Atmospheric-pressure cold plasma treatment of contaminated fresh fruit and vegetable slices: Inactivation and physiochemical properties evaluation. *The European Physical Journal D-Atomic, Molecular, Optical and Plasma Physics, 66*(10), 1–7.

Yun, H., Kim, B., Jung, S., Kruk, Z. A., Kim, D. B., Choe, W., & Jo, C., (2010). Inactivation of *Listeria monocytogenes* inoculated on disposable plastic tray, aluminum foil, and paper cup by atmospheric pressure plasma. *Food Control, 21*, 1182–1186.

Zhang, M., Oh, J. K., Cisneros-Zevallos, L., & Akbulut, M., (2013). Bactericidal effects of nonthermal low-pressure oxygen plasma on S. typhimurium LT2 attached to fresh produce surfaces. *Journal of Food Engineering, 119*(3), 425–432.

Ziuzina, D., Patil, S., Cullen, P. J., Keener, K. M., & Bourke, P., (2014). Atmospheric cold plasma inactivation of *Escherichia coli, Salmonella entericaserovar typhimurium* and *Listeria monocytogenes* inoculated on fresh produce. *Food Microbiology, 42*, 109–116.

CHAPTER 4

Refractance Window Drying and Its Applications in Food Processing

K. S. YOHA, S. R. PRIYADARSHINI, J. A. MOSES, and C. ANANDHARAMAKRISHNAN

Indian Institute of Food Processing Technology (IIFPT), Ministry of Food Processing Industries, Government of India, Thanjavur, Tamil Nadu, India, E-mail: anandharamakrishnan@iifpt.edu.in

ABSTRACT

The dehydration process is used to preserve the perishable foods by removing the moisture content and thereby increasing the shelf-life of the product. Refractance window (RW) drying is a novel pre-eminent dehydration technology which dehydrates the wet food products onto the thin Mylar film, which is transparent to infrared (IR) radiation and creates as a 'window' for drying. It enhances the self-limiting drying process in which the hot water transmits the heat energy into the wet product by forcing the product moisture to evaporate. Heating water transmits heat energy within itself by conduction, convection, and radiation. Among them, radiation is the most rapid process. In RW dryer, the circulating water temperature is maintained between 95 and 98°C. During drying, the hot water surface is covered by a Mylar film which blocks the evaporation of water and its heat loss. Hence, the product temperature remains at 60–72°C. At this reduced temperature, the pharmaceutical and bioactive products can be dried efficiently. This ideal technique promises to heat-sensitive food products by processing at low temperature and rapid drying with high retention of sensory, nutrients, and bioactive compounds in the end product. By applying this technique, dried products with free-flow in nature, lower water activity, and good storage stability can be achieved.

4.1 INTRODUCTION

Drying of food products plays a major role in the food processing sector. Dehydration promises food preservation and leads to value-added food products. The nutrition found in fresh foods is unparalleled. The objective of drying is to preserve the food products from microbial and enzymatic spoilage. Drying is not a new technology, since people preferred the drying method to stabilize the perishable foods many years ago. In last 4–5 decades, many R&D research has taken place to enhance the drying process for achieving a longer shelf-life of wet food products (Kroll et al., 1980). In the 18th century, the first artificial drying of food was introduced (Prescott et al., 1937). The concept of drying is to increase the shelf life of the food product beyond its average shelf-life by reducing the moisture content. Due to reduced water activity, the growth of microorganisms and enzymatic activities can be inhibited. Usually, thermal energy is used to remove the food moisture directly or indirectly (Chung and Chang, 1982). The dried form of a food product is easy to handle than that of wet form and also while packing it is most comfortable because of less weight and volume of the product (Sokhansanj and Jayas, 1987).

Drying uses about 12–20% of total energy, and it is considered as one of the energy-intensive operations in food processing (Raghavan et al., 2005; Moses et al., 2014). Energy management and its proper utilization is an essential task in industries. In order to improve the efficiency and efficacy of drying, it is essential to find ways to retain the nutrients in food products and also to make them consume less energy. Bioactive compounds in fruits and vegetables are considered as significant nutrients. Drying for a longer time may result in loss of nutrients. Conventional drying techniques require longer drying time; this long-term effect of temperature may lead to inferior quality products (Moses et al., 2013). The Refractance Window (RW) drying makes it possible to limit the loss of bioactive components through shorter drying time at low temperature. The RW drying has become the most eminent technology in food industries, especially because the equipment is relatively inexpensive and high quality dried products with retention of nutrients can be achieved through this technique. When compared to freeze dryer, the cost and energy consumption is too less, i.e., RW dryer consumes only one third to one half of that of a freeze dryer for drying the wet products. Figure 4.1 outlines the major advantages of RW drying.

RW drying is a novel dehydration technology, involves drying the wet food products onto the thin Mylar film, it is transparent to infrared (IR) radiation and creates as a 'window' for drying. Unique, simple design feature of this dryer promises cost-effective, energy efficient, and rapid reduction of moisture content with lesser drying time. Most of the dehydration techniques involve drying the products at high operational pressure and temperature, but this technique depends only on heat transfer modes, and there is a limited susceptibility to oxidation and free radical formation. The materials being dried are typically exposed to the dynamic changes upon heat transfer changing conditions of the drying system. In RW drying, dehydration occurs through the heat transfer modes along with the properties of IR and the refractance of light. This technique is ideal for heat-sensitive food products due to low-temperature processing and rapid drying. In RW dryer, the circulating water temperature is maintained between 95 and 98°C. During the drying process, the hot water surface is covered by a Mylar film which prevents the evaporation of water and its heat loss, during which only conduction occurs. When the wet food product is spread over the surface of the film, it allows IR energy from the hot water to pass through. At this point, all three modes of heat transfer occur, and this effective heat transfer into the product leads to rapid evaporation. Once the product gets dried, the IR "window" closes effectively, which limits the heat transfer to the product also from conduction.

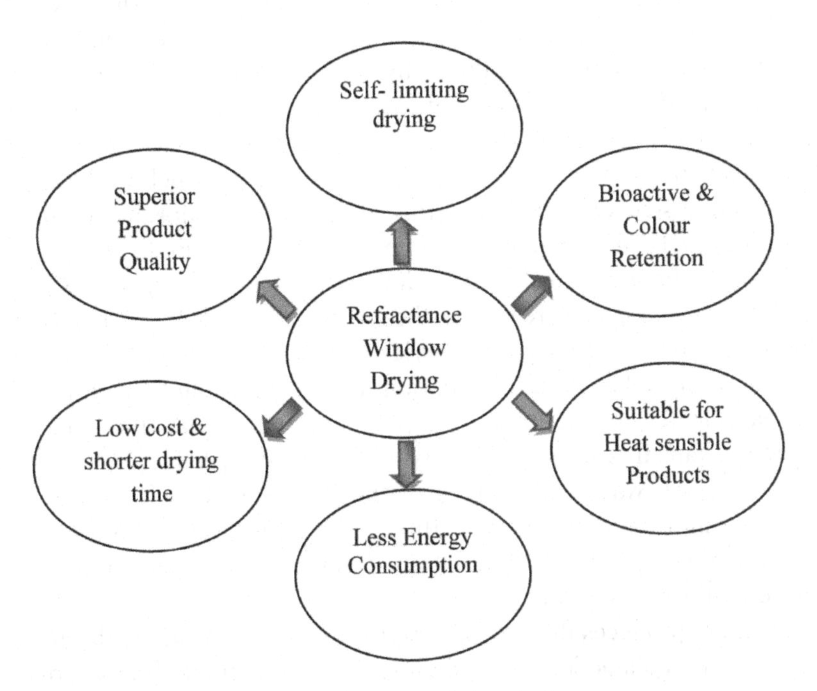

FIGURE 4.1 Major advantages of refractance window drying.

The contact film is a food grade insulated polyester film; hence, the product is protected from overheating. Even though the heating water temperature maintained between 95 and 98°C, the product temperature remains between 60 and 72°C due to evaporative cooling. At this low temperature, the pharmaceutical and bioactive products can be dried efficiently. When the product reaches the midpoint of drying, the free moisture gets evaporated. The dried product is discharged in the form of flakes, sheets, crystal powders, or small particles. As a result, it ensures the inherent sensory quality, color, flavor, aroma, nutrients, and bioactive retention in the end product. The dried products with free-flow crystal powder nature, lower water activity, and good storage stability can be achieved easily by this technique. In some instances, the product quality achieved by this technique is equivalent or better than that achieved in the freeze drying system. It is still challenging to identify the ideal path to improve the quality of dried products with retention of bioactives.

4.2 REFRACTANCE WINDOW DRYING SYSTEM

The R&D research is going on since several years for reducing the water content in food products and based on those technical developments, RW dehydration method was introduced by MCD Technologies, Inc., Tacoma, Washington, USA. In this system, heating water is the source of IR energy, which conveys the heat as IR radiation to the product through the Mylar film. Mylar film can be either moving via the conveyor belt in the continuous drying system or stationary in the batch system (Kudra and Mujumdar, 2009). Figure 4.2 depicts the batch-type RW dryer set-up. In this process, even though the hot water is maintained at the temperature 95–98°C, the actual product temperature above the Mylar film is below 70°C (Nindo and Tang, 2007). IR transparent Mylar film acts as a window which can transmit/absorb/reflect the IR radiation. By the contribution of this radiative heat transfer, the wet product applied on to the Mylar film gets dried. IR energy transmits through the Mylar film to the product in which the IR transmission wavelength range of the film matches with the absorption spectrum of water, which facilitates the rapid drying (Smith, 1994). Conduction, convection, and radiation all three modes of heat transfer are utilized by this drying system.

Hot water produces the thermal energy which is transmitted through the Mylar film by conduction and radiation. Due to the rapid transfer of heat energy, the water in the wet product gets evaporated quickly. The thermal energy transfer from the wet food product to the atmospheric air that is mediated by convection and also through evaporative cooling of the food product. At the state of thermal equilibrium, the evaporative cooling process occurs, which facilitates the self- limiting drying and also which plays a main role in energy consumption in the drying system. Once the product initiates to dry, the conductive heat transfer takes place immediately and in further stages of drying, heat transfer rate to the product gradually reduces by the reflection of IR energy into the hot water again. Product removal from the film is facilitated by cooling the dried food product below its glass transition temperature (T_g), while the product attains the glassy state, it becomes crystalline products or flakes.

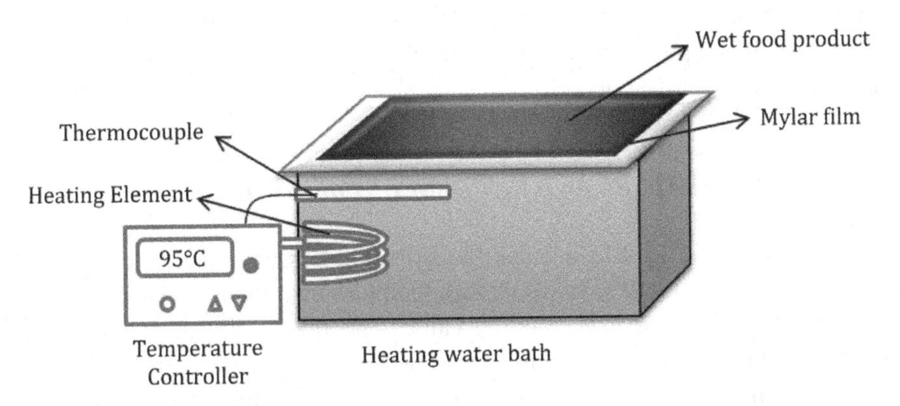

FIGURE 4.2 Refractance window dryer set-up.

4.3 CONCEPT OF REFRACTANCE WINDOW DRYING

The principle behind this RW dehydration phenomenon is the refractive index deviations at each stage of drying. When the wet food product is placed on to the Mylar film, the refractive index values of the water and the wet product becomes closer. Hence refractive index mismatch is less in this stage, which avoids the reflection of radiation and increases the transmission of radiative thermal energy towards the wet food product. The Mylar film acts as a "window" at this initial stage of drying, which allows the thermal radiation to pass through the film. According to the wet material dries, the refractive index of the product increases gradually, which creates refractive index mismatch as well as the window getting close in further stages of drying. Once the product gets dried, the air interface will be formed between the Mylar film and food product. This air interface formed because of increase in refractive index between hot water and atmospheric air; hence the thermal radiation gets reflected back into the water, which terminates the drying process (Ortiz-Jerez et al., 2015). While drying, the conduction mode of heat transfer dominates and at the end, the reflection of radiative thermal energy protects the product from overheating (Bolland, 2000; Vega-Mercado, 2001) and the evaporative cooling of food material facilitates self- limiting drying.

4.4 FACTORS INFLUENCING RW DRYING

In RW drying process, the drying depends on so many factors such as temperature and drying time, moisture content of the wet food material, the thickness of the pulp applied on to the film, optical properties of the Mylar film and its transmissivity, etc.

4.4.1 TEMPERATURE AND RESIDENCE TIME OF DRYING

Time-temperature profile decides the effectiveness and quality of drying. The conventional method involves high temperature and long residence time, which leads to losses in nutritional values and the quality of foods (Ratti, 2001). Usually, shorter drying time can be achieved under the high operational pressure and temperature. But in this case, low product temperature and relatively short drying time together with acceptable quality can be achieved. Ortiz-Jerez and Ochoa-Martínez (2015) observed the time-temperature profile of RW drying process, in which the product was dried at various temperatures, according to that the product temperature gets varied at each point of the product which was spread onto the film based on the removal of surface moisture by evaporation and finally the study reported that the quality of end-product greatly depends on temperature.

4.4.2 MOISTURE CONTENT

The decrease in the moisture content of food product enhances the quality of food in the drying process, while drying the reflectivity of the wet food product increases and the absorptivity decreases (Clarke, 2004; Nindo and Tang, 2007; Ratti and Mujumdar, 2006). The moisture sorption isotherm in the equilibrium state, directly explains the association between moisture content and water activity of product at a constant temperature (Moreira et al., 2008). The glass transition temperature and the sorption isotherm are the basic concepts which help to understand the role of water in food product stability (Kasapis, 2008). In RW dried samples, the moisture content was lower than freeze-dried samples (Baeghbali et al., 2010). Nindo et al., (2003b) reported the drying kinetics of asparagus puree using RW drying and concluded that, this technique could achieve the dried product with less moisture content at short residence time. Similar drying studies were conducted as well as the results were reported for carrot and strawberry puree (Abonyi et al., 2002).

4.4.3 THICKNESS OF PULP

The thickness of the wet food product applied on to the Mylar film in RW process has an impact on the residence time of drying. The wet product being dried initially follows a constant drying rate period, which depends on the thickness of the product. Zotarelli et al., (2015) reported that the drying rate was reduced around 40% while increasing the thickness of mango pulp from 2 to 3 mm. In this drying process, the transmitted radiation absorbed by the food products to be dried can be explained by Beer's law (Zotarelli et al., 2015). Higher product thickness implies less effective drying due to IR radiation, therefore the layer must be spread very thin to improve the effectiveness of drying. At low-temperature condition, the thick slices achieve larger diffusivity values, which lead to less shrinkage of slices, rather the thin slices undergo more shrinkage. Ochoa-Martinez et al., (2012) explained that the increase in thickness leads to an extension of drying time. This study reported that mango slices at 1mm thickness take 30 minutes, and 2 mm thickness takes 60 minutes for drying.

4.4.4 OPTICAL PROPERTIES OF CONTACT FILM

Optical transmissivity of the Mylar film influences the drying process. According to the optical properties of Mylar film, the transmission rate of thermal IR radiation can be estimated (Krimm, 1960). In RW process, three modes of the spectrum involve namely the emission spectrum by the hot water, the transmission spectrum of the Mylar film and the absorption spectrum of the wet food product. The thickness of the Mylar film affects IR radiation transparency. Thin Mylar film provides effective IR transparency when compared to thick films, but they have the poor mechanical strength to withstand the hot water pressure (Ratti and Mujumdar, 2006). The thickness and the chemical composition determine the transmissivity of the Mylar film. Ortiz-Jerez et al., (2015) reported that the total transmittance rate was increased ~128% while decreasing the thickness of the film from 0.2 to 0.1 mm at a constant temperature 90°C, i.e., the film thickness alone affects the drying. Transmissivity depends not only the thickness of the Mylar film; it also depends on the product moisture content. If the wet food product contains high moisture content, the absorption spectrum of the product is nearly similar to that of water. So the wet product absorbs more IR radiation at the wavelengths 3.0, 4.7, 6.0, and > 12 μm (Sandu, 1986; Zotarelli et al., 2015). It was reported that the wavelength ranges from 3 to 8 μm, and the water exhibits high absorptivity as well as high transmissivity (~90%) to IR radiation through Mylar film (Sandu, 1986). Therefore at these wavelengths, Mylar film shows better transmissivity to the IR radiation. Understanding the properties of different film materials is needed to know about the behavior to refract the thermal radiation. Usually, the polymer films have low transmissions at a range of 7–12 μm wavelength (Tsilingiris, 2003) and high transmissions at >12 μm wavelength (Ortiz-Jerez et al., 2015). Mylar® is the most suitable film for RW drying system having a thermal conductivity of 0.24 W/mK, the density of 1.39 g/cm^3 and specific heat of 0.28 Cal/g/°C at ambient conditions (Ortiz-Jerez and Ochoa-Martínez, 2015). Nindo and Tang (2007) explained that the refractive index at the film-wet pulp interface is reduced while the high moisture pulp is placed on top of the Mylar film and hence the radiative thermal energy can be easily transmitted to the product through the Mylar film. The refractive index of Mylar film is 1.64 (Tsilingiris, 2003) and the refractive indices of water and air are 0.33 and 1, respectively (Siegel and Howell, 1992).

4.5 THERMAL ENERGY TRANSFER AND ENERGY CONSUMPTION IN RW DRYING

Thermal efficiency in RW dryer can be improved by the circulation of hot water below the Mylar film; it mediates better heat transfer than the stagnant hot water. Evaporative cooling process of drying product plays a major role in energy consumption. The total energy consumed by RW dryer is correlated with the evaporative effects. In an industrial perspective, the energy requirement is a critical and decisive factor. RW dryer can achieve higher thermal efficiency with lower energy consumption. RW dryer required much lesser energy of about as compared to a spray dryer (Baeghbali and Niakousari,

2015). Similar results reported that RW dryer could be operated with half of the energy as required when compared to freeze dryer (Nindo and Tang, 2007). The amount of total heat transfer from the hot water to the wet food product can be calculated directly from the constant drying rate of the experiment. Zotarelli et al., (2015) conducted RW drying at constant drying rate period, in which the IR radiation emitted by the hot water was compared with the heat transfer fluxes from radiation, and also they assumed that the radiation emitted by the hot water was transmitted through the Mylar film and reaches the drying food product. The result of the study was obtained from the order of magnitude analysis, and the observed evaporated flux explained that 3% of the total heat energy was supplied to the wet food material. During drying in each stage, the temperature gets changed based on the thermal energy transfer, and also the thermal resistance will differ at each mode of heat transfer. Zotarelli et al., (2015) explained the detailed analysis of heat transfer mechanisms and the thermal resistance occurring at each stage of RW drying at a constant drying rate period.

4.6 BIOACTIVE RETENTION AND QUALITY OF PRODUCT

Bioactives are highly present in fresh foods, but the food products are dried to extend the shelf-life. During drying of such food products there may be significant losses in its nutritional value. RW drying technology maintains the product at a moderate temperature, which is helpful to achieve high retention of nutrients, vitamins, and bioactives in the final dried food product. This low-temperature short time drying technology enhances the product quality with attributes similar to that of freeze-drying and also inexpensive. β-Carotene, the precursor of Vitamin A and ascorbic acid (AA) (Vitamin C), are mostly found in fruits and vegetables, in RW drying technique these heat sensible components can be able to retain in the end products. One of the studies on asparagus found that RW drying technique is suitable for AA retention as well as total antioxidant retention when compared to freeze-dried product (Nindo, 2003b). AA content in the RW dried product is almost similar to that of fresh material, better to that of freeze-dried product and best when compared to all other thermal methods of drying. Abonyi et al., (2002) reported that the carotene loss in RW dried carrot puree was quite high but not significantly different from that of carotene content in freeze-dried carrot sample. But drum dried carrot puree has shown a severe nutrient loss in carotene content. Also, they compared the AA loss in strawberry puree, in which the loss of AA during RW and freeze-drying was 6.0%

and 6.4%, respectively. RW dried pomegranate juice has higher retention of anthocyanin content and color. Whereas in the spray drying process, the product color becomes darker, and also bioactive retention can't be achieved because of high working temperature (Baeghbali et al., 2010). The comparative drying study reported that when compared to raw purple potatoes, the RW dried purple potatoes shown the significant increase in total phenolic content (Nayak et al., 2011). Higher retention of anthocyanins and other bioactives were achieved in RW dried pomegranate pestils (Tontul and Topuz, 2017). A similar study reported that, 90% of anthocyanin content was retained in final RW dried haskap berry product (Celli et al., 2016). Raghavi et al., (2018) summarized the quality retention of final product through RW drying comparatively with other conventional drying techniques and highlighted that even at low temperature there is a scope for drying microbial products and other heat sensitive bioactives. Also it was proven that RW drying has significant potential for drying probiotics along with prebiotics. Yoha et al., (2019) reported that cell viability of probiotics was achieved 9.18 $\log_{10}$ CFU/g with 93% encapsulation efficiency using RW drying, it shown comparatively similar efficiency with freeze drying.

4.7 CONCLUSION

RW drying is a rapid, gentle drying technique which minimizes the product degradation caused by heat and oxidation while drying. It also prevents the flavor and aroma of the product and achieves the retention of color, nutritional attributes, and bioactives. RW dryer is the most economical, inexpensive equipment and easy to fabricate when compared to freeze dryers and spray dryers and also the energy consumption during the drying process is less. But it can achieve the quality of dried product equivalent to that of produced by freeze dryers and spray dryers. Due to drying at low-temperature RW drying is significantly suitable for heat sensible nutritional compounds in foods and herbs. Almost all the nutritional and sensory attributes can be retained in the end products. RW drying technique provides a better scope for the development of high value and nutritive-added food products. For some high sugar food emulsions spray dryer requires the addition of non-sugar carriers to dry, but in RW dryer we can dry without adding any filler materials as well as it can also able to dehydrate the highly viscous wet food materials. Even at low temperature, we can achieve the dehydration of the probiotic-containing foods with acceptable viability. These attributes of RW drying make it suitable for processing of high-value foods and nutraceuticals.

KEYWORDS

- **bioactive compounds**
- **dehydration**
- **IR transparent film**
- **low temperature**
- **refractance window drying**

REFERENCES

Abonyi, B. I., Feng, B. I., Edwards, C. G., & Tang, J., (2002). Quality retention in strawberry and carrot purees dried with refractance window system. *Journal of Food Science, 67*, 1051–1056.

Baeghbali, V., & Niakousari, M., (2015). Evaluation of a batch refractance window dryer in drying of some heat-sensitive foodstuff. *Iran. J. Food Sci. Technol., 13*(1), 185–192.

Baeghbali, V., Niakosari, M., & Kiani, M., (2010). Design, manufacture and investigating functionality of a new batch refractance window system. In: *Proceedings of 5th International Conference on Innovations in Food and Bioprocess Technology* (Vol. 7, p. 9).

Bolland, K. M., (2000). Refractance window™ drying: A new low temperature, energy efficient process. *Cereal Foods World, 45*(7), 293–296.

Celli, G. B., Khattab, R., Ghanem, A., & Brooks, M. S. L., (2016). Refractance window™ drying of Haskap Berrye preliminary results on anthocyanin retention and physicochemical properties. *Food Chem., 194*, 218–221.

Chung, D. S., & Chang, D. I., (1982). Principles of food dehydration. *J. Food Protection, 45*, 475–478.

Clarke, P., (2004). Refractance window TM-"down under." In: *Drying Proceedings of the 14th International Drying Symposium (IDS 2004)* (pp. 22–25). Sao Paulo, Brazil.

Kasapis, S., (2008). Recent advances and future challenges in the explanation and exploitation of the network glass transition of high sugar/biopolymer mixtures. *Critical Reviews in Food Science and Nutrition, 48*, 185–203.

Krimm, S., (1960). Infrared spectra of high polymers. In: *Fortschritte der Hochpolymeren-Forschung* (pp. 51–172). Springer, Berlin Heidelberg.

Kroll, K., Mujumdar, A. S., & Menon, A. S., (1980). Drying since the millenniums. In: Mujumdar, A. S., (ed.), *Drying* (Vol. 80, No. 2, pp. 485–494).

Kudra, T., & Mujumdar, A. S., (2009). *Advanced Drying Technologies*. CRC Press.

Moreira, R., Chenlo, F., Torres, M. D., & Vallejo, N., (2008). Thermodynamic analysis of experimental sorption isotherms of loquat and quince fruits. *Journal of Food Engineering, 88*, 514–521.

Moses, J. A., Karthickumar, P., Sinija, V. R., Alagusundaram, K., & Tiwari, B. K., (2013). Effect of microwave treatment on drying characteristics and quality parameters of thin layer drying of coconut. *Asian J. Food Agro. Ind., 6*(02), 72–85.

Moses, J. A., Norton, T., Alagusundaram, K., & Tiwari, B. K., (2014). Novel drying techniques for the food industry. *Food Engineering Reviews, 6*(3), 43–55.

Nayak, B., Berrios, J. D. J., Powers, J. R., Tang, J., & Ji, Y., (2011). Colored potatoes (*Solanum tuberosum* L.) Dried for antioxidant-rich value-added foods. *J. Food Process. Preserv., 35*(5), 571–580.

Nindo, C., Sun, T., Wang, S. W., Tang, J., & Powers, J. R., (2003b). Evaluation of drying technologies for retention of physical quality and antioxidants in asparagus (*Asparagus officinalis* L.). *LWT-Food Sci. Technol., 36*(5), 507–516.

Nindo, C. I., Feng, H., Shen, G. Q., Tang, J., & Kang, D. H., (2003a). Energy utilization and microbial reduction in a new film drying system. *J. Food Process. Preserv., 27*(2), 117–136.

Nindo, C. I., & Tang, J., (2007). Refractance window dehydration technology: A novel contact drying method. *Dry. Technol., 25*(1), 37–48.

Ochoa-Martínez, C. I., Quintero, P. T., Ayala, A. A., & Ortiz, M. J., (2012). Drying characteristics of mango slices using the refractance window™ technique. *J. Food Eng., 109*(1), 69–75.

Ortiz-Jerez, M. J., & Ochoa-Martínez, C. I., (2015). Heat transfer mechanisms in conductive hydro-drying of pumpkin (Cucurbita maxima) pieces. *Dry. Technol., 33*(8), 965–972.

Prescott, S. C., & Proctor, B. E., (1937). *Food Technology*. McGraw Hill Book Co., NY.

Raghavi, L. M., Moses, J. A., & Anandharamakrishnan, C., (2018). Refractance window drying of foods: A review. *J. Food Eng., 222*, 267–275.

Raghavan, G. S. V., Rennie, T. J., Sunjka, P. S., Orsat, V., Phaphuangwittayakul, W., & Terdtoon, P., (2005). Overview of new techniques for drying biological materials with emphasis on energy aspects. *Braz. J. Chem. Eng., 22*(2), 195–201.

Ratti, C., & Mujumdar, A. S., (2006). Infrared drying. In: Mujumdar, A. S., (ed.), *Handbook of Industrial Drying* (pp. 423–438). Marcel Dekker, New York.

Ratti, C., (2001). Hot air and freeze-drying of high-value foods–a review. *Journal of Food Engineering, 49*, 311–319.

Sandu, C., (1986). Infrared radiative drying in food engineering: A process analysis. *Biotechnol. Prog., 2*(3), 109–119.

Siegel, R., & Howell, J. R., (1992). *Thermal Radiation Heat Transfer*. Hemisphere Publishing Corporation, Washington DC.

Smith, T. M., (1994). Heat transfer dynamics. *TAPPI Journal, 77*, 239–245.

Sokhansanj, S., & Jayas, D. S., (1987). Drying of foodstuffs. In: Mujumdar, A. S., (ed.), *Handbook of Industrial Drying*. Marcel Dekker, Inc. NY.

Tontul, I., & Topuz, A., (2017). Effects of different drying methods on the physicochemical properties of pomegranate leather (pestil). *LWT-Food Sci. Technol., 80*, 294–303.

Tsilingiris, P. T., (2003). Comparative evaluation of the infrared transmission of polymer films. *Energy Convers. Manag., 44*(18), 2839–2856.

Vega-Mercado, H., Gongora-Nieto, M. M., & Barbosa-Canovas, G. V., (2001). Advances in dehydration of foods. *J. Food Eng., 49*(4), 271–289.

Yoha, K. S., Moses, J. A., & Anandharamakrishnan, C., (2019). Conductive hydro drying through refractance window drying–An alternative technique for drying of *Lactobacillus plantarum* (NCIM 2083). *Drying Technology,* 1–11.

Zotarelli, M. F., Carciofi, B. A. M., & Laurindo, J. B., (2015). Effect of process variables on the drying rate of mango pulp by refractance window. *Food Res. Int., 69*, 410–417.

CHAPTER 5

Enzyme Immobilization in Food and Agriculture: Principles, Practices, and Future Prospects

MITALI MADHUMITA[1] and PRAMOD K. PRABHAKAR[2]

[1]*Department of Agriculture and Food Engineering, Indian Institute of Technology, Kharagpur, West Bengal, India*

[2]*Department of Food Science and Technology, National Institute of Food Technology Entrepreneurship and Management, Sonipat, Haryana, India, E-mail: pramodkp@niftem.ac.in, pkprabhakariitkgp@gmail.com*

ABSTRACT

In this chapter, enzyme immobilization (EI) has been well discussed in order to get an insight on different EI techniques, economic status, properties, and potential application in agriculture and food industry. Immobilized enzymes are generally referred to as "enzymes physically confined or localized in a certain defined region of space with retention of their catalytic activities, and which can be used repeatedly and continuously." Due to high demand and various benefits of immobilized enzyme over soluble enzymes, they have lots of effective applications in food industries (meat, dairy, baking, starch conversion, brewing etc.), beverage industries (fruit juice, vegetable juice, beer, wine etc.), textile industry, pulp, and paper industry, detergent, and cosmetic industry, biofuel industry, medicinal, and pharmaceutical industry, etc. To achieve a well-balanced performance, high yield, good operational stability, and low mass transfer limitation, the enzyme should be immobilized for the use of above-discussed applications in an effective manner. There are several techniques that are adopted, including reversible methods (adsorption, ionic, and affinity binding) and irreversible methods (covalent binding, entrapment, cross-linking, and encapsulation) to immobilize the enzymes.

Recently valuable ongoing researches are going on and more focused on drug delivery, diagnosis, and treatment of diseases, tumor location analysis, bio-affinity chromatography, biosensors textile industry, etc. As a result, the demand for EI will improve for commercial development, consumer application, and large-scale application. The significance and efficiency of reactions utilizing immobilized biocatalysts are also improved. Lots of researches and projects should be focused for the future development of EI.

5.1 INTRODUCTION

Enzyme immobilization (EI) is a simplest and easier technique which is defined as imprisonment of enzyme molecules in a distinct phase in which the enzymes are immobilized and allow the exchange of medium and converted to final pure product. It can provide resistance to changes in conditions such as pH, temperature, ionic strength, etc. To maintain constant environmental conditions and to protect the enzyme against these changes, immobilization is very important. It allows proper suitable handling of procedure of enzyme preparations having two major benefits, and these benefits are: (1) the enzyme is separated from the product which is easier to get a well-founded and effective reaction technology (2) it reduces the cost for making the process effective. Due to these benefits of immobilized enzymes over their soluble form, there is a high increasing demand in the application of EI in industrial sectors. As enzymes are biological catalysts in nature, without consuming themselves, they promote the rate of reactions and used repeatedly in an immobilized or insolubilized form for as long as they remain active. There are several advantages in the immobilized state of an enzyme which includes convenient handling, more stability, and easier separation of the product by eliminating protein contaminants, repetitive use of enzymes, making the movement of the enzyme from one place to another easily, etc. Besides these, it also allows the development of a multi-enzyme reaction system that provides easier process control and reduces effluent disposal problems (Tischer and Wedekind, 1999). Having all these advantages, EI has a higher efficiency and adaptability. Thus, the enzymes, which are immobilized, are used in different sectors such as antibiotics, food, and beverages, transportation fuels, agrochemicals, chemical feedstock, monitoring devices, and disease diagnosis. Furthermore, they have huge applications in different industries, such as pharmaceutical, biomedical, biosensor, and bioreactor, etc. Because of the immobilization process, enzyme activity gets reduced, and these enzymes are subjected to mass transfer limitation.

5.2 ENZYME IMMOBILIZATION (EI): BACKGROUND, HISTORY, AND SIGNIFICANCE

There are several methods that have been developed on immobilization in the course of the last decades. The process of binding to carrier materials is a very suitable method for EI. Previously, in the 1950s, the use of binding enzymes onto solid materials by a human being was first prepared intentionally. The biofilms mostly occur all natural environments were first studied during the 1940s (Abdelmajeed et al., 2012). Generally, the biofilms consist of a number of layers of the cell which are placed in hydrated matrices and surface-attached microbial communities. The first enzyme, which was immobilized onto matrices like charcoal and aluminum hydroxide in late 1916, was invertase. The details of the development of immobilized enzymes from the year 1916 to 2016 are described in Tables 5.1 and 5.2.

Recently, several changes are brought in the field of biotechnology to promote the application of immobilized enzymes for various applications. A number of innovative researches are going on in the field of immobilization techniques such as biocatalytic reactions, biosensors, improved reactors, drug delivery, and tumor identification utilizing immobilized enzymes (Champagne, 2008). Immobilization techniques are used for these purposes such as (1) reuse of enzyme; (2) low cost and residence time; (3) easy separation of product; (4) continuous processing; (5) facilitates process control; (6) to get an optimized product yield; (7) enzyme isolation, stabilization, and purification is not needed; (8) multi-enzyme complex reaction; (9) cofactor regeneration in the native system; and (10) using mixed cultures. The immobilizations of enzymes are very precious due to its easy separation and multiple times of reusing the same reaction. Thus degradation becomes less by controlling reaction rate and time. It is very helpful to prevent the contamination of the product with enzyme molecules and decreases the purification costs. According to these benefits of immobilized enzymes, there is a high impact of the application on biotechnology and other sectors (Barrientos, 1994).

5.3 IMMOBILIZED ENZYMES: PROPERTIES, ADVANTAGES, AND DISADVANTAGES

Several factors, such as characteristics behavior of carrier material, nature, and a number of the enzyme and interactions between the support and enzyme, are responsible for affecting the properties of immobilized enzyme.

TABLE 5.1 Development of Immobilized Enzymes During the Year of 1916 to 2000

		Before 1970s
Early Phase	1916–1940	Different immobilizing matrices were discovered. Some of the examples are inorganic carriers such as glass, alumina, hydrophobic compound-coated glass, etc. Enzymes and biomolecules are attached with each other in the presence of these above inorganic carriers e.g., immobilization of invertase onto charcoal and aluminum hydroxide.
Underdeveloped Phase	1940–1965	Immobilization was done from physical adsorption to specific ionic adsorption. Advantages of this method are more specificity and durability. Different matrices used in this immobilization are phosphocellulose, DEAE-cellulose, cellulose, Amberlite XE-97, Dowex-2, and Dowex-50.
Developing Phase	1965–1970	In the developing phase, enzyme immobilization was done by adopting the covalent method. In this method, different chemical cross-linkers were used. e.g., Glutaraldehyde (natural and synthetic polymers)
		Between 1970–1990
Developed Phase	1970–1980	Enzyme immobilization was done by using different methods such as covalent, adsorption, entrapment, and encapsulation. Other new methods are developed like affinity binding and coordination binding.
Post Developed Phase	1980–1990	In this phase, immobilization was done with a designed high selectivity and specificity, which is useful in medicine, agriculture, and environment and industry sectors.
After 1990s to 2000		
Rational Design of Immobilized Enzymes	1990–2000	New advanced immobilization techniques were developed by using robust immobilized enzymes having the similar catalytic ability and higher operational stability. For example, the development of cross-linked enzyme crystals (CLEC).

(*Source*: Adapted from Nisha et al., 2012)

TABLE 5.2 Development of Immobilized Enzymes from the Year of 2000 to 2016

	During 2000–2016
2000	Enzymes were not immobilized onto any carrier till the year of 2000. Firstly, Penicillin G acylase enzyme was aggregated by such methods like chemical cross-linking; forming in soluble cross-linked enzyme aggregates (CLEAs).
2001	In this year, the catalase enzyme produced from *Bacillus* SF was immobilized by using covalent linkage via glutaraldehyde onto silanized alumina which treats the textile bleaching effluents.
2002	Immobilized laccase was used for remediation of polluted soil that contaminated with 2,4-dichlorophenol (DCP)
2006	In this year, mitochondrial creatine kinase and type I hexokinase enzymes were developed. These were immobilized on biological membranes which are applicable for any protein with known three-dimensional structures
2008	A new method was developed, i.e., Leukemia Inhibitory Factor (LIF) which present on mouse embryonic stem cell (mESC) was immobilized into maleic anhydride copolymer thin-film.
2009	A lactose nano-probe was developed by immobilizing β-galactosidase isolated from pea seeds onto gold nanoparticles.
2010	Protease enzyme is immobilized sequentially onto polystyrene sulfonate followed by covalent linkage.
2011	The matrix metalloproteinase-8 (MMP-8) was immobilized onto epoxy activated silica and packed into a reactor which is used for the online screening of drugs. Useful for the treatment of heart diseases, multiple sclerosis, osteoarthritis, and various other inflammatory conditions.
2012	Surface plasmon resonance (SPR) was found to be very useful in studying biomolecular interactions, which is based on immobilization of a number of biomolecules including proteins, enzymes, and antibodies.
2013	The endoproteinase Glu-C from *Staphylococcus aureus* was developed that immobilized onto a magnetic bead. Bead is made of cellulose activated with divinyl sulfone. Another enzyme, acetyl coenzyme, A-carboxylase was immobilized onto the tip of atomic force microscope (AFM) and used as nanobiosensors for various detection processes.
2014	Immobilization of tannase (Tan410) was taken from soil metagenomic library onto different matrices: mesoporous silica SBA-15, chitosan, calcium alginate, and Amberlite IRC 50. This is used for the removal of tannins from green tea infusion after characterization.
2015	A geosensor is synthesized by using immobilized DNA onto gold nanoparticles and can be used for early diagnosis of various genetic diseases.
2016	Nowadays, computer modeling has been used to screen various matrices to immobilize inulinase. It helps in understanding interaction mechanisms and gives structural properties of the bound enzyme with the matrix

(*Source*: Adapted from Nisha et al., 2012)

Due to the result of the interaction, the behavior of immobilized enzymes differs from dissolved enzymes. Several properties such as pH, specific activity, kinetic property, selectivity, and stability significantly change upon EI. These changes in properties occur in the limited enzyme system due to the changes of the protein, reaction of enzyme, and substrate (Sharma et al., 2008). There is a less change in physical immobilization methods (entrapment and encapsulation) than chemical methods for the catalytic behavior of enzyme. This is due to the native conformation with no covalent modification. As a consequence, the operational stability of the immobilized enzyme increases, which is based on temperature and time. The reason behind this increment is due to the enzyme loading that causes controlled diffusion (Nisha et al., 2012). Figure 5.1 describes a brief interaction between immobilized enzyme and carrier having several properties such as chemical, biochemical, mechanical, and kinetic properties.

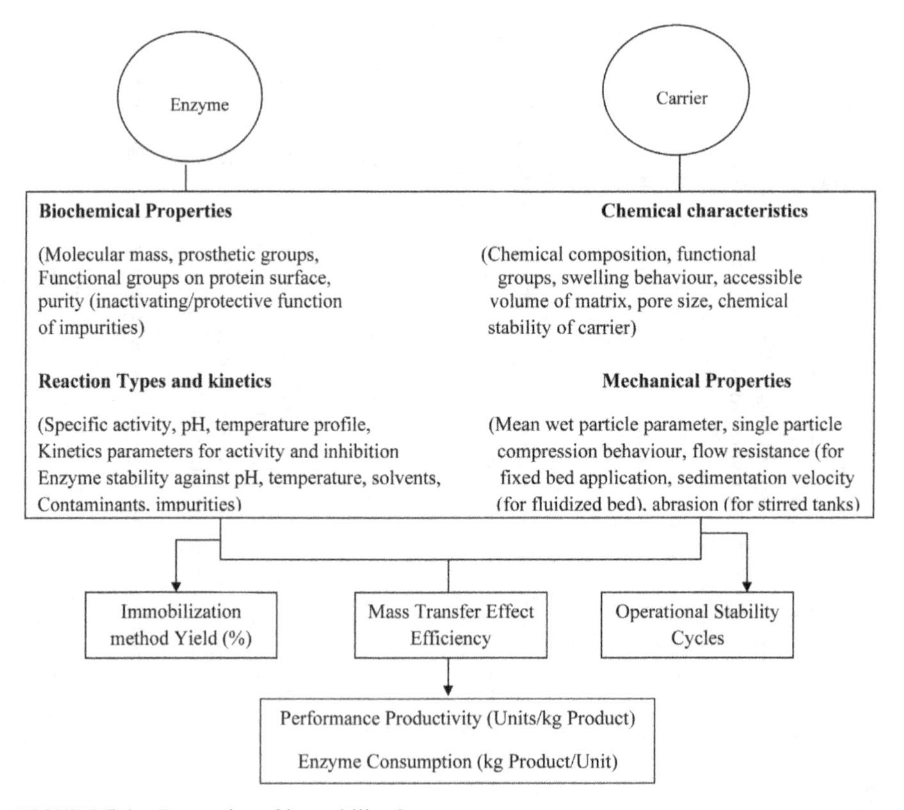

FIGURE 5.1 Properties of immobilized enzymes.

Due to the interactions between matrices and carriers, there is a change in its microenvironment during EI (Guisan et al., 1997). This interaction is based on the physical and chemical properties. The important parameters such as pH, temperature, viscosity, etc. are affected during EI. The change in pH depends on the type of immobilization method, physical, and chemical properties of the matrix. The electrons shift the optimum pH towards the basic side and acidic side when the matrix is positively and negatively charged, respectively. This change in pH is due to the surface charge density of the matrix. At a high loading of enzyme and intrinsic specific activity, the optimum pH is obtained. As a result, the concentration of released product molecules is increased. Another parameter, temperature also depends on the type of immobilization method and properties of the matrix. This parameter increases when substrate molecules gain some kinetic energy and reach the active site rapidly. Optimum temperature is obtained at the higher side of the temperature, which depends on the type of matrix and the interactions between the enzyme and the matrix. Also, due to the diffusion barriers, kinetic parameters are changed during EI. The three important parameters (Immobilizing matrix, Immobilization conditions, and Reaction time) determine the enzyme stability, which can increase or decrease during immobilization (Sharma et al., 2008).

The advantages of immobilized enzymes are:

1. Separation of the enzyme from the reaction mixture and re-use of these enzymes for other reactions;
2. Providing the ability to control reaction times;
3. Provides pure products;
4. Protection from degradation and deactivation;
5. Enhanced stability;
6. Cost efficiency;
7. Use as controlled release agents;
8. Develops a multienzyme reaction system.

The disadvantages of immobilized enzymes are:

1. Cost of carriers;
2. Changes in properties;
3. Affects the stability and activity of enzymes;
4. Mass transfer limitation;
5. Problems with cofactors and regeneration;
6. Problems with the multienzyme system.

5.4 ENZYME IMMOBILIZATION (EI) TECHNIQUES

Immobilized enzymes are the enzymes either covalently bound or adsorbed onto the surface of an insoluble support or incorporated in a gel matrix. It is very precarious that the substrates and the products move freely in and out of the phase to which the enzyme molecules are confined. So, the movement of immobilized enzymes is completely or severely restricted. It is not necessary for immobilized enzyme molecules to make them immobile. Several EI techniques are developed in which enzyme molecules can move freely within their phase, and these techniques are entrapment, membrane confinement, etc. In other methods, like adsorption and covalent bonding, they cannot move freely. The catalytic reaction of enzymes occurs in three different states (molecules in individual form, aggregates with other entities, and as attached to surfaces) under mild conditions with a very high degree of substrate specificity. Due to these reactions, the formation of by-products is decreased. To immobilize the enzymes onto surfaces and to achieve a high-performance biocatalyst, four ways are developed. These four ways include that (1) a suitable reaction is performed prior to binding to activate the enzyme; (2) the carrier should be modified and activated; (3) to conciliate between carrier and enzyme, a bi or multifunctional coupling agent is used; and (4) to generate a protein with "(bio) specific" groups, the enzyme is modified by recombinant DNA. So that it can absorb onto special carriers using (bio) affinity binding. The immobilization method is classified into different groups, which are shown in Figure 5.2, and all the images of EI techniques are represented in Figure 5.5.

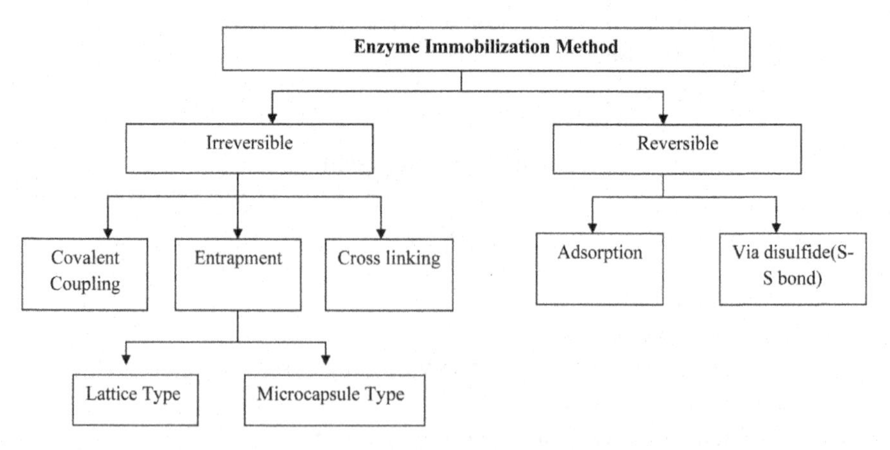

FIGURE 5.2 Schematic representation of different methods of enzyme immobilization.

5.4.1 IRREVERSIBLE IMMOBILIZATION

The main goal of the irreversible immobilization is that without destroying the biological activity of the enzyme, biocatalyst cannot be separated easily from the support once it is attached with it. The most irreversible EI methods are covalent coupling, entrapment or microencapsulation, and cross-linking.

5.4.1.1 COVALENT COUPLING

Covalent coupling is the binding which activates polymers and commonly used on a laboratory scale. Immobilized enzymes are covalently bound to stationary carriers, which are shown in Figure 5.5(A). Variety of carriers used in this immobilization is extensive, and such carriers are glass, silica, sephadex, cellulose, nylon, and polystyrene. The binding forces generated between the enzyme and carriers are very strong. So that enzyme cannot easily be separated from carriers in the presence of high concentrated substrates. For the selection of covalent immobilization of an enzyme, two main factors, such as the formation of covalent bonds by using a type of functional groups on the protein and their physical and chemical characteristics are considered.

Advantages:

1. Firmly binding between enzyme and carrier;
2. Protein is not lost in the solution.

Disadvantages:

1. Laborious and expensive;
2. Selection of conditions is more difficult for covalent binding than other carrier binding methods.

5.4.1.2 ENTRAPMENT

This EI technique is invented by the Snamprogetti's Laboratories (now named Eniricerche). In entrapment, the enzymes are simply trapped inside the polymer matrix. They are not directly attached to the support surface, but. Based on the incorporation of an enzyme into the lattice of a polymeric group, enzymes are held or entrapped within it, but the product is separated. There is no bond formation between the enzyme and polymer carrier due to the breakup of the polymeric mass (Abdelmajeed et al., 2012). Under optimal conditions, the efficiency of entrapment exceeds 90%, which means almost the entire enzyme present in the initial emulsion is eventually entrapped. The process of entrapment method is described in Figure 5.5(B). It can be classified into two types of entrapment. These are:

a. **Lattice Types Entrapment:** In this type of entrapment; enzymes are placed within the spaces of a cross-linked water-insoluble polymer. Some of the examples of synthetic polymers are polyacrylamide, polyvinyl alcohol, and natural polymer, i.e., starch.
b. **Microcapsule Type Entrapment:** Technically, this type of entrapment is called encapsulation, which encloses the enzyme within a semi-permeable polymer membrane and shown in Figure 5.5(C). Encapsulation of enzymes into semi-permeable membranes is very useful in the medical field. Example: Liposome, aminoacylase, etc. (Figure 5.3).

Procedure:

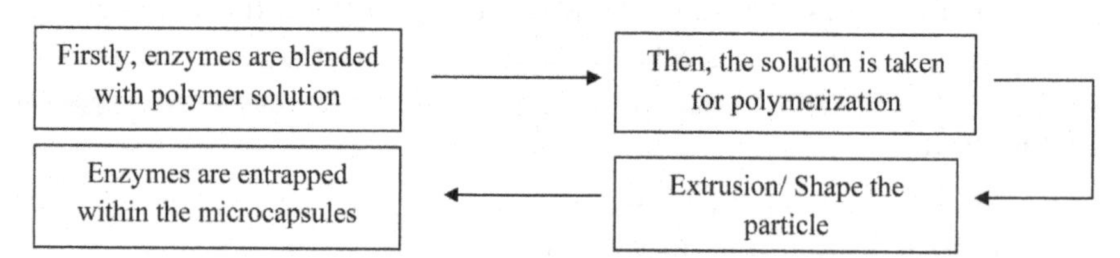

FIGURE 5.3 Process of enzyme immobilization by entrapment.

Advantages:

1. Simple and low cost;
2. High yield;
3. Easy recovery of reaction product;
4. No change in intrinsic enzyme properties;
5. Minimize enzyme requirement and matrices.

Disadvantages:

1. Possibility of the leakage of enzyme;
2. Only small sized substrate can be used;
3. Only low molecular weight substrates can diffuse rapidly in the enzyme.

5.4.1.3 CROSS-LINKING

The method of cross-linking is based on covalent cross-linking of enzymes with appropriate bifunctional agents on an insoluble support, which is shown in Figure 5.5(D). It is achieved by intermolecular cross-linking of enzymes by using bi or multi-functional reagents. These reagents are glutardialdehyde, glutaraldehyde, glyoxal, diisocyanates, hexamethylene diisocyanate, diazobenzidine, toluene diisocyanate, etc. Multiple numbers of covalent bonds are formed between the enzyme and the matrix. Also, a large, 3-D network is formed by attaching the enzyme molecules to each other (Abdelmajeed et al., 2012).

Advantages:

1. Simple and can easily be separated from the reaction mixture;
2. Mixture can be reused (3) Suffers limitations caused by diffusion.

Disadvantages:

1. Enzyme may denature due to the use of polyfunctional reagents in cross-linking;
2. High cost;
3. Irreversible nature.

5.4.2 REVERSIBLE IMMOBILIZATION

The main aim of reversible immobilization of enzymes is that it immobilizes labile enzymes and can be applicable in bioanalytical systems. Under gentle conditions, enzymes can be detached due to the support binding. Also, this support can be regenerated and re-loaded with fresh enzyme when the enzymatic activity decays (Gupta and Mattiasson, 1992).

5.4.2.1 ADSORPTION

Adsorption method is very simple and easy to immobilize the enzymes. In this method, enzymes are adsorbed physically on an active adsorbent surface by contacting an aqueous solution of the enzyme with an adsorbent. The insoluble materials, on which enzymes can be adsorbed, are ion exchangers, organic polymers, silica gels, collagen, and even glass, etc. A film of the adsorbate is created on the surface of the adsorbent by this process. This technique was first used by Nelson and Griffin in 1916 to produce immobilized invertase using charcoal as adsorbent (Figure 5.4).

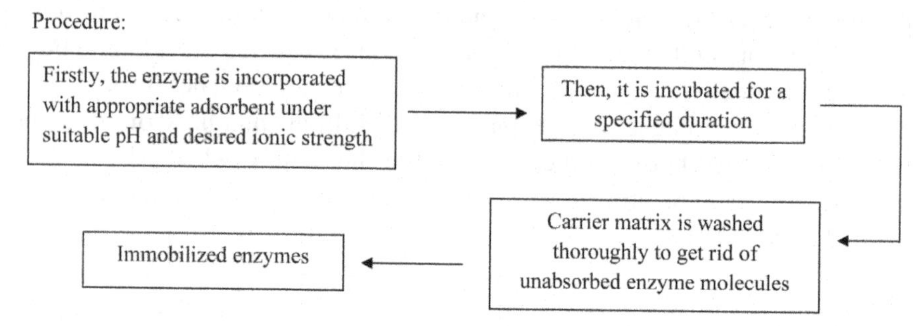

FIGURE 5.4 Process of enzyme immobilization by adsorption.

a. **Nonspecific Adsorption:** In this type of adsorption, the enzymes are attached to the matrix by non-covalent linkages including hydrogen bonding, van der Waals forces, or hydrophobic interactions without any pre-activation of support (Messing, 1976; Woodward, 1985) which is shown in Figure 5.5 (F). The process of adsorption involves the weak interaction between enzyme and carrier. The enzyme may be leaked out due to the effect of temperatures, pH, and ionic strength, presence of substrate and also for the weak interaction.
b. **Ionic Binding:** This method is reversible and carried out under mild conditions. But it is very difficult to find the conditions under which the enzyme remains strong and active. Figure 5.5 (E) shows that the enzymes are bound very strongly through salt linkages. Depending on the application, it is very useful to shift the optimal conditions of a certain enzyme towards more alkaline or acidic conditions (Guisan et al., 1997).
c. **Hydrophobic Adsorption:** It is based on the chromatographic principle in which a chemical bond is formed. And, it depends on several factors like pH, salt concentration, temperature, etc. The strength of interaction of enzyme and carrier depends both on the hydrophobicity of the adsorbent and protein. The regulation of hydrophobicity depends on the degree of substitution of the support and the size of the hydrophobic ligand molecule (Caldwell et al., 1976; Solomon et al., 1987).

d. **Affinity Bonding:** This immobilization method involves optimized variables such as pH, temperature, nature, and concentration of the adsorbent, ionic strength, the concentration of enzyme, etc. The matrices used in this method are either organic or inorganic in nature. Some of the examples of matrices used in this method are ceramic, starch, alumina, activated carbon, cellulose, gelatin, etc. Without removing any non-adsorbed enzyme, the enzyme is directly attached to the adsorbent (Figure 5.5(G)).

Advantages:

1. Simple method and cheap;
2. No reagents are needed and easily carried out;
3. Less disruptive to the enzyme;
4. Easy to separate from the reaction mixture and reused;
5. Helps in purification and EI without any conformational change.

Disadvantages:

1. Desorption of the enzyme occurs rather often;
2. Non-specific.

5.4.2.2 FORMATION OF DISULPHIDE BOND

Between matrix and enzyme, a stable covalent bond is formed. This bond can be broken only by the reaction with dithiothreitol (DTT) medium under mild conditions. The activity yield of this method is high and provides an appropriate thiol-reactive adsorbent with high specificity by changing the pH (Figure 5.5).

5.5 CARRIER MATERIALS USED IN ENZYME IMMOBILIZATION (EI)

The substances used for the immobilizing of enzymes are termed as "carriers/support." The interaction between the enzyme and carrier provides an immobilized enzyme with the above-specified properties, including chemical, biochemical, and mechanical properties. To get an effective EI, the support materials should have some required features which are summarized in Figure 5.6.

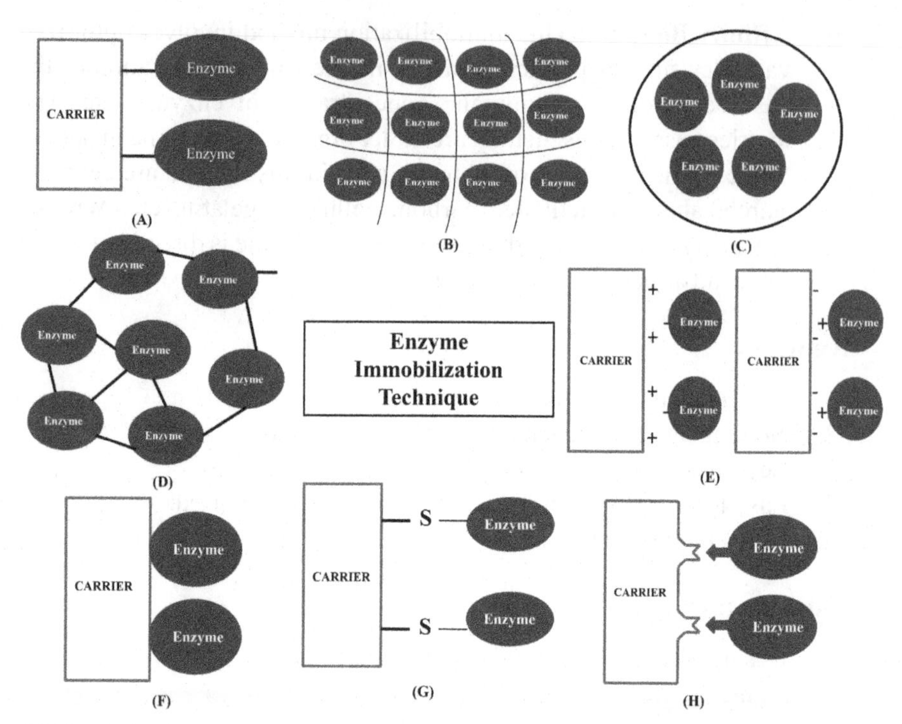

FIGURE 5.5 Schematic representation of enzyme immobilization technique: (A) covalent coupling, (B) entrapment (C), microencapsulation, (D) cross-linking, (E) ionic binding, (F) adsorption, (G) affinity binding, and (H) disulfide binding.

The role of carrier materials is the most important factor in the EI. So the selection of these materials is mandatory for immobilization. Three main factors which should be considered in the selection of carriers are the cost, availability, and stability. Different types of carriers are used in EI, which are divided into organic, inorganic, and hybrid materials. In severe reaction conditions, the carrier has to protect the enzyme structure to retain a high catalytic activity. The physicochemical parameters of the carrier which influence the reactor configuration are form, shape, density, porosity, pore size distribution, operational stability and particle size distribution of the supporting matrix (Wong, 2013). There are several types of carrier materials used in EI. Generally, carrier materials are classified into two types, such as chemical and morphological. Chemical carrier is again divided into inorganic and organic carriers. And the morphological carrier is divided into porous and non-porous material. Synthetic and biopolymer are grouped under an organic carrier. The ideal carrier matrices used for EI should be physically

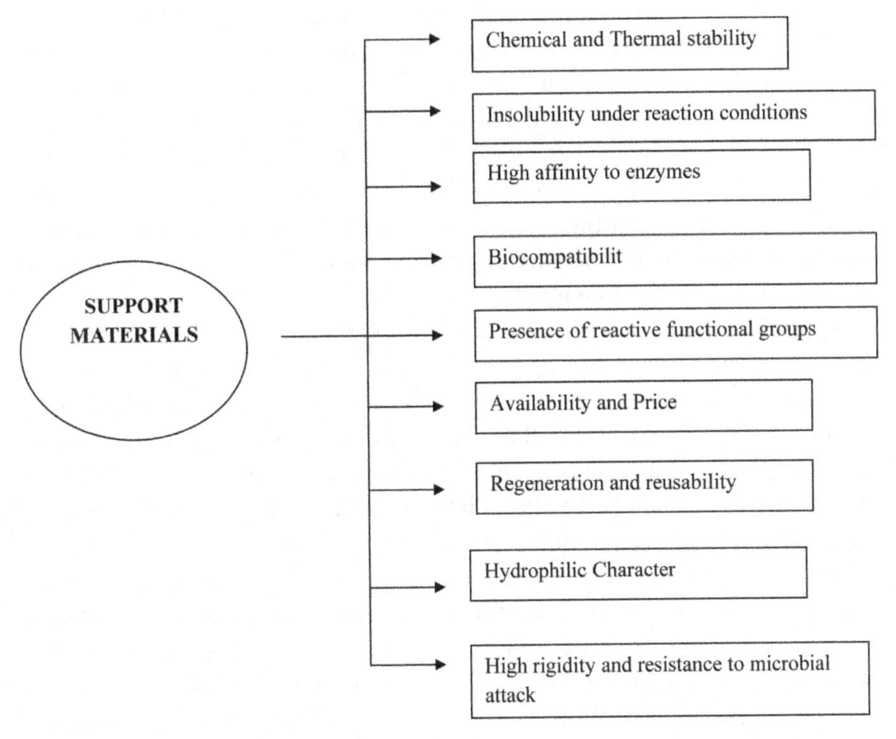

FIGURE 5.6 Features of support materials used for the enzyme immobilization.

very strong and stable. The carriers used in the EI by the chemical method are divided into both organic and inorganic carriers. Inorganic carriers have several benefits such as good thermal stability, mechanical properties, and resistance against microbial attack. Some of the examples of inorganic carriers, that used in EI are glass, silica gel, alumina, hydroxyapatite, nickel/nickel oxide, titania, zirconia, etc. Silica-based supports, as an inorganic carrier, can be easily fabricated. It provides a desirable morphology, pore structures, and microchannels to allow the interaction between enzyme and carrier. These are the most used for industrial and research purposes (Blanco et al., 2004; Vianello et al., 2000). Besides the above-mentioned carriers, magnetic particles, mesoporous materials, nanoparticles, ceramic materials, carbon nanotubes, graphene, and graphene oxide (GO) are also used as support materials for enzymes. Magnetic particles are the supporting materials having high mechanical stability, low porosity, and large surface area for easy modification (Fan et al., 2003). The main advantages of nanoparticles are to require a large surface area for enzyme binding, increase immobilization yield, and minimize diffusional limitations (Jia et al., 2003; Kim et al., 2006). The most important benefits of both ceramic materials and carbon nanotubes exhibit good mechanical stability and high resistance to temperature, pressure, and chemicals (Hong et al., 2015; Pedrosa et al., 2010). Due to the presence of several features, enzymes like lipase or peroxidase can be immobilized on GO surfaces by physical methods such as adsorption, covalent binding, or entrapment. And these features include biodegradability, two-dimensional structure, high surface area, good thermal, and chemical stability (Tseng et al., 2014).

Organic carriers mainly classified into synthetic polymer and biopolymers. Synthetic polymers are highly hydrophilic and stable in nature. Diffusion limitation is the main drawback of this polymer (Boller, 2002). Generally, biopolymers are the polysaccharides which are insoluble in water, and these are cellulose, starch, agarose, carrageenans, and chitosan, etc. (Velde et al., 2002). On the other hand, the organic materials such as electrospun materials and polymeric membranes are used in single particle form. Biocatalytic systems have more valuable applications in industrial-scale processes due to their performance, higher efficiency, and stability. Also, electrospun support materials have useful properties, including biocompatibility, nontoxicity, biodegradability, high mechanical strength, and hydrophilicity. Therefore, it can be used as suitable matrices for several types of biocatalysts (Dai et al., 2016; Liang et al., 2007). Polymeric membranes having good porosity and large surface area, it can allow efficient enzyme attachment and facilitates the immobilization of biomolecules on the surface of the support (Dwevedi, 2016).

5.6 ECONOMIC IMPORTANCE OF ENZYME IMMOBILIZATION (EI)

From the literature survey, it is observed that for industrial enzymes, the economy is increased to $US3.3 billion in 2010 and then exceeds to around $4.5 to $5 billion by 2015 at a growth rate of 7–9%, annually. For non-industrial markets such as in pharmaceutical, diagnostic, and research applications, it is estimated around $2.4 billion in 2010. In case of biocatalysis, the value of immobilized enzymes is $160 million in 2010 and then increase to $230 million by 2015 (Cornelius et al., 2011; Harris and Chrispeels, 1975).

Most of the enzymes are cheaper and much more expensive. So they should be utilized with an effective way of having a great social impact in every sector. Enzymes are not directly used by the processes having catalytic molecules. When they are used in a soluble form, they cannot be economically recovered for re-use due to retaining some activity after the reaction. So, generally, they are becoming wasted and contaminated the product (He et al., 2015). Therefore to eliminate this wastage, a simple and economical method must be used by which the productivity can be improved and the enzyme can be separated from the reaction product using a two-phase system that is described before. By this process, several benefits are raised, and these are the prevention of contaminating the product, minimization of cost and labor, possible effluent handling problems, etc. The demand of enzymes from various sources (food products, detergents, beverages, etc.) including plant; microbes, etc. and the demand of consumer goods must be increased. Lots of innovative researches are going on to get a high and effective yield, which can be helpful for plant and microbial enzymes. Instead of microbial enzymes, plant enzymes increase the industrial applications and production of economically useful compounds.

5.7 IMPORTANCE OF IMMOBILIZED PLANT ENZYMES IN AGRICULTURAL INDUSTRY

Most of the enzymes such as carbohydrases, proteases, phytases, lipases, etc. are found in plants, and these are beneficial to the various applications used in industries. Plant enzymes are highly region and stereospecific by comparing with other enzymes. Due to the high demand of plant enzyme, they can be used in pharmaceuticals companies, synthesis of various drugs, oxidation, hydroxylations, acetylations, (de)esterification and isomerizations, etc. Among the plant parts, seeds having starch and protein are the richest source of enzymes. These seeds are hydrolyzed by lipase, protease enzyme (Morkunas et al., 2013). Enzymes degrade the storage products in the endosperm, which promote the growth of seedling. Several plant enzymes such as α- and β-Amylase, phytase, protease, lipase, transglutaminase, cellulose, tyrosinase, pectinase, lipoxygenase, α- and β-galactosidase, etc. can produce pure products. α- and β-amylase enzymes break down starch into simple sugars, which helps to germinate the seeds and in the growth of the seedlings (Champagne, 2008; Witcher et al., 1998). These are used in bread, detergents, and clothing industries. Also, it helps in the production of high-fructose corn syrup (HFCS). Phytase is alkaline in nature and found in pollen grains and seed part of the plant (Nakajima et al., 2000). It has several benefits, such as used in the diets of simple-stomached animals and corn wet milling. A protease is a group of enzymes which hydrolyzes peptide bonds to protein and involves in several activities such as digestive purposes, growth, and development, extracellular modeling or remodeling tissues, defense mechanisms, senescence, cell death, etc. Transglutaminase is alkaline in nature and catalyzes the formation of an isopeptide bond between a free amine group and the acyl group. It plays an important role during photosynthesis and used in the food industry, meat processing, bakery industry, etc. Lipase is widely available from natural sources due to their low cost and easy purification. It has several applications in the dairy, detergent, paper, pesticide, and biodiesel industries (Herrero et al., 2005). Cellulose enzyme is produced by fungi, bacteria, protozoa that generate cellulolysis. Mostly, it is found in all plant species and very helpful for human bodies by controlling sugar blood level, preserve, and lower the cholesterol level. Tyrosinase mostly found plant and animal tissue which have the ability to catalyze melanin and other pigments from tyrosine by oxidation, and it plays an important role in immune response, several medicinal applications such as melanin synthesis for therapeutic purposes (Kieleczawa et al., 1992). Pectinase is an enzyme found in plant cell wall that helps in breaks down pectin. It helps in growth and development of plant, fruit ripening, and seed germination (Sun and Henson, 1991). The benefits of this enzyme are proper maceration of fruit pulp, used as a thickener and gelling agent in the preparation of jellies jams, etc. It is very helpful for the health benefits of human being by increasing digestibility. It also helps in the growth and development of plant. Lipoxygenase is a family of iron-containing enzyme and available in plants, fungi, and animals. It helps to catalyze the oxygenation of polyunsaturated fatty acids to form fatty acid hydroperoxides. It is used as a lubricant, emulsifier, and plasticizer. It is also applicable in bread making, aroma production. α- and β-galactosidase enzymes are obtained from various sources such as microorganisms, plants, and animals. It helps in hydrolysis of lactose in milk and whey.

To achieve a high productivity and low-cost labor, various techniques are developed for conducting EI. Plant enzymes have better economical advantages due to their cheaper rate, easier production, and usage. A systematic analysis should be developed for designing strong immobilized plant enzymes, which has a valuable benefit for upcoming applications. There are several limitations for soluble plant enzymes due to the exposure to several non-natural conditions (high pH, temperature, concentration, etc.). EI is the method to overcome this problem according to the enzyme's selection, duration, and activity. Due to easier product separation and less laborious, immobilized enzymes are found to be having more catalytic efficiencies compared to soluble enzymes. Some of the plant immobilized enzymes are β-amylase (produced from fenugreek onto chitosan beads coated with polyvinyl chloride and polyvinyl pyrrolidone), α-galactosidase (produced from chickpea seeds onto graphene nanosheets), β-galactosidase (produced from pea onto gold nanoparticles and also from chickpea onto grapheme alkylamine glass beads), immobilized chlorophyllase (produced from *Brassica* spp. onto magnetic alginate beads), immobilized alkaline protease (produced from brinjal onto mesoporous matrices), immobilized lipase (produced from *Pachira* aquatic onto calcium alginate beads coated with polyvinyl alcohol) and urease (produced from soybean seeds onto chitosan beads). By using these matrices, the physicochemical properties are improved. The immobilizing efficiencies of both β-galactosidase and urease were found to be 67.12 and 77% corresponding to the protein of 50 and 500 μg, respectively (Dwevedi, 2016). Recently, computer modeling, a new technique is developed, which predicts the efficacy of given matrix and gives the information about the changes during EI (Jacob, 2009). According to Kishore et al., 2012 and Yang et al. (2014), they found that immobilized soybean lipoxygenase has immobilized onto DEAE-cellulose with 30.6% efficiency and rice peroxidize immobilized onto poly-5-carboxy-indole having high thermostability, respectively. The factors such as number, position, and nature of bond acting between enzyme and matrix determine the loss or gain in enzymatic activity on immobilization. Having catalytic properties and better economical impact, immobilized plant enzymes have a greater beneficial effect on enzymatic reactions (Singh et al., 2012).

5.8 IMPLICATION OF ENZYME IMMOBILIZATION (EI) IN PHARMACEUTICALS

The use of immobilized enzymes in pharmaceuticals is very much extensive and used for curing several diseases (Herrero et al., 2003). Stability and growth are the two major benefits of the EI in pharmaceutical industry. The enzymes provide biological and functional systems when combined with the biomaterials. Pharmaceutically enzymes are defined as the prodrugs that target a specific biological reversible or irreversible reaction for the treatment of diseases (Yan et al., 2014). Some of the pharmaceutical enzymes are cysteine proteinases, asparaginase, streptokinase, urokinase, deoxyribonuclease, hyaluronidase, pegademase, glucocerebrosidase, etc. In the pharmaceutical industry, production of 6-aminopenecillinic acid using immobilized penicillin amidase is the major application of immobilized enzymes which helps in the deacylation of the side chain of either penicillin G or penicillin V. The use of immobilized enzymes in pharmaceuticals is classified into two categories: biosensors and bioreactors. Biosensor possesses several advantages such as reliability, sensitivity, accuracy, easy handling, low cost, etc. whereas the advantages of bioreactor are blood detoxification, inborn metabolism errors, cancer treatment, removal of waste metabolites, etc. Figure 5.7 describes the application of biological enzyme. The pharmaceutical enzymes such as superoxide dismutase, asparaginase, urease, lactate oxidase, heparinase, and alcohol oxidase, etc. are used in bioreactor (Guo et al., 2006; Zhang et al., 2007).

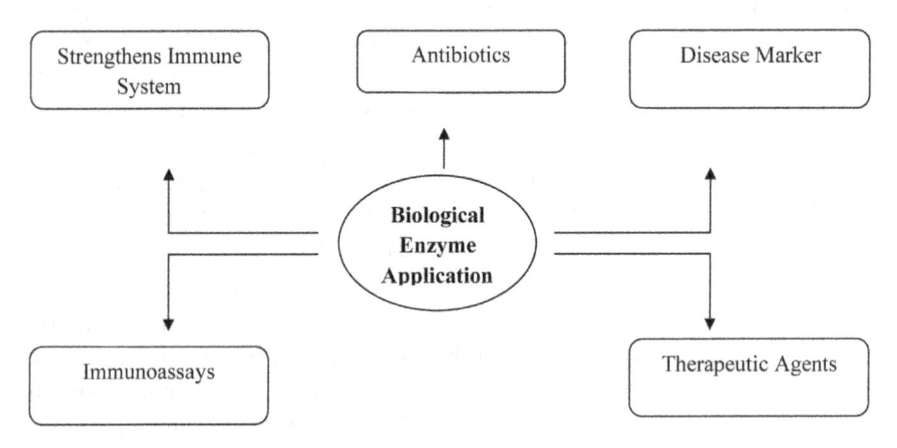

FIGURE 5.7 Applications of various biological enzymes.

5.9 APPLICATION OF ENZYME IMMOBILIZATION (EI) IN FOOD PROCESSING

Enzymes are the proteins which have been used deliberately in food processing sector since ancient times to make a variety of food products such as bakery products, fermented alcoholic beverages, dairy products, fish, meat products, etc. Traditionally, these enzymes are produced by the extraction and fermentation processes from plant, cultivable micro-organisms, and animal sources. These food enzymes have powerful catalytic functions which have a number of distinct advantages. The advantages include high yield, less processing time, catalytic efficiency, less residual production, control enzyme activity, improve sensory characteristics, reaction rate, etc. High cost and low stability are the main disadvantages of food enzyme.

5.9.1 MEAT PROCESSING

Distribution of tenderizing protease evenly into meat pieces is a challenging method in meat processing. Biochemical consistency and structure of meat are the factors which affect the tenderization. This tenderization effect is executed by the protease enzyme present in the muscle of meat. Several methods are used, such as dipping or marinating of meat pieces in proteolytic enzymes solution. Besides this method, proteolytic enzyme solution is also directly injected into meat pieces. Enzymes used in the meat industry and their activities are described in Table 5.3.

TABLE 5.3 Enzymes Used in Meat Industry and Their Activity

Meat Industry	Enzymes Used	Activity
	Proteolytic enzyme	Increase the tenderization effect and modify texture (Melendo et al., 1996) Improve chewability, digestibility, flavor, nutritional value, etc. Reduce bitterness
	Transglutaminase	Improve texture, flavor, and shelf life of product (Romero de Avila et al., 2010)
	Lipases	Improves flavor in production of sausage (Miklos et al., 2011)
	Phytase	Reduce cost and environmental contamination (Lim and Lee, 2009)

5.9.2 DAIRY INDUSTRY

Milk, highest producer in India, is a wholesome natural food. The demand of dairy products is increasing worldwide day today. Due to surplus availability of milk, nowadays the liquid milk is converted to different types of value-added product by using different biochemical and food enzymes. Table 5.4 gives details about the enzymes used in milk and cheese industries.

5.9.3 BAKERY INDUSTRY

Baking is the processing method to produce the baked products like bread, biscuits, cookies, cake pastries, etc. where wheat flour and enzyme substrate are taken as important ingredients. There is a high demand in bakery industry

TABLE 5.4 Enzymes Used in Milk and Cheese Industry and Their Activity

	Enzymes Used	Activity
Milk and Cheese Industry	**Proteases**	Hydrolyze caseins during cheese production and stabilizes micelle formation preventing coagulation Modify texture, flavor, and solubility properties of milk Accelerates cheese ripening
	Rennet	Oldest application for cheese production in the milk industry
	Rennin	Most widely used to coagulate milk Hydrolyzes the Phemet bonds and releases a strong hydrophilic C-terminal fragment (Ageitos et al., 2006)
	Lipases	Helps in breakdown of milk fat, modify the length of fatty acid chains, and improves flavor.
	Lactose	Hydrolyzed lactose four times sweeter than non-hydrolyzed Improves scoop ability and creaminess of the product
	Peptidases	Improves sensory characteristics (taste and aroma) of cheese due to the production of peptides, amino acids, fatty acids, deaminases, decarboxylases, aminases, transaminases, etc.
	β-galactosidase	Helps in catalyzation of formation of galacto-oligosaccharides (Grosová et al., 2008) Helps in hydrolysis of whey lactose to glucose and galactose to avoid intolerance
	Lysozyme	An antimicrobial enzyme that limits the growth of *Clostridia* in aged cheese. Causes swelling of the cheese shape
	Nisin	Useful in non-thermally processed dairy products

for using food enzymes, such as peptidases, lipases, xylanases, pullulanases, pentosanases, cellulases, glucose oxidase, lipoxygenase, etc. Generally, enzymes from three sources are used in baking. These are endogenous enzymes which are used in flour and exogenous enzymes which are added in the dough. To get a nutritionally rich and qualitative product, following food enzymes are used by replacing chemical additives completely or partially. By using these enzymes, the rheological properties and qualities (volume, color, flavor, aroma, structure, tenderness, and shelf life) of final bakery product increases. Enzymes used in bakery industry and their activities are given in Table 5.5.

TABLE 5.5 Enzymes Used in Bakery Industry and Their Activity

	Enzymes Used	Activity
Bakery Industry	**Amylases**	Breakdown of starch for improving human nutrition Helps in hydrolyzation of starch into glucose, maltose, and dextrin Helps in increasing the loaf volume and reducing the aging rate of starch (Gupta et al., 2003)
	Lipoxygenase	Increases the strength, stability, viscosity, and rheological properties of the flour
	Glucose oxidase	As a conditioner, it helps in the preparation of wheat flour dough and also helps in oxidation of the protein matrix to make dough strong and more elastic
	Amyloglucosidase	Saccharification
	Lipase	Increases the volume, softness, strength, elasticity, and reduces bread aging
	Proteases	Used to improve dough machinability, crumb texture, color, flavor, and water absorption Reduces mixing time

5.9.4 FRUIT JUICE INDUSTRY

The enzymes present in fruit juice basically contain pectolytic enzymes such as polygalacturonases, pectinesterases, pectin lyases, cellulases, hemicellulases, proteases, beta-glucanase, etc. These enzymes can increase the yield of solid recovery during the processing of fruit juice. The enzymes used in the fruit juice industry and their activities are given in Table 5.6.

5.9.5 WINE PRODUCTION

Wine is an alcoholic beverage which is the result of fermentation of grape juice, and in this fermentation processes, enzymes play an important role. The most widely used enzymes and their activities are presented in Table 5.7.

During the purification of enzymes, the taken enzymes are denatured. Therefore, the immobilization technique of food enzyme is adapted to make the enzyme stable. The enzymes which are immobilized, they have a great importance in food processing sector. Several applications of EI are available such as production of syrups, production of baker's yeast by immobilized β-galactosidase which helps in hydrolysis of lactose in whey.

TABLE 5.6 Enzymes Used in Fruit Juice Industry and Their Activity

	Enzymes Used	Activity
Fruit Juice Industry	**Pectolytic enzyme**	Increase juice yield by pressing and filtration Promote the removal of suspended matter Prevents precipitation
	Amylase	Converts starch to glucose and clarification
	Cellulases	Lowering viscosity and maintains the texture
	Laccase	Increases the susceptibility of browning during storage
	Limoninase	Causes bitterness in citrus juices

TABLE 5.7 Enzymes Used in Wine Industry and Their Activity

	Enzymes Used	Activity
Wine Industry	**Pectinolytic enzymes**	Increases the yield, volume, and reduce the pressing time during grape crushing (Romero-Cascales et al., 2008)
	Maceration enzymes (Commercial enzyme preparations with cellulosic, hemicellulosic, and pectinolytic activities)	Improves the efficiency, quality, stability of wine and increases the extraction of pigments (Kaur et al., 2004)

Currently, some immobilized enzymes are being used in commercial food processing operations. These are the production of L-aspartic acid, production of 5'-ribonucleotides, isomaltulose production, sucrose hydrolysis, and aspartame synthesis by using amylase, 5'-phosphodiesterase, isomaltulose synthase, invertase, and thermolysin enzymes, respectively in their immobilized forms. Table 5.8 (Chaudhary et al., 2015) gives some food processing substrates by using their respective immobilized enzymes.

5.10 SUMMARY AND FUTURE PROSPECTS

Immobilization technology is used not only for bio-catalysis but also for various applications used in various industries. The stability and catalytic efficiency of immobilized enzyme would be greater if the size of the matrix is very small. Implementation of immobilized enzymes for several applications such as food processing industry, medicinal, and pharmaceutical industry has now been accepted after a long fight of experimentations and researches. For present and future requirements, specific types of enzymes and their applications must be available. Currently, new innovative work has been patented in case of cancer treatment by using immobilized enzymes. Also, the research approaches regarding EI should continue to improve yield and profit. Therefore the availability of immobilized enzymes with suitable price should be available easily in an efficient manner. Also, various researches are going on drug delivery, tumor location analysis, and biosensors with the use of immobilized enzymes. Some projects and researches should be going on the development of bioreactors by which the efficiency of reactions will improve.

TABLE 5.8 Application of Immobilized Enzymes with Respective Substrate

Enzyme	Immobilization support	Food substrate
β- galactosidase	Organic and inorganic support	Lactose removal from milk
Pectinase	Anion exchange resin	Pectin solution
Pectin lyase	Alginate beads	Esterified protein
Tyrosinase	Polyacrylic acid carbon nanotubes	Red wine
Lipase	Calcium alginate beads	Oil and grease
Trypsin	Cellulase	B-lactoglobulin
Laccase	Silica gel	Wine, fruit juice processing
Glucoamylase	Chitin	Hydrolyzed starch
Amyloglucosidase	Bone powder	Whey, skimmed milk, lactose

5.11 CONCLUSIONS

From, the above precisely discussed chapter, we concluded that EI is one of the most promising techniques that widely used in several industries such as food industry, pharmaceutical industry, bioremediation, detergent industry, textile industry, etc. to achieve maximum efficiency. To achieve several technological advantages, the method and application of EI in above-described sectors should be adopted. Nowadays, food enzymes are widely used by replacing the chemical method to get a quick and efficient performance. Immobilization permits the repetitive use of enzymes and hence can lower the cost and provides operational stability to enzymes. Several researches and projects are to be focused to overcome the problems related to immobilization techniques.

KEYWORDS

- **bio-catalyst**
- **biosensor**
- **enzyme immobilization**
- **food**
- **immobilization methods**

REFERENCES

Abdelmajeed, N. A., Khelil, O. A., & Danial, E. N., (2012). Immobilization technology for enhancing bio-products industry. *African Journal of Biotechnology, 11*(71), 13528–13539.

Ageitos, J. M., Vallejo, J. A., Paza, M., & Villa, T. G., (2006). Fluorescein thiocarbamoylkappa-casein assay for the specific testing of milk-clotting proteases. *Journal of Dairy Science, 89*, 3770–3777.

Barrientos, L., Scott, J. J., & Murthy, P. N., (1994). Specificity of hydrolysis of phytic acid by alkaline phytase from lily pollen. *Plant Physiology, 106*, 1489–1495.

Blanco, R., Terreros, P., Fernandez-Perez, M., Otero, C., & DiazGonzalez, G., (2004). Functionalization of mesoporous silica for lipase immobilization-characterization of the support and the catalysts. *Journal of Molecular Catalysis B: Enzymatic, 30*, 83–93.

Boller, T., Meier, C., & Menzler, S., (2002). Eupergitoxirane acrylic beads: How to make enzymes fit for biocatalysis. *Organic Process Research & Development, 6*, 509–519.

Brena, B., Pombo, P. G., & Viera, F. B., (2013). *Immobilization of Enzymes: A Literature Survey. Molecular Biology* (3rd edn., p. 1051). Springer Science Business Media New York.

Caldwell, K., Axén, R., Bergwall, M., & Porath, J., (1976). Immobilization of enzymes based on hydrophobic interaction. I. Preparation and properties of a beta-amylase adsorbate. *Biotechnology Bioengineering, 18*, 1573–1588.

Champagne, P., (2008). Bioethanol from agricultural waste residues. *Environmental Progress and Sustainable Energy, 27*, 51–57.

Chaudhary, S., Sagar, S., Kumar, M., Sengar, R. S., & Tomar, A., (2015). The use of enzymes in food processing: A review. *South Asian Journal of Food Technology and Environment*, 190–210.

Cornelius, S., Witz, S., Rolletschek, H., & Möhlmann, T., (2011). Pyrimidine degradation influences germination seedling growth and production of Arabidopsis seeds. *Journal of Experimental Botany, 62*, 5623–5632.

Dai, Y., Yao, J., Song, Y., Liu, X., Wang, S., & Yuan, Y., (2016). Enhanced performance of immobilized laccase in electrospun fibrous membranes by carbon nanotubes modification and its application for bisphenol A removal from water. *Journal of Hazardous Material, 317*, 485–493.

Dwevedi, A., (2016). Basics of Enzyme Immobilization. In: *Enzyme Immobilization* (pp. 21–44). Springer International Publishing: Switzerland.

Fan, J., Lei, J., Wang, L., Yu, C., Tu, B., & Zhao, D., (2003). Rapid and high-capacity immobilization of enzymes based on mesoporous silicas with controlled morphologies. *Chemical Communication, 17*, 2140–2141.

Grosova, Z., Rosenberg, M., & Rebros, M., (2008). Perspectives and applications of immobilized β-galactosidase in food industry-a review. *Czech Journal of Food Science, 26*, 1–14.

Guisan, J. M., Alvaro, G., Rosell, C. M., & Fernandez, L. R., (1994). Industrial design of enzymic processes catalyzed by very active immobilized derivatives: utilization of diffusional limitations (gradients of pH) as a profile table tool in enzyme engineering. *Applied Biotechnology and Biochemistry, 20*, 357–369.

Guisan, J. M., Penzol, G., Armisen, P., Bastida, A., Blanco, R., Fernández-Lafuente, R., & García-Junceda, E., (1997). Immobilization of enzymes acting on macromolecular substrates. In: Bickerstaff, G. F., (ed.). *Immobilization of Enzymes and Cells* (pp. 261–275).

Guo, T., Zhang, L., Liu, W., & Zhou, Z., (2006). A novel solution to power problems in implanted biosensor networks. *Medicine and Biology Society, 1*, 5952–5955.

Gupta, M., & Mattiasson, B., (1992). Unique applications of immobilized proteins in bioanalytical systems. *Methods of Biochemical Analysis, 36*, 1–34.

Gupta, R., Gigras, P., Mohapatra, H., Goswami, V. K., & Chauman, B., (2003). Microbialα-amylase: A biotechnological perspective. *Process Biochemistry, 38*, 1599–1616.

Harris, N., & Chrispeels, M. J., (1975). Histochemical and biochemical observations on storage protein metabolism and protein body autolysis in cotyledons of germinating mung beans. *Plant Physiol*ogy, *56*, 292–299.

He, M., Zhu, C., Dong, K., Zhang, T., Cheng, Z., Li, J., & Yan, Y., (2015). Comparative proteome analysis of embryo and endosperm reveals central differential expression proteins involved in wheat seed germination. *BMC Plant Biology, 15*, 1–17.

Herrero, G. F., Escribano, J., & García-Carmona, F., (2005). Betaxanthins as substrates for tyrosinase. An approach to the role of tyrosinase in the biosynthetic pathway of betalains. *Plant Physiology, 138*, 421–432.

Hong, G., Diao, S., Antaris, A. L., & Dai, H., (2015). Carbon nanomaterials for biological imaging and nanomedicinal therapy. *Chemical Reviews, 115*, 10816–10906.

Jacob, N., (2009). Pectinolytic enzymes. In: *Biotechnology for Agro-Industrial Residues Utilization* (pp. 383–396). New York: Springer.

Jia, H., Zhu, G., & Wang, P., (2003). Catalytic behaviors of enzymes attached to nanoparticles: The effect of particle mobility. *Biotechnology and Bioengineering, 84*, 406–414.

Kaur, G., Kumar, S., & Satyanarayana, T., (2004). Production, characterization and application of a thermostable polygalacturonase of a thermophilic mold *Sporotrichum thermophile* Apinis. *Bioresource Technology, 94*, 239–243.

Kieleczawa, J., Coughlan, S. J., & Hind, G., (1992). Isolation and characterization of an alkaline phosphatase from pea thylakoids. *Plant Physiol*ogy, *99*, 1029–1036.

Kim, J., Grate, J. W., & Wang, P., (2006). Nanostructures for enzyme stabilization. *Chemical Engineering Science, 61*, 1017–1026.

Kishore, D., Talat, M., Srivastava, O. N., & Kayastha, A. M., (2012). Immobilization of *β*-galactosidase onto functionalized graphene nano-sheets using response surface methodology and its analytical applications. *PLoS ONE, 7*, 1–13.

Liang, D., Hsiao, B. S., & Chu, B., (2007). Functional electrospun nanofibrous scaffolds for biomedical applications. *Advanced Drug Delivery Reviews, 59*, 1392–1412.

Lim, S. J., & Lee, K. J., (2009). Partial replacement of fish meal by cottonseed meal and soybean meal with iron and phytase supplementation for parrot fish *Oplegnathusfasciatus. Aquaculture, 290*, 283–289.

Melendo, J. A., Beltrán, J. A., Jaime, I., Sancho, R., & Roncalés, P., (1996). Limited proteolysis of myofibrillar proteins by bromelain decreases toughness of coarse dry sausage. *Food Chemistry, 57*, 429–433.

Messing, R. A., (1976). Adsorption and inorganic bridge formations. *Methods in Enzymology*, 148–169.

Miklos, R., Xu, X., & Lametsch, R., (2011). Application of pork fat diacylglycerols in meat emulsions. *Meat Science, 87*, 202–205.

Morkunas, I., Formela, M., Marczak, L., Stobiecki, M., & Bednarski, W., (2013). The mobilization of defense mechanisms in the early stages of pea seed germination against Ascochytapisi. *Protoplasma., 250*, 63–75.

Nakajima, M., Nakayama, A., Xu, Z. J., & Yamaguchi, I., (2000). Gibberellin induces alpha-amylase gene in seed coat of Ipomoea nil immature seeds. *Bioscience Biotechnology Biochemistry, 68*, 631–637.

Nisha, S., Arun, K. S., & Gobi, N., (2012). A review on methods, application and properties of immobilized enzyme: Chemical science review and letters. *Chemical Science Review and Letters, 1*(3), 148–155.

Pedrosa, V. A., Paliwal, S., Balasubramanian, S., Nepal, D., Davis, V., Wild, J., Ramanculov, E., & Simonian, A., (2010). Enhanced stability of enzyme organophosphate hydrolase interfaced on the carbon nanotubes. *Colloids and Surfaces: B Biointerface, 77*, 69–74.

Romero de Ávila, M. D., Ordónez, J. A., De la Hoz, L., Herrero, A. M., & Cambero, M. I., (2010). Microbial transglutaminase for cold-set binding of unsalted/salted pork models and restructured dry ham. *Meat Science, 84*, 747–754.

Romero-Cascales, I., Fernandez-Fernandez, J., Roz-Garcia, J. M., Lopezz-Roca, J. M., & Gomez-Plaza, E., (2008). Characterization of the main enzymatic activities present in six commercial macerating enzymes and their effects on extracting color during winemaking of Monastrell grapes. *International Journal of Food Science and Technology, 43*, 1295–1305.

Sharma, M., Kumar, V., & Pundir, C. S., (2008). *Indian Journal of Biotechnology, 7*, 328–332.

Singh, P., Prakash, R., & Shah, K., (2012). Effect of organic solvents on peroxidases from rice and horseradish: Prospects for enzyme-based applications. *Talanta, 97*, 204–210.

Solomon, B., Hollaander, Z., Koppel, R., & Katchalski, K. E., (1987). Use of monoclonal antibodies for the preparation of highly active immobilized enzymes. In: Mosbach, K., (ed.), *Methods in Enzymology* (Vol. 135, pp. 160–170).

Sun, Z., & Henson, C. A., (1991). A quantitative assessment of the importance of barley seed α-amylase, β-amylase, debranching enzyme, and α-glucosidase in starch degradation. *Archives of Biochemistry and Biophysics, 284*, 298–305.

Tischer, W., & Wedekind, F., (1999). Immobilized enzymes: Methods and applications. *Topics in Current Chemistry, 200*, 95–125.

Tseng, C., Liao, C., Sun, Y., Peng, C., Tzen, J. T. C., Guo, R., & Liu, J., (2014). Immobilization of Clostridium cellulolyticum. *Journal of Agriculture and Food Chemistry, 62*, 6771–6776.

Van de Velde, F., Lourenço, N., Pinheiro, H., & Bakker, M., (2002). Carrageenan: A food-grade and biocompatible support for immobilization techniques. *Advanced Synthesis and Catalysis, 344*, 815–835.

Vianello, F., Zennaro, L., Di Paolo, M. L., Rigo, A., Malacarne, C., & Scarpa, M., (2000). Preparation, morphological characterization and activity of thin films of horseradish peroxidase. *Biotechnology and Bioengineering*, *68*, 488–495.

Witcher, D. R., Hood, E. E., Peterson, D., Bailey, M., Bond, D., Kusnadi, A., et al., (1998). Commercial production of β-glucuronidase (GUS): A model system for the production of proteins in plants. *Molecular Breeding*, *4*, 01–12.

Wong, D. W. S., (2013). *Food Enzymes: Structure and Mechanism* (p. 390). Dordrecht: Springer Science & Business Media.

Woodward, J., (1985). Immobilized enzymes: Adsorption and covalent coupling. In: Woodward, J., (ed.), *Immobilized Cells and Enzymes: A Practical Approach* (pp. 3–17).

Yan, D., Duermeyer, L., Leoveanu, C., & Nambara, E., (2014). The functions of the endosperm during seed germination. *Plant Cell Physiol.*, *55*, 1521–1533.

Yang, C. H., Yen, C. C., Jheng, J. J., Wang, C. Y., Chen, S. S., Huang, P. Y., Huang, K. S., & Shaw, J. F., (2014). Immobilization of *Brassica oleracea* chlorophyllase 1 (BoCLH1) and *Candida rugosa* lipase (CRL) in magnetic alginate beads: An enzymatic evaluation in the corresponding proteins. *Molecules*, *19*, 11800–11815.

Zhang, M., Haider, M. R., Huque, M. A., Adeeb, M. A., Rahman, S., & Islam, S. K., (2007). A low power sensor signal processing circuit for implantable biosensor applications. *Smart Materials and Structure*, *16*, 525–530.

CHAPTER 6

Aonla: Composition, Medicinal Properties, Processing, and Food Applications

PARVEEN KUMARI and B. S. KHATKAR

Department of Food Technology, Guru Jambheshwar University of Science and Technology, Hisar, Haryana, India,
E-mail: bskhatkar@yahoo.co.in

ABSTRACT

Emblica officinalis (aonla) is native to tropical India and Southeast Asia, commonly known as Indian gooseberry, amla, nelli, amalaki. It belongs to family Eubhorbiaceae. Nutritional, commercial, and medicinal significance of aonla fruit makes it popular all over the world, and it is commonly recommended in both in Unani and Ayurvedic systems of medicine. Aonla is an excellent source of ascorbic acid (AA) (300–900 mg/100 g), amino acid and minerals along with phytochemicals such as polyphenols, tannins, emblicol, linoleic acid, corilagin, phyllemblin, and rutin. Technological importance of aonla is mainly attributed to the phytochemical part, which contributes to antioxidant properties. Aonla has amazing remedial qualities and believed to increase defense against various diseases viz., ulcers, gastrointestinal disorders, heart troubles, and diabetes. Additionally, aonla is useful in enhancing memory, lowering cholesterol, and also act as an antimicrobial agent. The embolic fruit is a rare example of an edible material, which is rich in tannins and AA. Aonla fruit is acidic, cooling, diuretic, and laxative in nature, and the dried aonla fruit is used in treating hemorrhage, diarrhea, and dysentery. Amazing antibacterial and astringent properties of aonla help in preventing infection and also in healing of ulcers. Various aonla extracts have expectorant, purgative, spasmolytic, anti-tumor activities, hypoglycemic, and strong antibacterial activity. Table value of aonla fruit is very limited due to its sour and astringent taste. Processing of aonla into different kind of products like preserve, juice, pickle, candy, chutney, RTS beverages, and dried powder adds value to the aonla fruits.

6.1 INTRODUCTION

Fresh fruits are the natural staple food having highly beneficial natural effect on the human system. India is the second largest producer of fruits in the world after China. *Emblica officinalis* (aonla) is native to tropical India and Southeast Asia, commonly named as Indian gooseberry, amla, nelli, amalaki (Barthakar and Arnold, 1991). Aonla, member of family Euphorbiaceae, is one of the most important non- traditional fruit having immense potentiality for cultivation on wasteland. Aonla being a minor fruit crop gained importance among the farmers, and scientists as its production is highly remunerative without care expenses and has commercial significance. India ranks 1st in aonla area and production (Kore et al., 2013). Aonla varieties are divided into three categories on the basis of maturity status (Pathak, 2002) (Table 6.1).

TABLE 6.1 Classification of Aonla Varieties Based on Maturity Status

Variety	Season	Group
Banarasi, NA-9, NA-10, and Krishna	Mid-October–mid-November	Early maturing
Francis, NA-7, Kanchan, and NA-6	Mid-November–mid-December	Median maturing
Chakaiya, Bhawanisagar	Mid-December–mid-January	Late maturing

Aonla fruits are fleshy, yellowish green in color having six vague perpendicular furrows enclosing seeds. Nutritional, commercial, and medicinal significance of aonla fruit makes it popular all over the world (Goyal et al., 2007) and it is commonly recommended in both in Unani and Ayurvedic systems of medicine. Aonla is an excellent source of ascorbic acid (AA) (300–900 mg/100 g), amino acid and minerals along with phytochemicals such as polyphenols, tannins, emblicol, linoleic acid, corilagin, phyllemblin, and rutin (Ghorai and Sethi, 1996; Jain and Khurdiya, 2004; Murthy and Joshi, 2007; Baliga and Dsouza, 2011) (Figure 6.1).

These days an interesting trend in the use of "food as a medicine" is emerging. Medicinal and therapeutic properties of aonla are well documented, and it is considered a wonder fruit for health conscious people (Anon, 1969; Pathak, 2003). Technological importance of aonla is mainly attributed to the phytochemical part, which contributes to antioxidant properties. The embolic fruit is a rare example of an edible material, which is rich in tannins along with AA (Kalra, 1988). Table value of aonla fruit is very limited due to its sour and astringent taste. The demand for processed products of aonla is steadily increasing in domestic as well as in global market. Processing of aonla into different kind of products like preserve, juice, pickle, candy, chutney, RTS beverages, dried powder, etc. is done (Tondon et al., 2003). In recent time, due to many folds increase in area and production of aonla, its processing, and value addition has increased (Goyal et al., 2007).

FIGURE 6.1 Aonla fruit.

6.2 PHYSICO-CHEMICAL AND NUTRITIONAL COMPOSITION OF AONLA FRUIT

6.2.1 PHYSICAL PROPERTIES

Processing of fruits into different products depend on their physicochemical properties as these affect final texture, appearance, and nutritional quality of the end product. In general, for the judgment of nutritional profile, proximate composition of a food commodity is evaluated. Physical parameters includes fruit weight, seed weight, pulp weight, pulp: seed ratio, volume, specific gravity, fruit width, length, diameter, texture of fruit in terms of hardness. Various research studies were conducted by different researchers to evaluated physico-chemical properties of aonla fruit. Data pertaining to physical properties of aonla fruit is given in Table 6.2. Yellowish green color and round to oblate shape of aonla fruit is reported by Ganchari et al., (2010); Dahiya and Dhawan (2001); Ghorai and Sethi (1996) and Kalra (1988). Hazarika et al., (2009) reported varietal difference in physical parameters such as fruit weight, seed weight, pulp weight, pulp: seed ratio, volume, and specific gravity of 14 types from Jorhat region of Assam. Varietal difference exists in hardness and toughness of aonla fruits.

TABLE 6.2 Physical Properties of Fresh Aonla Fruit

Parameter	Quantity
Weight (g)	14.27–49.97
Length (mm)	25.12–41.10
Width (mm)	29.38–44.85
Geometric Diameter (mm)	33.7–42.9
Hardness (g)	253.13–583.27

Source: Parveen and Khatkar (2015); Kumar and Singh (2013); Ganchari et al. (2010); Hazarika et al. (2009); Goyal et al. (2007).

6.2.2 NUTRITIONAL COMPOSITION

The fruits are rich in AA and tannins. Tannins present in aonla are known to retard the oxidation of vitamin C (Srivastava and Ranjan, 1967). Aonla fruit contained 160 times higher AA content and approximately three times higher protein content than apple. AA is an important nutritional component for human health. Consumption of aonla fruit as staple food and in processed form is increasing due to its high ascorbic content. Amino acid such as glutamic acid, proline, alanine, lysine, and aspartic acid were reported higher in aonla fruit compared to apple (Barthakur and Arnold, 1991). Aonla is a good source of carotene, thiamine, riboflavin, carbohydrate, and minerals (Pareek and Kaushik, 2012). Nutritional composition of fresh aonla fruit is reported in Table 6.3. Moisture content is an important parameter in assessing the quality of fresh fruits and vegetables.

Mishra et al., (2009) compared wild and Chakaiya variety with respect to physicochemical characteristics and reported that AA and total polyphenol content as gallic acid equivalent (GAE) were higher in wild variety than variety Chakaiya. Phenol content of aonla cultivar varied in range 162.2 to 175.7 mg/100 g. Study by Singh et al., (2009) reported good AA content, total sugar, titratable acidity, and TSS in different aonla cultivars evaluated for physicochemical properties under Chhattisgarh conditions. Singh et al., (1987), studied physico-chemical properties of different cultivars of aonla. Highest AA and starch content was observed in variety Banarasi. Maximum acidity was observed in variety Harpharori while variety Desi showed maximum total soluble solid. The chemical composition of fresh fruit of aonla with respect to moisture (80–81%), protein (0.5–0.8%), fat (0.1–0.2%), crude fiber (3.1–3.4%), starch (3.4%), sugars (5–15%), minerals (0.7%) and vitamins (400–600 mg/100 g) has been reported by various researchers (Srivastava, 1960; Tripathi et al., 1988; Dahiya and Dhawan, 2001; Goyal et al., 2008; Khan, 2009).

TABLE 6.3 Nutritional Composition of Fresh Aonla Fruit

Parameter	Quantity
Moisture (%)	80–87
Pulp (%)	89.8–94.8
Stone (%)	3.5–7.2
Fiber (%)	0.8–2.1
Carbohydrate (%)	12.5–21.5
Protein (%)	0.4–0.6
Fat (%)	0.1–0.12
Total Soluble Solid (%)	9.6–14.3
Acidity (%)	1.6–2.4
Sugar	
Total (%)	3.5–4.9
Reducing (%)	2.1–2.9
Pectin (%)	2.4–3.4
Vitamins	
Ascorbic acid (mg/100 g)	500–1500
Vitamin A (I.U.)	10.5–65.3
Thiamine (mg/100 g)	0.04–0.06
Riboflavin (mg/100 g)	0.5–1.0
Nicotinic acid	0.20–0.30
Minerals	
Total (%)	0.40–0.61

TABLE 6.3 *(Continued)*

Parameter	Quantity
Iron (mg/100 g)	10.8–17.6
Calcium (mg/100 g)	0.04–0.08
Phosphorus (mg/100 g)	1.5–2.5
Sulfur (mg/100 g)	4.4–16.0
Phenols	
Total (%)	2.4–3.1
Gallic acid (mg/g)	0.44–3.60
Ellagic acid (mg/g)	0.15–0.80
Ascorbic acid (% w/w)	0.40–2.36

Source: Parveen and Khatkar (2015); Singh (2012); Sawant et al. (2011); Jaiswal et al. (2007); Scartezzine et al. (2006).

Dahiya and Dhawan (2001) reported appreciable amount of total sugar (7.43g/100 g) and AA (454.40mg/100 g) and significant minerals content such as calcium (14.91g/100 g), iron (0.62mg/100 g) and phosphorus (11.51mg/100 g) in fresh aonla fruit.

6.3 PHYTOCHEMICAL COMPOSITION OF AONLA FRUIT

Researchers all around the world have made various attempts from time to time to extract phytochemicals present in aonla. Studies mainly emphasized on evaluation of therapeutic value of phytochemicals present in different aonla extract. Alkaloids, tannins, phenolic compounds, and amino acids were phytochemical extracted from aonla, which may be responsible for anti-pyretic and analgesic activities (Perianayagam et al., 2004). Total polyphenolic content in methanolic extract of aonla fruits from different regions of China ranged from 81.5 to 120.9 mg GAE/g and flavonoids content varied in range 3.7–18.7 mg catechin equivalents (CE)/g. Mishra et al., (2015) separated gallic acid, ellagic acid, emblicanin A, and Bby TLC. Study concluded that emblicanin A and B had 7.86 and 11.20 times more DPPH scavenging activity than AA and 1.25 and 1.78 times more than gallic acid, respectively. Hydrolyzable tannins present in *Emblica officinalis* are credited for antioxidant activity of extract which were found to have AA-like action. Total phenolic content of extract was 7.02 mg GAE/ml of extract, and their % DPPH radical scavenging activity was found to be 42.69%.

Ethanolic and aqueous extract of *Emblica officinalis* were evaluated for their antipyretic and analgesic activity against brewer's yeast induced hyperthermia in rats. Study by Perianayagam et al., (2004) revealed that both extracts showed potent antipyretic and analgesic activity and preliminary phytochemical screening of extracts showed presence of different compounds such as alkaloids, tannins, phenolics, amino acids and carbohydrates which might be responsible for their antipyretic and analgesic activity. Preliminary phytochemical screening of petroleum ether extract of *Phyllanthus emblica* by Elangovan et al., (2015) reported presence of tannins, phenolic compounds, flavonoids, and some other compounds such as alkaloids, carbohydrate, glycosides, steroids, saponins.

Preliminary phytochemical screening of aqueous extract of aonla powder showed presence of glycosides, tannins, flavonoids, saponins, alkaloids, and inorganic constituents and durva extract was also reported to have alkaloids, tannins, flavonoids, and saponins (Kapure et al., 2015). Qualitative and quantitative analysis of *Emblica officinalis* leaf extract by Krishbaiah et al., (2009) revealed that extract of aonla contained alkaloids (0.24%), Tannins (11.2%), saponins (1.1%), flavonoids (0.55%) and phenols (0.037%).

Alcoholic extract of aonla fruit has a protective effect on cerebral ischemic–reperfusion injury. The protective effect of the aonla extract was attributed to phytochemicals like vitamin C, gallic acid, polyphenols, and phyllemblin (Panchal et al., 2009). Fruit extract reported to have hypolipidemic (Anila and Vijayalakshmi, 2002), antidiabetic (Sabu and Kuttan, 2002) and anti-inflammatory activities (Asmawi et al., 1993) and inhibited retroviruses such as HIV-1 (El-Mekkawy et al., 1995), tumor development (Jose et al., 2001) and gastric ulcer (Bandyopadhyay et al., 2000). Study by Ghosal et al., (1996) concentrations of different phytochemical suchemblicanin A (37%), emblicanin B (33%), punigluconin (12%), pedunculagin (14%), rutin (3%) and gallic acid (1%) in the extract were established by HPTLC (high performance thin layer chromatography), using authentic markers.

Study by Dwivedi and Noorani (2011) revealed that aqueous and methanolic extract of aonla fruits possessed potent antihelmintic activity in adult earthworm's pheritima posthuma. Separation and purification of different compounds in methanolic extract of aonla was carried out using RP-HPLC (reverse-phase high-performance liquid chromatography). Mobile phase used was consisted of solvent A (0.1% trifluoroacetic acid) and solvent B (acetonitrile) at a flow rate of 1.0 ml/min. Peaks were studied at 280 nm. Compound identified were *kaempferol 3-β-D-glucopyranoside, kaempferol, quercetin, isocorilagin, and geraniin respectively* by spectral methods using ^{1}H and ^{13}C nuclear magnetic resonance spectroscopy, UV-spectroscopy, mass spectroscopy and comparison with literature. Study was carried out by Filipiak-Szok et al., (2012) for determination of phenolic acid and quercetin derivatives in Indian aonla and European gooseberry using HPLC with diode array detector. Mobile phase used was 2% acetic acid (solvent A) and methanol (solvent B) at a flow rate of 1 ml/min. In both amla and gooseberry extract predominant compound was gallic acid 5.37 mg/100 g and 3.21 mg/100 g on dry mass basis respectively followed by caffeic acid. Among flavonoids, rutin was prominent 3.11 mg/100 g and 2.12 mg/100 g of dry matter basis.

A new method based on combination of HPTLC and diode array detector was developed by Pozharitskaya et al., (2007) for identification of individual compound in methanolic extract of *Emblica officinalis*. Different compounds separated from aonla extract by TLC (thin layer chromatography) were free gallic acid (1.85%), ellagic acid (6.45%), emblicanin A (13.12%) and B (9.98%).

Isolation and purification study of compounds, present in methanol aonla extract and responsible for anti-oxidative properties by Liu et al., (2008) using RP-HPLC and identification was done by spectral methods (^{1}H and ^{13}C nuclear magnetic resonance spectroscopy, UV-spectroscopy, mass spectroscopy and by literature comparison). Ethyl acetate fraction has maximum phenolics. Phenolics: geraniin, quercetin 3-β-D-glucopyranoside, kaempferol 3-β-D-glucopyranoside, isocorilagin, quercetin, and kaempferol identified in study. Being rich source of antioxidant these compounds could be used in foodstuffs and traditional medicine. Different compounds isolated and purified from *Phyllanthus emblica* were gallic acid, ellagic acid, mucic acid, 1,4-lactone 3-o-gallate, isocorilagin, chebulanin, chebulagic acid, and mallotusinin. Results revealed that all phenolics showed excellent radical scavenging activity, fair inhibition capacity of lipid oxidation, and potency to chelate Fe^{2+}. Mallotusinin was first time reported in aonla and also showed good antioxidant activity and antiproliferative ability (Luo et al., 2011). Vitamin C content in aonla was determined by enzymatic method, spectrophotometric method, and HPLC method in the study by Raghu et al., (2007). Results of enzymatic method showed presence of vitamin C. Conventional colorimetric method quantified AA content while HPLC method helped in isolation, identification, and quantification of vitamin C content in aonla fruit.

Majeed et al., (2009) developed a new method of HPLC for estimation of vitamin C content present in aonla juice and extract. Seven different fractions were observed from freeze-dried aonla juice/extract using preparative HPLC system. Separation was done on C-18 column with mobile phase: solvent A (0.1% formic acid) and solvent B (methanol) at a flow rate of 35 ml/min. In HPLC analysis, peak 1 was found to have same retention time as AA, and further NMR (nuclear magnetic resonance) study was done for confirmation of AA.

In HPLC profiling compounds such as caffeic acid, coumaric acid, syringic acid, and myricetin were reported in both pulp and seeds while gallic acid was observed in pulp. Response surface methodology was applied by Mishra et al., (2015) for optimizing condition for maximum extraction of phenolics from aonla. Results revealed that ethanol concentration and pH have significant effect on extraction of phenolics from aonla seed powder. Optimum conditions for extraction were: ethanol 88.1%, temperature 35.9°C, and pH 4.4. Ethyl acetate fraction was subjected to HPLC for fractionation into six different compound viz. gallic acid, catechin, caffeic acid, quercetin, p-coumaric acid, vanillic acid, and chlorogenic acid. Gas chromatography-mass spectroscopy (GC-MS) was also used for determination of ethyl acetate extract fractions. Extraction of amla and durva was carried out by three different methods: maceration, percolation, and soxhlet apparatus. Soxhlet apparatus extraction was found to be the best among all three methods.

6.4 MEDICINAL PROPERTIES OF AONLA FRUIT

Aonla (*Emblica officinalis)* is a gift of nature to mankind. Aonla enjoys a good status over other fruits because of its varied valuable properties. It is an indispensable part of the Ayurvedic and Unani system of medicine. Aonla has amazing remedial qualities and believed to increase defense against various diseases viz., ulcers, gastrointestinal disorders, heart troubles, and diabetes. Additionally, aonla is useful in enhancing memory, lowering cholesterol, and also act as an antimicrobial agent (Khan, 2009). Aonla is also recommended for the treatment of age disorders such as forgetfulness, confusion, fatigue, and loss of strength. Aonla is regarded as "the best among rejuvenative herb," "useful in relieving cough and skin diseases," and "the best among the sour fruits" as per Ayurveda system. In Unani medicine, aonla is recognized as a tonic for heart and brain and also protect eyesight. It is a major ingredient in several medicinal preparation, including Triphala and the famous Chyavanprash which improve physical and mental well being (Nandhkarni, 1976; Manjunatha et al., 2001). Aonla fruit is acidic, cooling, diuretic, and laxative in nature, and the dried aonla fruit is used in treating hemorrhage, diarrhea, and dysentery. Amazing antibacterial and astringent properties of aonla help in preventing infection and also in healing of ulcers (Rakesh and Dhawan, 2006). Various aonla extracts have expectorant, purgative, spasmolytic, anti-tumor activities, hypoglycemic, and strong antibacterial activity (Jamwal et al., 1959; Jayshri and Jolly, 1993; Jose et al., 2001).

Alcoholic extract of aonla has protective action against cerebral ischemia-reperfusion injury in rat and its aqueous extract possess antipyretic, laxative, tonic properties along with antibacterial activity (Vinayagamoothy, 1982; Panchal et al., 2009). Saponins, flavonoids, tannins, and phenols compounds present in aonla are the potential inhibitors of most common respiratory pathogens such as *Staphylococcus aureus*, *Klebsiella pneumoniae*, *Streptococcus pyogens* (Javale and Sabins, 2010). *Emblica officinalis* possessed potent antimicrobial activity against common respiratory pathogens and can be used in treatment of various diseases caused by these organisms (Javale and Sabins, 2010). Strong antibacterial potential of *Emblica officinalis* extract was reported against *Escherichia coli*, *S. paratyphi A*, *S. paratyphi B*, K. ozaenae, *Proteus mirabilis*, *Pseudomonas aeruginosa*, *Klebsiella pneumoniae*, and *Serratia marcescens* in the study by Saeed and Tariq (2007). Antimicrobial activity data of ethanolic aonla extract against *Staphylococcus aureus*, *Pseudomonas*, and *Klebsiella pneumonia* revealed that gram-positive bacteria were more susceptible against aonla extract as compared to gram-negative, which might be due to complex structure of later ones (Abdulhafza, 2013).

Aonla fruit is used for the treatment of diarrhea, jaundice, and inflammations (Deokar, 1998). The pulp of the fruit is smeared on the head to dispel headache and dizziness (Perry, 1980). Leaves and fruits of aonla are utilized by rural populations as a remedy for the treatment fever and inflammation, and its juice is consumed as tonic, especially during winter season. Therapeutic value of aonla fruit and its various extracts is given in Table 6.4.

Free radicals are reported to play an important role in pathogenesis of neurodegenerative disorders, cancer, liver cirrhosis, cardiovascular diseases, atherosclerosis, cataracts, diabetes, and inflammation (Aruoma, 1998). Aonla extract, by antioxidant action, prevent LDL (Low-Density Lipoprotein) oxidation, thereby inhibited endothelial dysfunction and formation of atheromatous plaques (Antony et al., 2006). Significance reduction reported in total cholesterol and triglyceride due to antioxidant properties of low molecular weight hydrolyzable tannins present in aonla extract as well as increase in HDL by Antony et al., (2008). Conjugated ring structures and hydroxyl groups of phenolic compounds is credited for potential to function as antioxidants by scavenging superoxide anion (Robak and Dryglewski, 1988), lipid peroxy radicals (Torel et al., 1986)

TABLE 6.4 Therapeutic Value of Aonla Fruit

Phytochemical	Health benefits	References
Gallic acid, ellagic acid, and ascorbic acid	Reduced oxidative stress (concentration dependent), inhibited α-amylase, and α-glucosidase activity with antiglycation activity. Prevent LDL (low-density lipoprotein) oxidation in human under *in-vitro* condition	Nampoothiri et al., 2011
Gallic acid and ellagic acid	Inhibited UVB-induced photoaging in human skin fibroblast via its strong ROS scavenging ability	Dar et al., 2010
Tannic acid	Reduced NO levels, erythrocytes membrane lipid peroxidation, activities of Na^+/K^+ and Mg^{2+} ATPase and fluorescent anisotropic values in alcohol induced biochemical and biophysical changes in erythrocyte membrane of rat	Reddy et al., 2009
Tannins and quercetin	Partial recovery of pathological manifestation and suppressed tumor-forming potential of 2-AAF (2, acetylaminofluorene) in rat liver, inhibited expression of several cell proliferation markers	Sultana et al., 2008
Crude ethanolic extract	Reduce vacuolar degeneration around centrilobular zone, fatty liver, and glycolysis induced by ethanol in rat	Pramyothin et al., 2006
Crude methanolic extract	Ulcer protective and healing effect in ethanol, aspirin, cold-restrained stress, pylorus-ligation, acetic-acid induced ulcers	Sairam et al., 2002
Crude aqueous extract	Inhibit growth of L 929 cells in culture, reduced solid tumor induced by DLA (Daltons lymphoma ascites cells) cells, inhibit activity of enzyme Cdc25 phosphatase (conc. dependent)	Jose et al., 2001
	Reduced elevated level of serum and lipid peroxidize, glutamate-pyruvate transaminase and alkaline phosphatase induced by CCl_4 (carbon tetrachloride) administration	Jose et al., 2000
	Inhibited hepatocarcinogenesis induced by N-nitrosodiethylamine (concentration dependent)	Jenna et al., 1999
Gallic acid, Rutin, Emblicanin A and B,	Restore activity of cardiac superoxide dismutase, catalase, and glutathione dismutase reduced by IRI (ischemia-reperfusion) in rats	Bhattacharya et al., 2002
Punigluconin, Pedunculagin	Lowered serum cholesterol, TG (triglyceride), Phospholipid, LDL level in cholesterol-fed rabbits	Mathur et al., 1996

and singlet oxygen (Husain et al., 1987), and stabilizing free radicals implicated in oxidative processes through hydrogenation or complexing with oxidizing species (Shahidi and Wanasusdara, 1992).

Studies by Ghosal et al., (1996) and Bhattacharya et al., (1999) of low molecular weight (<1000) hydrolysable gallotannins (EOT) comprising emblicanin A, emblicanin B, punigluconin, and pedunculagin, separated from fresh juice or solvent extracts of *Emblica* fruits, supported significant antioxidant potential of extracts. Aqueous extract of aonla reported to inhibit lipid peroxidation induced by radiation in rat liver microsome due to its antioxidant properties (Khopde et al., 2001). Methanolic and acetone extract of Triphala exhibited good antioxidant potential compared to their chloroform extract (Singh et al., 2010). Methanolic extract of aonla fruit of different varieties have better antioxidant properties compared to ethanolic and ethyl acetate extract (Kumari and Khatkar, 2016).

Ethyl acetate extract of aonla also suppressed inducible nitric oxide synthase (iNOS) and cyclooxygenase (COX)-2 expressions in aorta of aging rats by inhibiting NF-kappa B activation. Thus aonla could be considered as antioxidant for anticipation of age-related renal disorders (Yokozawa et al., 2007). Total antioxidant activity of aqueous extracts of aonla, spirulina, and wheat grass observed were7.78, 1.33 and 0.278 mmol/l, respectively, at concentration of 1 mg/ml. Significant cytoprotective activity and inhibition of t-BOOH (tertiary butyl hydroperoxide) induced ROS (reactive oxygen species) generation of was observed in both alcoholic and aqueous extract of aonla (Shukla et al., 2009). Strong antioxidant activity reported in methanolic extract of aonla fruit as compared to commercial compounds such as butylated hydroxyl anisole (BHA) and quercetin. So aonla might be considered as prospective source of antioxidants (Liu et al., 2008).

Aonla extract orally administrated to diabetic rats improved their body weight gain and significant increase observed in various oxidative stress indices of the serum with this decreased levels of albumin. Due to their antioxidant properties, aonla can be used for improving glucose metabolism in diabetes (Rao et al., 2005). Antioxidant activity of the aonla extract was attributed to gallic acid and tannic acid and also exhibited significant protection to DNA against oxidative damage (Kumar et al., 2006).

6.5 PROCESSING AND FOOD APPLICATIONS OF AONLA FRUIT

Aonla fruit is highly perishable in nature; processing is required to increase shelf life and for value addition, especially during glut period. Processing not only give better returns to producers but also reduces post-harvest losses. Aonla preserve, candy, pickles, jam, juice, and squash are major traditional products of aonla which are widely used by consumer as a health food and as a natural source of vitamin C and polyphenols.

6.5.1 AONLA PRODUCTS

Aonla chutney prepared by different researchers and composition varied from sweet salty to pungent taste depending on sensory score. Ready to eat aonla chutney having composition: aonla pulp–65.59%, salt–8%, green chili–3.3%, garlic paste–2.11% and coriander leaves paste–18.89% prepared by Mishra et al., (2010). Chutney having 50° Brix could be stored up to 2 months, and the chutney with 25° Brix had only 10 days shelf life while chutney of 35° Brix could be stored up to 20 days at refrigeration temperature. Aonla chutney have better characteristics than other similar products, as it had higher polyphenol content with great antioxidant activity and it could be stored at refrigeration temperature without significant change in its green color using class one preservatives. Aonla chutney prepared using three different cultivars (NA-6, NA-7, and Chakaiya) of aonla and other ingredients used were: sugar, salt, ginger, hot spices, red chili, and glacial acetic acid. Chutney stored at ambient condition was acceptable for 150 days. Chutney prepared from NA-7 reported the most acceptable in terms of sensory properties and AA, total sugar, non-reducing sugar content, and pH.

Pure aonla based carbonated beverage and aonla based carbonated beverage blended with aloe vera juice prepared by Thorat, (2007). Results indicated that Aloe vera juice increased quality of carbonated beverage by increasing stability of biochemical parameters like AA content, good sensory characteristics. Indian gooseberry juice was blended with other juices such as apple, lime, pomegranate, Perlette grape, and Pusa Navarang grape juice to prepare ready to serve (RTS) beverage and comparison of physico-nutritional quality of blended juice was done with the control juice. Aonla juice was reported to increase AA content of other juices, and also its acceptability was increased by alteration in taste. RTS prepared by blending aonla juice with Pusa Navarang grape juice (20:80) was reported to be best in terms of overall sensory quality and vitamin C content (Jain and Khurdiya, 2004). Thermal processing of aonla juice was conducted at different time-temperature combination for the destruction of yeast and maintaining maximum AA content. Treatment of aonla juice at 90°C for 1 minute found to be the best as it destructed yeast completely with minimum loss in AA content (Jain et al., 2003).

Aonla squash recipe standardized by Gajanana et al., (2007). Aonla squash containing 30% aonla juice, 5% lime juice, 2% ginger with 40° Brix was reported the best in terms of organoleptic characteristics with respect to color, flavor, taste, appearance, and overall acceptability. Aonla squash was prepared using Aonla juice, sugar, and citric acid (CA) and response observed were viscosity, taste, and overall acceptability. Aonla squash was observed to have better retention of phytochemical: Gallic acid (191.4 mg/100 g), ethyl gallate (0.21 mg/100 g), ellagic acid (160.4 mg/100 g) and quercetin (0.50 mg/100 g) as well as sensory characteristics. AA content of squash was 115.6 mg/100 g. Product retained fair amount of nutrients and also found acceptable organoleptically (Kumari and Khatkar, 2016). Singh et al., (2013) investigated suitable blending ratio for preparation of aonla based beverages. Fruits such as mango, jamun, jackfruit, and guava were blended in aonla for preparing blended RTS, nectar, and squash. Study concluded that squash containing 25% aonla pulp and 75% mango pulp scored maximum sensory score with prominent and attractive color. RTS prepared using aonla (80%), and mango (20%) showed better nutritional quality with better consumer appeal. Jaiswal et al., (2008) prepared squash using seven cultivars of aonla (Kanchan, Krishna, Francis, Chakaiya, NA-6, NA-7, and NA-10) by standard protocol. Cultivar NA-7 was reported to be most suitable for squash preparation in terms of nutritional and organoleptic quality parameters followed by NA-6. Aonla squash was stored for 1 year at ambient condition and evaluated at interval of 30 days. Results indicated that gradual decrease in AA content and sensory score of aonla squash while increase was noted in total polyphenol content during storage.

Aonla candy was prepared using the segments of aonla separated by boiling in water and then steeping in sugar syrup of increasing concentration (50 to 70°B). Ginger and cardamom flavor were added in syrup of 70°B, and after drying, candies were stored. Candies prepared from cultivar Krishna were found to be the best in terms of sensory and biochemical characteristics (Nayak et al., 2012). Srivastava and Genitha, (2012), standardized recipe for preparation of candy from aonla fruit segments using different blanching (microwave, steam, and water) treatments. Texture of aonla segments blanched in microwave was better compared to water and steam blanched segments. To overcome the loss of vitamin C at each stage of processing crystallized candies were prepared using high sugar concentration. Blanching in microwave was reported to be best among all treatments as it retained maximum AA content. Study was conducted by Vikram et al., (2014) to see impact of varieties and honey coating on shelf life of aonla candy. Variety NA-6 was found superior over others with respect quality parameters studied. Data revealed that AA content, pH, and sensory score showed decrease while TSS, acidity, and browning increased during storage. Study concluded that honey-coated candy was superior over sugar coated candy with respect to all quality parameters studied.

Shredded aonla were given three different (with salt, alum, and calcium carbonate) treatments for laddu preparation using sugar syrup of 80°B fortified with cardamom powder. Resultant laddu by all three methods were rich in antioxidants, protected against stress, and were healthcare (Borole et al., (2012). Dehydrated RTE (Ready to eat) aonla shreds were prepared by Sagar and Kumar (2006) by blanching fruits for 10 min in boiling water bath. After shredding, segments were dipped into sugar syrup (78°Brix) and then dehydration was done in cabinet dryer. Packing of shreds was done in 200 g low-density polyethylene, 200 g high-density polyethylene, and 150 g polypropylene and stored at room temperature (17–33.5°C) and low temperature (7 ± 2C) for period of six months. Moisture content of aonla shreds was 5%. Shreds prepared were good in nutritional aspects as well as in sensory characteristics.

Mishra et al., (2010) optimized composition of aonla bar using the factorial design. Ingredients like aonla pulp (125 g), glucose (12% of aonla pulp) and SMP (8% of aonla pulp) were kept constant while the level of sugar, pectin, and sodium alginate were varied from 90 to 110 g, 6 to10 g and 0 to 4 g, respectively. Sugar level below 90 g failed to give appropriate sweetness while higher level than 110 g produced sticky and harder aonla bar. Product prepared using 10% pectin was also hard in texture while product with 4% sodium alginate gave after taste after setting. Aonla bar prepared by 6% pectin showed poor setting characteristic while at 10% level product was found to have hard texture. The best-optimized formulation for aonla bar was 125 g aonla pulp, 100 g sugar, 12% glucose, 8% SMP (Skim Milk Powder), 8% pectin and 2% sodium alginate and product was found better in terms of calcium, phosphorus, and calorific value. Mixed fruit slab was prepared by mixing aonla pulp and mango pulp in 1:1 ratio and total soluble solid content was adjusted to 30%. Fruit slab contained good amount of β-carotene (4.12 mg/100 g) and AA (76.87mg/100 g). Product might be in an innovative intermediate moisture food having better sensory and nutritional qualities (Verma and Chopra, 2010).

6.5.2 FOOD APPLICATIONS OF AONLA

Aonla chyawanprash was prepared by using aonla pulp mixed with different types of herbs and spices in a definite ratio and then concentrating resultant mix in microwave oven. Results showed that final product retained maximum natural vitamin C along with better color, appearance, and shelf stability (Beharilal, 2007). Instant oil less pickle was prepared from two different varieties of aonla Desi and Chakaiya. The overall quality of pickle made pre treated segment of Desi variety was found better than Chakaiya variety (Premi et al., 2002). The preserve made from blanched aonla fruits treated with 2.5% pectic enzyme solution at 40C for 4 hours and then treated with syrup, resultant preserve was compared with the conventionally made preserve and was found best in all respects. Absorption of sugar was also higher in the preserve made by this process (Mehta and Tomar, 1979). Triphala is formulated by mixing amalaki, haritaki, and bibhitaki in equal proportions. Therapeutic value of Triphala was due compounds such as AA, carotene, gallic acid, ellagic acid, tannins, flavonoids, phyllembic acid, chebulagic acid (Irani and Khaled, 2015).

Wine was prepared from aonla fruits using *Saccharomyces cerevisiae* by Soni et al., (2009). Results showed that alcohol content varied from 14 to 16% depending on fermentation conditions and ingredients used. Storage of wine for 1 month in oak barrel improved quality by reducing level of undesirable components such as isoamyl alcohols, n-butanol, n-propanol, and isobutanol. Aonla wine can be used as good health drink with phenolic compound and AA as antioxidant and alcohol as stimulant.

Singh et al., (1999) prepared aonla preserve with variation in initial sugar syrup concentration (40, 50 and 60° Brix) at different temperature. Study on the basis of chemical composition revealed that sugar syrup of 60° Brix at 40°C was the best for the preparation of aonla preserve. Recipe for aonla syrup preparation was standardized by Gajanan et al., (2007) using aonla juice, lime, ginger juice, and sugar. Aonla syrup having 55% aonla juice, 10% lime juice, 4% ginger juice with 68° Brix, reported to be best in terms of color, flavor, appearance, and overall acceptability. Aonla and carrot preserve was prepared by Sethi and Anand, (1983) to see retention of nutrients. Study revealed that 48.3% AA and 18.5% tannins of aonla leached out in the syrup. Aonla fruit, along with syrup, was recommended for the health benefits in terms of minerals and nutrients. Aonla and carrot preserve available in market were evaluated for their nutritional as well as for microbial status by Sethi and Anand (1982). Microbial contamination was observed more in carrot preserve than in aonla preserve.

To overcome the astringency of aonla juice, an effort was done by Garg (2010) by blending with bitter gourd juice to prepare RTS drink. Aonla juice and bitter guard juice were blended in the ratio of 75:25 to prepare RTS drink having 20% juice, 15° Brix and 0.3% acidity. Product was found to have good nutritional and sensory qualities. Aonla juice RTS beverage was prepared by Shere et al., (2008) by adding sugar with lime juice and flavoring agents used were black salt and jeera extract. On the basis of sensory characteristics, drink was found acceptable. Investigation was conducted by Bahadur et al., (2009) for preparation of aonla juice from different cultivars of aonla with appropriate level of preservative dose to extend storage life. Juice obtained from Banarasi variety and preserved with 750 ppm of KMS was found to be the best among other juices in terms of organoleptic properties. Carbonated health drink was prepared by Throat et al., (2007) with 10% aonla juice and 1.2% ginger juice. Juice extracted by cold method was found suitable for fresh consumption while for preparation of beverage juice was extracted by hot method. Health drink beverage blended with aloe juice (2%) was found most acceptable.

A study was conducted by Kothari and Bhatnagar, (2010) for preparation of aonla blended spiced fruit juice. Pineapple and sweet lime juices were blended with aonla juice in different ratio individually. Blend of aonla-pineapple (15:85) with 0.35% black salt and aonla–sweetlime (15:85) with 0.5% black salt was found most acceptable on the basis of sensory qualities. Blends were reported to be superior in terms of AA content than pure sweet lime and pineapple juices and also secured higher sensory score.

Alcoholic, self carbonated beverage was prepared from amla carrot juice by blending in different proportions, and final TSS was adjusted to 16° Brix by adding sucrose. Pasteurization was done for 15 sec at 82°C, after cooling juice was inoculated with *Saccharomyces cerevisiae* at the rate 0.5% and incubated at 20±5°C for 36 hours. Results of organoleptic revealed that low alcoholic self carbonated aonla - carrot (50:50) was the best. Beverage contained 0.3% w/v Alcohol content, 0.36% acidity and CO_2 0.90 bar (Sahota et al., 2009). Aonla juice and bael juice were added to jamun juice for increasing vitamin C and micronutrient level. Vitamin C content of aonla juice blended with jamun juice, increased from 27 mg/100 g to 256 mg/100 g and sensory panelist gave 6.5 out of 9 for overall acceptability of juice, and it could be satisfactory alternative to harmful soft drinks (Gupta, 2015).

Biscuits prepared by incorporating these natural extracts of aonla, drumsticks leaves and raisins were found better in terms of sensory characteristics and chemical analysis compared to control and also showed higher antioxidant activity than BHA treated biscuits. After six months of storage, peroxide and acid value were lower than BHA control and BHA treated biscuits (Reddy et al., 2005).

6.6 CONCLUSION REMARKS

Aonla being a rich source of antioxidants and phenolic compounds it can be used as functional food for health improvement. Medicinal aspects of aonla are mainly related to their phytochemicals. Aonla is a rich source of AA and phytochemicals which accounts of its nutraceutical properties. Gallic acid, chebulanin, ellagic acid, mallorusinin, isocorilagin, Emblicannin A, Emblicannin B, Punigluconin, and Pedunculagin are the major phytochemicals present in aonla fruit that have apparent antiproliferative and potent antioxidant activity. Phenolic compounds such as quercetin, quercetin 3-β-D-glucopyranoside, kaempferol 3-β-D-glucopyranoside, kaempferol, and isocorilagin in ethyl acetate extract of aonla exhibited strong antioxidant and radical scavenging potential. Further studies on isolation and identification techniques of more bioactive phytochemicals from crude extract of aonla fruits and their utilization in value-added food products can help in better understanding of aonla fruit's medicinal value and its commercial applications.

KEYWORDS

- **aonla**
- **disorder**
- **medicinal**
- **nutritional**
- **phytochemical**
- **polyphenol**
- **product**

REFERENCES

Abdulhafza, N.N. (2013). Study of phytochemical composition and antibacterial activity of *Emblica officinalis* (amla) fruit extract. *Al-Anbar Journal of Veterinary Science* **6(1):**107-113.

Anila, L. and Vijayalakshmi, N.R. (2002). Flavonoids from *Emblica officinalis* and *Mangifera indica*- effectiveness for dyslipidemia. *Journal of Ethnopharmacology* **79(1)**:81-87.

Anonymous. (1969). Wealth of India [Raw materials], Council of Scientific and Industrial research, New Delhi.

Antony, B.; Benny, M. and Kaimal, T.N.B. (2008). A pilot clinical study to evaluate the effect of *Emblica officinalis* extract (Amalmae™) on markers of systemic inflammation and dyslipidemia. *Indian Journal of Clinical Biochemistry* **23(4)**: 378-381.

Aruoma, O.I. (1998). Free radical, oxidative stress and antioxidants in human health and disease. *American Journal of Oil Chemical Society* **75(2)**:199-212.

Asmawi, M.Z.; Kankaanranta, H.; Moilanen, E. and Vapaatalo, H. (1993). Anti-inflammatory activities of *Emblica officinalis* Gaertn. Leaf extracts. *Journal of Pharmacy and Pharmacology* **45(6)**: 581-584.

Bahadur,V.; Mishra, D.; Singh, D.B.; Roshan R.K. and Singh, A. (2009). Effect of varieties and preservative levels of shelf life of aonla juice. 10th *International Controlled and Modified Atmosphere Research Conference,* 4-7. Antalya, Turkey. Pp. 80.

Baliga, M.S. and Dsouza, J.J. (2011). Amla (*Emblica officinalis* Garten), a wonder berry in the treatment and prevention of cancer. *European Journal of Cancer Prevention* **20(3)**: 225-239.

Bandyopadhyay, S.K.; Pakrashi, S.C. and Pakrashi, A. (2000). The role of antioxidant activity of *Phyllanthus emblica* fruits on prevention from indomethacin induced gastric ulcer. *Journal of Ethnopharmacology* **70(2)**: 171-176.

Barthakur, N.N. and Arnold, N.P. (1991). Chemical analysis of the emblic (*Phyllanthus emblica* L.) and its potential as a food source. *Scientia Horticulturae* **47(1)**: 99-105.

Battacharya, A.; Chatterjee, A.; Ghoshal, S. and Bhattacharya, S.K. (1999). Antioxidant activity of active tannoid principles of *Emblica officinalis* (amla). *Indian Journal of Experimental Biology* **37(7)**:676-680.

Beharilal, A.; Singhal, R.; Singh, J. and Jindal, J.M. (2007). Preparation of nutritive aonla chyawanprash by the use of microwave radiation. *Beverage and Food World* **34**:38-40.

Bhattacharya, S.K.; Bhattacharya, A.; Sairam, K. and Ghosal, S. (2002). Effect of bioactive tannoid principles of *Emblica officinalis* on ischemia-reperfusion-induced oxidative stress in rat heart. *Phytomedicine* **9(2)**:171-174.

Borole, A.V.; Thakur, P.P.; Zakiroddin, Md. and Bansode, V.V. (2012). Studies on preparation of aonla laddu. *Beverage and Food World* **39**: 49-51.

Dahiya, S.P. and Dhawan, S.S. (2001). Physico-chemical characteristics of Aonla (*Emblica officinalis* Garten) cv. Chakaiya. *Indian Food Packer* **55(6)**: 133-136.

Dar, A. M.; Peerzada; Kumar, K.S.N.; Afzal, Z.M. and Abdullah, T.S. (2010). Effect of *Emblica officinalis* (fruit) against UVB-induced photo-aging in human skin fibroblasts. *Journal of Ethnopharmacology* **132(1)**:109-114.

Deokar, A.B. (1998). Medicinal Plants Grown at Rajegaon, 1st Edition. DS Manav Vikas Foundation, Pune, pp. 48–49.

Dwivedi, G. and Noorani, A.A. (2011). Anthelmintic activity of *Emblica officinalis* fruit extract. *International Journal of Pharma:Research and Development–online (IJPRD)* **3(1)**:50-52.

Elangovan, N.M.; Dhanarajan, M.S. and Elangovan, I. (2015). Preliminary phytochemical screening and HPTLC finger printing profile of leaf extracts of *Moringa oleifera* and *Phyllanthus emblica. International Research Journal of Pharmaceutical and Bioscience* **2(2)**:32-40.

El-Mekkawy, S.; Meselhy, M.R.; Kusumoto, I.T.; Kadota, S.; Hattori, M. and Namba, T. (1995). Inhibitory effects of Egyptian folk medicines on human immune-deficiency virus (HIV) reverse transcriptase. *Chemical and Pharmaceutical Bulletin* **43(4)**:641-648.

Filipiak-Szok, A.; Kurzawa, M. and Sztyk, E. (2012). Determination of anti-oxidant capacity and content of phenols, phenolic acids and flavonols in Indian and European gooseberry. *Chemical Papers* **66(4):**259-268.

Gajanana, K.; Rokhade, A.K.; Patil, P.B. and Kulkarni, M.S. (2007). Standardisation of recipe for preparation of aonla squash. *Beverage and Food World* **34:**55-56.

Ganachari, A.; Thangavel, K.; Ali, S.M.; Nidoni, U. and Ananthacharya (2010). Physical properties of Aonla fruit relevant to the design of processing equipments. *International Journal of Engineering Science and Technology* **2(12):**7562-7566.

Garg, A. (2010). Studies on processing and development of carbonated spiced RTS drink from bitter gourd and amla fruit. *Processed Food Industry* **13(11):**23-26.

Ghorai, K. and Sethi, V. (1996). Varietal suitability of amla ('Desi' and 'Banarasi'.) fruits for storage and preservation. *Indian Food Packer* **50(1)**:11-18.

Ghosal, S.; Tripathi, V.K. and Chauhan, S. (1996). Active constituents of *Emblica officinalis*: The chemistry and antioxidant effects of two new hydrolysable tannins: Emblicanin A and B. *Indian Journal of Chemistry* **35(B)**:941-948.

Goyal, R.K.; Kingsly, A.R.P.; Kumar, P. and Walia, H. (2007). Physical and mechanical properties of Aonla fruits. *Journal of Food Engineering* **82(4)**:595-599.

Goyal, R.K.; Patil, R.T.; Kinglsly, A.R.P.; Walia, H. and Kumar, P. (2008). Status of post harvest technology of aonla in India- a review. *American Journal of Food Technology* **3(1)**:13-23.

Gupta, P. (2015). Preparation and standardization of jamun RTS supplemented with amla juice and bael juice. *Online International Interdisciplinary Research Journal* **5 (3)**:165-174.

Hazarika, B.N.; Deka, B.C.; Choudhury, S. and Sarma, B. (2009). Studies on variability in physico-chemical characters of different aonla accessions from Jorhat region of Assam. *Indian Journal of Horticulture* **66(2):**190-192.

Irani, R. and Khaled, K. (2015). A review on polyherbal formulation of Triphala- the Ayurvedic wonder. *Journal of International Academic Research for Multidisciplinary* **3 (4)**:407-418.

Jain, S.K. and Khurdiya, D.S. (2004). Vitamin C Enrichment of Fruit Juice Based Ready-to-Serve Beverages through Blending of Indian Gooseberry (*Emblica officinalis* Gaertn.) Juice. *Plant Foods for Human Nutrition* **59(2)**:63–66.

Jaiswal, R.; Singh, G. and Singh, A.K. (2007). Studies on physico-chemical characteristics of aonla *(Emblica officinalis* Gaertn.) fruits. *Progressive Agriculture* **7 (1/2)**:90-92.

Jaiswal, R.; Singh, G. and Singh, A.K. (2008). Evaluation of aonla *(Emblica officinalis Gaertn.)* cultivars for squash making. *Progressive Agriculture* **8(1):**29-31.

Jamwal, K.S.; Sharma, I.P. and Chopra, L. (1959). Pharmacological investigations on the fruits of *Emblica Officinalis*. *Journal of Scientific and Industrial Research* **18**: 180-181.

Javale, P. and Sabnis, S. (2010). Antimicrobial properties and phytochemical analysis of *Emblica officinalis. Asian Journal of Experimental Biological Science Special* **1**:96-100.

Jayshri, S. and Jolly, C.I. (1993). Phytochemical antibacterial and pharmacological investigations on *Monordica chiranlia* and *Emblica officinalis*. *Indian Journal of Pharmaceutical Sciences* **1**: 6-13.

Jeena, K.J.; Joy, K.L. and Kuttan, R. (1999). Effect of *Emblica officinalis*, *Phyllanthus amarus* and *Picrorrhiza kurroa* on N-nitrosodiethylamine induced hepatocarcinogensis. *Cancer letters* **136(1)**:11-16.

Jose, J.K.; Kuttan, G. and Kuttan, R. (2001). Antitumor activity of *Emblica officinalis*. *Journal of Ethnopharmacology* **75(2):**65-69.

Kalra, C.L. (1988). The chemistry and technology of amla (*Phyllanthus emblica*)- a resume. *Indian Food Packer* **38(4)**:67-82.

Kapure, P.L.; Makade, K.P.; Sanap, M.D.; Gandhi, S.J.; Ahirrao, R.A. and Pawar, S.P. (2015). Various extraction method and standardisation parameter of amla and durva. *Pharma Science Monitor* **6(2):**61-72.

Khan, K.H. (2009). Role of *Emblica officinalis* in medicine – a review. *Botany Research International* **2(4):**218-228.

Khopde, S. M.; Priyadarsini, K. I. ; Mohan, H. ;Gawandi, V. B. ; Satav, J. G.; Yakhmi, J. V.; Banavaliker, M. M.; Biyani, M. K. And Mittal, J. P. (2001). Characterizing the antioxidant activity of amla *(Phyllanthus emblica*) extract. *Current Science* **81 (2):**185-190.

Kore , V. T.; Devi , H. L. and Kabir , J. (2013). Packaging, storage and value addition of aonla, an underutilized fruit, in India. *Fruits* **68(3)**:255–266.

Kothari, C. and Bhatnagar, V. (2010). Development and quality evaluation of aonla blended spiced fruit juice. *Beverage and Food World* **37(4)**:29-32.

Krishnaiah, D.; Devi, T.; Bono, A. and Sarbatky, R. (2009). Studies on phytochemical constituents of six Malaysian medicinal plants. *Journal of Medicinal Plants* **3(2):**067-072.

Kumar, G. S.; Nayaka, H.; Dharmesh, S. M. and Salimath, P. V. (2006). Free and bound phenolic antioxidants in amla (*Emblica officinalis*) and turmeric (*Curcuma longa*). *Journal of Food Composition and Analysis* **19(5)**: 446–452.

Kumar, S. and Singh, I.S. (2013). Physico-chemical studies of various cultivars of aonla fruits. *Progressive Horticulture* **45(1):**80-82.

Kumari, P. And Khatkar, B.S. (2016). Assessment of total polyphenols, antioxidants and antimicrobial properties of aonla varieties. *Journal of Food Science and Technology* **53(7):**3093-3103.

Kumari, P. and Khatkar, B.S. (2016). Process standardisation and phytochemical of contents aonla squash. *International Journal of Advanced Research in Engineering Technology and Science* **3(6):**95-100.

Liu, X.; Cui, C.; Zaho, M.; Wang, J.; Luo, W.; Yang, B. and Jiang, Y. (2008). Identification of phenolics in the fruit of Emblica (*Phyllanthus emblica* L.) and their antioxidant activities. *Food Chemistry* **109(4)**:909-915.

Liu, X.; Zhao, M.; Wang, J.; Yang, B. and Jiang, Y. (2008). Antioxidant activity of methanolic extract of emblica fruit (Phyllanthus emblica L.) from six region in China. *Journal of Food Composition and Analysis* **21(3)**:219-228.

Luo, W.; Zaho, M.; Yang, B.; Ren, J.; Shen, G. and Rao, G. (2011). Antioxidant and antiproliferative capacities of phenolics purified from *Phyllanthus emblica* L. Fruit. *Food Chemistry* **126(1)**:277-282.

Majeed, M.; Bhat, B.; Jadhav, A.N.; Srivastava, J.S. and Nagabhushanam, K. (2009). Ascorbic acid and tannins from *Emblica officinalis* Gaertn. Fruits-a revisit. *Journal of Agricultural and Food Chemistry* **57(1)**:220-225.

Manjumatha, S.; Jaryal, A.K.; Bijlani, R.L.; Sachdeva, U. and Gupta, S.K. (2001). Effect of chyawanprash and vitamin C on glucose tolerance and lipoprotein profile. *Indian Journal of Physiology and Pharmacology* **45(1)**:71-79.

Mathur, R.; Sharma, A.; Dixit, V.P. and Verma, M. (1996). Hypolipidaemic effect of fruit juice of *Emblica officinalis* in cholesterol-fed rabbits. *Journal of Ethanopharmacology* **50(2)**:61-68.

Mehta, G.L. and Tomar, M.C. (1979). Studies on the simplification of preserve making II Amla (*Phyllanthus emblica* L.). *Indian Food Packer* **33(5)**: 27-30.

Mishra, P.; Dutta, N. and Mahanta, C.L. (2015). Partial extraction and identification of phenolics in amla (*Emblica officinalis*) seed coat powder. *Journal of Food Science Technology* 52(11):6990-7001.

Mishra, P.; Mishra, P. and Rai, G.K. (2010). Process and product standardisation for the development of amla bar. *Beverage and Food World* **37**:58-60.

Mishra, P.; Srivastava, V.; Verma, D.; Chauhan, O.P. and Rai, G.K. (2009). Physico-chemical properties of Chakiya variety of Amla (*Emblica officinalis*) and effect of different dehydration methods on quality of powder. *African Journal of Food Science* **3(10)**: 303-306.

Murthy, Z.V.P. and Joshi, D. (2007). Fluidized bed drying of Aonla (*Emblica officinalis*). *Drying Technology* **25(5)**: 883-889.

Nadkarni, K.M. (1976). Dr KM Nadkarni's Indian Meteria Medica. Bombay, India: the Karnataka printing press and the popular press Ltd. pp.480-484.

Nampoothiri, S.V.; Prathapan, A.; Cherian, O.L.; Raghu, K.G.; Venugopalan, V.V. and Sundaresan, A. (2011). In vitro antioxidant and inhibitory potential of Terminalia bellerica and *Emblica officinalis* fruits against LDL oxidation and key enzymes linked type 2 diabetes. *Food and Chemical Toxicology* **49(1)**:125-131.

Nayak, P.; Tondon, D.K. and Bhatt, D.K. (2012). Study on changes of nutritional and organoleptic quality of flavoured candy prepared from aonla (*Emblica officinalis* G.) during storage. *International Journal of Nutrition and Metabolism* **4(7):**100-106.

Panchal, N.M.; Patel, M.B.; Patel, J.V.; Anand, I.S. and Patel, C.N. (2009). Protective effect of alcoholic extract of amla (*Emblica officinalis*) fruits on cerebral reperfusion injury in rats. *Journal of Young Pharmacists* **1(3):**213-219.

Pareek, S. and Kaushik, R.A. (2012). Effect of drying methods on quality of Indian gooseberry (*Emblica officinalis* Gaertn.) powder during storage. *Journal of Scientific Research* **71(11)**:727-732.

Parveen, K. and Khatkar, B.S. (2015). Physico-chemical properties and nutritional composition of aonla *(Emblica officinalis)* varieties. *International Food Research Journal* **22(6)**: 2358-2363.

Pathak, R.K. (2002). *Emblica officinalis.* ICAR (Indian Council of Agricultural Research). *Food and Agriculture Organisation.*

Pathak, R.K. (2003). Status report on Genetic Resources of Indian Gooseberry in South- East Asia. pp.4-18.

Perianayagam, J.B.; Sharma, S.K.; Joseph, A. and Christina, A.J.M. (2004). Evaluation of anti-pyretic and analgesic activity of *Emblica officinalis* Gaertn. *Journal of Ethnopharmacology* **95(1)**: 83-85.

Perry, L.M. (1980). Medicinal Plants of East and South East Asia: Attributed Properties and Uses (Vol 620). MIT Press, Cambridge. London.

Pozharitskaya, O.N.; Ivanova, S.A.; Shikov, A.N. and Makarov, V.G., (2007). Separation and evaluation of free radicalscavenging activity of phenol components of *Emblica officinalis* extract by using an HPTLC–DPPH9 method. *Journal of Separation Science* **30**:1250–1254.

Pramyothin, P.; Samosorn, P.; Poungshompoo, S. and Chaichantipyuth, C. (2006). The protective effect of *Phyllanthus emblica* Linn. extract on ethanol induced rat hepatic injury. *Journal of Ethnopharmacology* **107(3)**:361-364.

Premi, B.R.; Sethi, V. and Bisaria, G. (2002). Preparation of instant oil-less pickle from Aonla (*Emblica officinalis gaertn.*). *Indian Food Packer* **56**:72-75.

Raghu, V.; Platel, K. and Srinivasan, K. (2007). Comparison of ascorbic acid content of *Emblica officinalis* fruits determined by different analytical methods. *Journal of Food Composition and Analysis* **20(6)**:529-533.

Rakesh and Dhawan, S.S. (2006). Processing techniques of aonla beverages. *Beverage and Food World* **33**:49-50.

Rao, T.S.; Kumari, K.K.; Netaji, B. and Subhokta, P.K. (1985). Auurveda Siddha. *Journal of Research* **6**:213-224.

Reddy, V.D.; Padmavathi, P.; Paramahamsa, M. and Varadacharyulu, N. (2009). Modulatory role of *Emblica officinalis* against alcohol induced biochemical and biophysical changes in rat erythrocyte membrane. *Food and chemical Toxicology* **47(8)**:1958-1963.

Robak, J. and Dryglewski, R.J. (1988). Flavonoids are scavengers of superoxide anion. *Biochemical Phramacology* **37(5)**:83-88.

Sabu, M.C. and Khuttan, R. (2002). Anti-diabetic activity of medicinal plants and its relationship with their antioxidant property. *Journal of Ethnopharmacology* **81(2)**:155-160.

Saeed, S. and Tariq, P. (2007). Antibacterial activities of *Emblica officinalis* Gaertn. and *Coriandrum sativum* (Euphorbiaceae) against Gram negative urinary pathogens. Pakistan *Journal of Pharmaceutical Science* **20(1):** 32-5.

Sagar, V.R. and Kumar, R. (2006). Preparation and storage of ready–to-eat dehydrated gooseberry (aonla) shreds. *Journal of Food science and Technology* **43(4):**349-352.

Sahota, P.; Pandove, G.; Jairath, S. And Phutela, R.P., (2009). Preparation of low alcoholic, self carbonated beverage from carrot and amla juices. *Beverage and Food World* **36**:18, 23.

Sai Ram, M.; Neetu, D.; Yogesh, B.; Anju, B.; Dipti, P.; Pauline, T.; Sharma, S.K.; Sarada, S.K.S.; Ilavazhagan, G.; Kumar, D. and Selvamurthy, W. (2002). Cyto-protective and immunomodulating properties of amla (*Emblica officinalis*) on lymphocytes: an *in- vitro* study. *Journal of Ethnopharmacology* **81(1)**: 5-10.

Sawant, L.; Prabhakar, B.; Mahajan, A.; Pai, N. and Pandita, N. (2011). Development and validation of HPLC method for quantification of phytoconstituents in *Phyllanthus emblica. Journal of Chemical and Pharmaceutical Research* **3(4):**937-944.

Scartezzini, P.; Anrognoni, F.; Raggi, M.A.; Poli, F. and Sabbioni, C. (2006). Vitamin C content and antioxidant activity of fruit and of the ayurvedic preparation of *Emblica officinalis* Gaertn. *Journal of Ethanopharmocology* **104(1-2)**:113-116.

Sethi, V. and Anand J.C. (1982). Physico-chemical and microbiological quality of carrot and amla preserve. *Indian Food Packer* **36**:38-43.

Sethi, V. and Anand J.C. (1983). Retention of nutrients in carrots and amla preserves. *Indian Food Packer* **37**:64-67.

Shahidi, F. and Wanasusdara, J.P.K.P.D. (1992). Phenolic antioxidants. *Critical Review in Food Science and Nutrition* **32(1)**:67-103.

CHAPTER 7

Traditional Technology of Fish Preservation in Northeast India

R. K. MAJUMDAR

College of Fisheries, Central Agricultural University (I), Lembucherra, Tripura–799210, India

ABSTRACT

The northeastern region of India (NEI) extends between latitudes 21° 57′ and 29° 30′ North, and between longitudes 89° 46′ and 97° 30′ East, lie deep in the lap of easternmost Himalayan hills comprising of eight states of Arunachal Pradesh, Assam, Manipur, Meghalaya, Mizoram, Nagaland, Tripura, and Sikkim. It is one of the most thinly populated regions of India, with about 45.5 million people (as per 2011 census), contributing about 4% of the whole country's population and the population density ranges from 14–340 Km^{-2}. There are about 166 numbers of ethnic groups inhabiting NEI, and each tribe is well distinguished from others by their inherent culture and lifestyle. A lot of indigenous knowledge systems pertaining to agriculture, medicine, food, and natural resources management exist in the region from time immemorial. The technology of processing and preservation of fish is unique and indigenous to the northeastern states, evolved by compulsion of people. The geographical location, heavy rainfall, etc. factors necessitated many age-old indigenously fermented fish products such as *Shidal*, *Ngari*, *Hentak*, *Lonailish*, *Tungtap*, *Numsing*, etc. to be a common practice for surplus fish harvest. The traditional fish processing units of NEI, by, and large, are rural, seasonal, labor-intensive, informal, and capital deficient. The indigenous fish processing techniques were developed in homes, normally inherited from generation to generation, and a little improvement, if any, were solely based on the observations of the practitioners. One of the most commercially important product, i.e., Shidal, whose technology is very old and probably originated in the former undivided India (now

Bangladesh) and believed to came into existence at least before the British Era in NEI, i.e., before 1824. Hentak and Numsing are indigenous fermented fish pastes of NEI, where plants are also used along with fish. There is only salt fermented fish product is Lonailish, which is prepared exclusively from Indian shad (*Tenualosa ilisha*), a high-fat fish, and extensively consumed in NEI and Bangladesh mainly due to its typical flavor, aroma, and texture. In general, there exists a traditional belief that all such ethnic fish products improve the immunity against seasonal illness in forest dominated NE India. From the nutritional point of view, the ethnic fermented fish products of northeast India have two important properties such as 'high anti-oxidative activity' and 'prevalence of lactic acid bacteria (LAB).' Except few, so far no significant scientific intervention has yet been initiated for improvement and validation of such age-old indigenous fish processing technologies. These traditional fish processing methods deserve genuine scientific intervention for their improvement and standardization, and this may reveal the intellectual richness of our indigenous people in terms of their ability to use microbes, although unknowingly, for preparing products for varied purposes in addition to food and beverages.

7.1 INTRODUCTION

Since ancient times, humans have exploited the fermentation process for the development of different food products to satisfy their taste. The earliest archaeological record of fermentation was an alcoholic drink, made from fruit, rice, and honey, dates from 7000 to 6600 B.C. in China. The earliest evidence of food fermentation dates back as far as 6000 B.C. in the Fertile Crescent—and practically every civilization included at least one fermented food in its culinary heritage to enrich their diet. In the early era, the significance of food fermentation to the people was for improving shelf-life and storing properties of foods. From the history of Asian civilizations it was revealed that there was a wide variety of fermented foods, such as Japanese *natto* (soybeans), Vietnamese *mám* (seafood), Chinese *douchi* (black beans), *Lao pa daek* (fish sauce), Korean *banchan* (side dishes), and all of these contributed an essential component of their daily cuisine. Food fermentation appears to be the oldest example of 'biotechnology,' and varieties of tastes, flavors, and textures are supplemented to the foods and additionally make food easily digestible. Fermentation prevents food from spoilage, thus enhancing shelf-life and also improves the nutritional quality as well. This also develops congenial physicochemical characteristics leading to favorable sensory properties. Fermentation, thus, tends to become a very promising technology, in any aspects whether traditional methods are 'improved' or modified to develop new products or whether knowledge of the effects of certain fermentations leads to new foods or food ingredients.

The prospect of food fermentation is enormous, and it offers a wealth of possibilities from food preservation to the development of diversified food products, and most significantly, the fermented foods have been found to possess therapeutic properties. The food safety unit of World Health Organization (WHO) has attached high priority for the research in food fermentation, due to its potential to improve the food safety by controlling the growth and activity of pathogens in foods. Fermentation technology has adapted itself to social demands. Although, fermented, and traditionally cured fish products are very popular in NEI, however, research and development on these are meager. The traditional fish processing units of NEI, by, and large, are rural, seasonal, labor-intensive, informal, and capital deficient. Commonly, fermented foods, including fish are sold and consumed in the areas where they are produced. The indigenous fish processing techniques were developed in homes, normally inherited from generation to generation, and a little improvement, if any, were solely based on the observations of the practitioners. Moreover, hardly there is any interest from the producer to know the role of microbes or physical and chemical changes in the products during fermentation. Some characteristics of the final products such as appearance, color, odor, and taste are modified based on the consumers' requirement, and process variations are made in the ingredients or conditions. Nevertheless, most processes are conducted on a trial-and-error basis without any quality control. Quality of the products primarily depends on the experience of the processor.

The therapeutic and nutritional properties of many fermented foods originating from various countries have been "experienced," and thereby the process has been continuing for several generations. The possible benefit to good health from fermented foods was first elucidated by the famous Nobel prize-winning Russian Bacteriologist, Elie Metchnikoff in 1910. In the primary phase, it was observed that the Bulgarians had an average life-span of 87 years, and that was exceptional for the early 1900s, and it was also a fact that four out of every thousand lived past 100 years of age. In the next phase, the investigation by the scientific community got momentum to these beliefs after the publication of the book "prolongation of life" by Metchnikoff in 1908. In this book, the author recognized that the mystery behind the long life of Bulgarian people was the consumption of high amounts of fermented milk, which he described as the controlling agent for intestinal putrefaction. Following this, several studies have been oriented to observe the lifestyle of a number of populations and found that the use of fermented food was very common. Although an increasing number of recent scientific articles is emphasizing the positive effects of fermented foods on various aspects of health and disease through different mechanisms.

The northeastern region of India (NEI), lie deep in the lap of easternmost Himalayan hills in North-East part of India, connected to rest of India by merely 20 km of wide land. The region is known for its vast natural resources and a cauldron of different people and cultures. States like Arunachal Pradesh, Nagaland, Manipur, Tripura, Mizoram, Meghalaya, Assam, and Sikkim are included in the NEI. The region shares over 2,000 km of border with Bhutan, China, Myanmar, and Bangladesh. Its total geographical area is 2,55,168 sq km with a population of 40 million (as per 2010 census), of which the Brahmaputra valley in Assam alone houses almost half of its population. People of all religions like Hinduism, Islam, Buddhism, and Christianity live here. NEI is home to a varied number of ethnic groups (almost 166), whose existence can be traced back to the pre-historic times. Each tribe is having a unique cultural identity, especially in food, dress, lifestyle, and religious beliefs. From a historical point of view, tribes of NEI can be largely associated with the ethnic groups of Indo-Mongoloids, Tibeto-Burmese, and proto-Austrioloids which represent the Asio-Austric culture on Indian hilly regions. This presumption is also supported by the looks as well as the traditions which are followed by the ethnic communities of NEI. Therefore, the tribes of NEI provide a cultural bridge between India on one side and South-East Asia, China, and Inner Asia and Burma on the other by ethnic and linguistic angles. The intrinsic part of its rich culture and tradition of the region are manifested by numerous forms of art and festivals. The food culture of NEI is indigenous, and it differs vastly from the rest of the country in its taste as well as flavors. In general, inhabitants of the region are non-vegetarian, and fond of spices in their dish.

The northeastern region is a great treasure of cultural, social, and linguistic variability, besides rich in plant diversity, and these are conserved by tribal people. Indigenous knowledge systems pertaining to agriculture, medicine, food, and natural resource management are the wealth of the region. The famous culture named 'jhum' cultivation, which indigenous people are habituated to live and survive with the forest, ensure a range of ethnic foods rich in nutrition and compatible to culture and ethnicity of tribes. From ancient times, rural women of this region have selected many wild plants and non-vegetarian foods through trial and error for satisfying their hunger. Moreover, different fermented and non-fermented foods are used in various combinations and even with traditional vegetables to meet the food and nutritional security of the region. The traditional foods consumed by ethnic people of the region are intimately associated to virtually all aspects of their socio-cultural, spiritual life, health, and seasonal illness. But, the new generation in the same community has witnessed a rapid change over the past 50 years, and this has been mainly due to the intervention of modern crop varieties and culture methods, materialistic life, and current trends towards increasing use of commercially processed foods.

The fish fermentation technology in the northeastern states evolved by compulsion of people. Considering the fact that NEI, being the highest rainfall area of the world, simple sun drying of fish is not congenial to preserve the fish for a lean period. High humid atmosphere and frequent rainfall particularly during the peak fishing seasons (i.e., from May to September) prolongs the drying of fish under the open sun. The well known 'Beel Fisheries' as well as accumulation of water in abundant low lying areas during the rainy season, offers an ideal habitat for the breeding of weed fishes such as *Puntius* spp., 'Darkina' (*Esomus danricus*), 'Mola' (*Amblypharyngodon mola*). The necessity of preserving fish for lean period and also to utilize the huge catch of weed fishes during monsoon motivated the fishermen to search for a method through which they could preserve the heavy catches of such less valued weed fishes for consumption and sale in the dry seasons (November to April) when there was scarcity of raw fresh fish in the market. Furthermore, those perishable produce could not have been transported to the distant markets where they could get a higher price compared to their local markets, due mainly to the non-availability of ice and good road communications. In such a situation, probably after several trial and errors, the Shidal and other fish fermentation technologies were originated to cater their needs.

Northeast has a number of indigenous fermented fish products such as Shidal, Ngari, Hentak, Lonailish, Tungtap, Numsing, etc. Fermented fish appears to be a staple diet, especially for the tribal people and those of non-tribal who lives in hilly areas. Fermentation of fish without salt is unique for the NEI, and in this respect, these differ from Southeast Asia, where fish is fermented with salt. This indicates the fact that the technology of salt-free fermentation originated long before the men started using salt (sodium chloride). Even though the only salt-fermented fish product of NEI is Lonailish, probably originated following the fermentation technology of Southeast Asia. Although the process of such fermented fish products is simple, but most of the critical steps are optimized by experiences for getting a yield of different qualities. Traditional practices are still prevalent without much scientific intervention and methods standardization. This chapter is a compilation of all available information regarding different traditional practices of fish preservation prevailing in the northeast sector of India, based on my own research, personal communication and from the available literature in this respect.

7.2 SHIDAL

One of the most popular and commercially important indigenously processed fish product is Shidal. Shidal is a salt-free fermented fish product and popularly called as 'seedal,' 'sepaa,' 'hidal,' 'verma' and 'shidal' in Tripura, Assam, Mizoram, Arunachal Pradesh, and Nagaland. Amongst the northeastern states, presently Assam and Tripura being the major producer of Shidal. This is a very ancient technology and probably originated in the erstwhile undivided India (now Bangladesh) in the period believed to be least before the British Era in northeastern states of India, i.e., before 1824. According to the Indian history, people of this region were unaware about the use of salt before its introduction by the British. Even though, after British Era, the salt was considered as a highly valued and scarce commodity and people used to use a substance known as 'khar,' made from banana or papaya plant as an alternative of salt.

Therefore, the question of preserving fish with a huge quantity of costly salt might be the financial burden to the fishermen. This may possibly be one of the reasons to preserve fish in a unique way without using salt. 'Beel Fisheries,' a kind of weed infested shallow water bodies and gets dried up fully or partially during winter, were abundant in the plains of Bangladesh and adjoining northeastern part of India. Such water bodies serve as an excellent habitat for weed fish such as *Puntius* spp. (Figure 7.1). With the beginning of rainy season, such weed fishes propagate naturally and form a good fishery coinciding the decreasing of the water level with the onset of winter. This may be the possible explanation for exclusive use of *Puntius* sp. for Shidal production. But non-availability of adequate amount of *Puntius* spp. in the northeast India for Shidal production compelled the Shidal manufacturer to import dried *Puntius* spp. from Gujarat, Lucknow, Chennai, West Bengal, Bangladesh, etc., and with time, the price of Shidal has been increased considerably in the markets. In order to provide Shidal at a cheaper price, the traditional producer tried Shidal production with other varieties of fish by trial and errors method. And lastly, an estuarine fish, i.e., *Setipinna phasa* was found to be suitable for Shidal production. In West Bengal, *Setipinna phasa* is caught in huge quantity in the mid to later part of the monsoon period, which is dried with or without salt, since there is less demand of this fish in fresh condition. Shidal producer of NEI import dried *S. phasa* from West Bengal. The cost of production of Shidal from phasa (Figure 7.2) fish is less compared to Shidal from *Puntius* spp.

FIGURE 7.1 (See color insert.) Punti shidal.

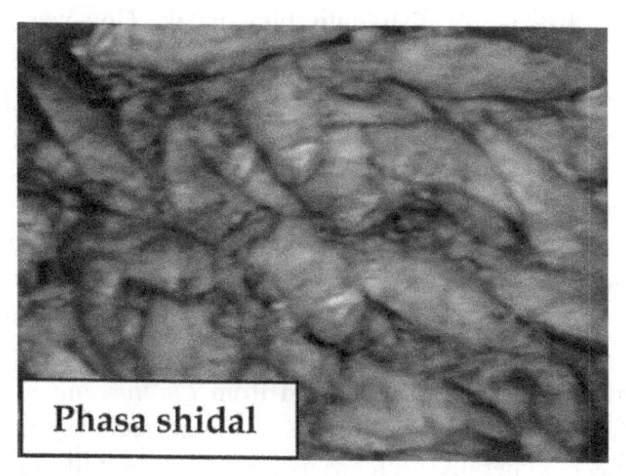

FIGURE 7.2 (See color insert.) Phasa shidal.

A strong and attractive flavor of Shidal seems to be responsible for its wide popularity even beyond the boundary of NEI. Traditionally, Shidal is prepared from *Puntius* spp., a weed fish used to be captured in huge quantity during the receding phase of monsoon. The product is dull white in color; solid, bilaterally compressed and little pasty with the shape of the fish remains almost unchanged except little disintegration near belly and caudal portion. The color of the product gradually turns to slight brownish to deep brownish on continuous exposure to air. The odor of Shidal is so strong, that it infuses the air in and around the storage and gives the area a characteristic smell of Shidal. The quality deterioration of Shidal becomes very fast when the products are taken out from the fermenting container and exposed to air. Major Shidal production centers are present at Assam (Nagaon, Dhubri, Goalpara, and Kachar districts), Manipur (Imphal city) and Tripura (West Tripura district). The ideal time for Shidal production is from November to February. Dried salt-free *Puntius* species, a major part of it are imported from other states and neighboring country Bangladesh, are usually used for the preparation of Shidal. The different steps in shidal production are discussed below.

1. **Matka and Its Oil Processing:** The earthen container usually used for Shidal production is locally called as 'Matka.' It is a pear-shaped container, available in different sizes, but most common size has the dimension such as the diameter of neck 8 inches, the diameter of middle expanded part 24 inch and height 36 inches having a capacity of 40 kg. Since earth made matkas are breakable, can be used for several batches of fish until they break. However, best quality matkas are made from very fine black soil, and such matkas absorb very less amount of oil during oil processing and also provide very less air permeability. Producers' also believe that older or re-used matkas yield better quality product and also the cost of production is lower compared to new matkas, because of the fact that older matkas absorb very less amount of oil during oil processing and they also provide very good air sealability. Prior to use, matkas are smeared with oil with an idea to seal the micropores present in its wall to make it almost non-permeable to air and vapor. The preferred oil for oil smearing is the oil extracted from *Puntius*, but presently, their availability is scarce. Therefore, the large scale commercial Shidal producers generally prefer vegetable oil, especially mustard oil for this purpose.

   Both the inner and outer walls of the matka are smeared with oil efficiently followed by drying in the sun. This process of oil smearing and subsequent drying in the sun is repeated for 7 to 10 times in case of new matka, until they become fully saturated with oil and unable to absorb any more oil even after a fresh drying. After oil processing, matka is ready for filling of fish for fermentation

(Figure 7.3). Whereas, only 2 to 5 times of oil smearing and subsequent drying is required in case of used matkas.

FIGURE 7.3 (See color insert.) Oil-smeared matkas.

2. **Water Soaking and Drying of Dried Fish:** Raw material of shidal, i.e., dried punti fish is first cleaned and sorted and occasionally further dried in the sun for 3 to 5 days to remove moisture from the fish to maximum possible extent and also to drive away from the maggots, if any. Special care is taken during sorting to remove broken fish and fish with a sign of infestation. Dried fishes are soaked in water while washing using bamboo baskets (locally called 'tukri'). Water soaking-cum-washing is preferably done in running water like a river, so that the scales, bones, broken parts, etc. are moved along with the water current. Although, cemented wash tanks with proper drainage facility are also used in some commercial Shidal production centers, where there is no river nearby. This soaking-cum-washing step is continued for 5 to 15 minutes. This step is very critical in Shidal technology. Normally the washing time is decided on the basis of initial moisture content of the dried fish, and most importantly, the duration of the fermentation period is dependent on this step. Water absorption is higher and quicker due to the previous drying of fish. Immediately following water soaking, wet fish are spread over the bamboo mattress or cemented floor, preferably in a well-ventilated room with bamboo fencing, and kept overnight for partial drying. Evening hours are considered as the right time for water soaking and washing, because the drying period of water soaked fish of 8 to 10

hours passes in the absence of any flies and birds. The partially dried fishes are soft textured with dry surface and are ready for filling in the matka (Figures 7.4 and 7.5).

FIGURE 7.4 **(See color insert.)** Water soaking.

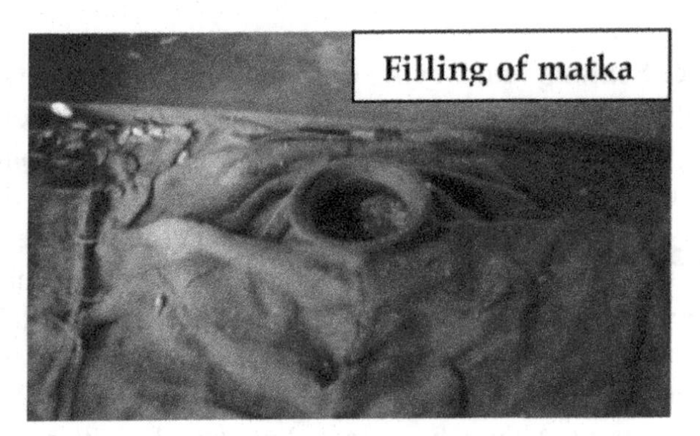

FIGURE 7.5 **(See color insert.)** Filling of matka.

3. **Filling of Matka:** During this step, oil processed matka is first placed in a pit dug in the ground in such a way that one-third of the belly remains buried in the ground. The matka is fixed to the pit very tightly gathering dug out soil in the surroundings, ensuring that it stands exactly at the vertical position and can withstand the pressure during filling of fish with compaction. In order to avoid any raw material getting contaminated with the soil underneath, clean gunny bags are spread surrounding the matka. While filling, partially dried fish are spread in layers, and uniform pressure is applied with bare hand or feet (in case of largemouth matka) between each layer.

Sometimes wooden stick is also used along with hand or feet for almost airtight packing. About 35–37 kg of dried fish is required to fill one 40 kg capacity matka.

FIGURE 7.6 (See color insert.) Filled matkas.

4. **Sealing of Filled Matka:** The matka, when filled up to the neck, is primarily sealed with a cover paste. The cover paste is made with the residues obtained while cleaning and sorting of dried fish. The usual thickness of the cover layer is 2–2.5 inch. Any broadleaf or newspaper is placed over the wet seal of cover paste, and finally, the matka is sealed by a layer of wet mud made from clay soil. The mud layer is checked on and often for about a week to repair any crack immediately by wet mud again. Sometimes, the mud layer is covered with a thin polyethylene sheet to avoid entry of any insects through the unnoticed cracks.

After completion of filling and final sealing, matkas are lifted to the surface and left undisturbed under shade for maturation. The fermentation shed is a big room in the ground with roof and sides made with rough bamboo fencing, so that air, sunlight, and even some rainwater are allowed inside the room. The usual period of maturation is 4–6 months, but it may be extended to one year. Also, some manufacturers start selling of the products after 2 months of fermentation. Moreover, filled matkas are also sold in this state in the wholesale markets. The yield of Shidal varies from 40–42 kg from each matka (Figure 7.7).

FIGURE 7.7 (See color insert.) Fermentation shed.

7.2.1 NUTRITION PROFILE OF SHIDAL

A complete nutritional evaluation of punti Shidal was first reported by Majumdar et al., (2009). The moisture, ash, protein, and lipid content of Shidal collected from markets of Tripura were recorded as 18.84%, 16.3%, 38.93%, and 16.73%, respectively. The pH and total titratable acidity (TTA) were also determined and found to be 6.9 and 1.66, respectively. The protein nitrogen and non-protein nitrogen (NPN) was found to be 46% and 54%, respectively of the total nitrogen. Although, they reported a higher value of TVBN content (509 mg%) of Shidal, but such high concentration of TVBN usually does not manifest any ammonia-like odor in the product. This, according to them, may probably be due to masking of ammonical odor by the characteristic strong odor of Shidal. The free alpha-amino-nitrogen content (79.54 mg%) of Shidal was found to be in the moderate range. The peroxide and thiobarbituric acid value were determined as 18.1 milliequivalent O_2 / kg fat and 0.41 mg malonaldehyde/kg respectively, and the explanation was given that the fermentation was carried out in an anaerobic environment in salt-free condition. Glutamic acid, aspartic acid, leucine, alanine, and lysine were found to constitute the major amino acids, while phenylalanine was the dominant aromatic amino acid. Some amino acids such as tyrosine, histidine, arginine, and tryptophan were found to be very low in amount and proline was not detected, and reason was given that this may be due to possibility of formation of derivatives of amino acids such as amines and gluconeogenic substances during fermentation. Amongst saturated fatty acids, palmitic acid (25.83%) was detected as highest, followed by stearic acid (10.81%). The most prominent monoenoic acids were oleic acid (25.33%) and palmitoleic acid (4.59%). The *n*–6 PUFAs in Shidal were composed mainly of linoleic acid (10.77%) and a very low amount of arachidonic acid (0.71%). The *n*–3 PUFAs were constituted by linolenic acid (4.18%), EPA (0.41%) and DHA (0.57%) along with two others (C 18:4 and C 22:5). The *n*–3/*n*–6 ratio was found to be 0.51.

7.3 NGARI

Popular fermented fish product of Manipur is known as Ngari (Figure 7.8). 'Iromba,' a special culinary preparation is made using Ngari, and it serves a daily compulsory side dish with cooked rice of people of all communities of Manipur. Although, production of Ngari is restricted to the valley region of Imphal (Manipur) and surrounding areas, but commercial-scale production centers are very less. October to January is the favorable period for production of Ngari, when huge quantity of weed fishes, mainly *Puntius* spp. are caught in the low lying areas including paddy lands. Ngari is prepared exclusively from the punti fish (*Puntius sophore*), which are available locally and also imported from Brahmaputra valley of Assam and Bangladesh. In Manipuri language, the sundried salt-free punti fish is known as 'Phoubu.' But for commercial production of Ngari, dried punti fish are procured in bulk from far states like Andhra Pradesh and West Bengal. Due to lack of any large scale fish processing industry in Manipur, the production and marketing of Ngari are confined to some of the rich household. The processing of fish for Ngari is almost similar to Shidal with some differences. The Ngari technology can be grouped into two major types. One method that has been used since ages is known as the indigenous method, and the other one is the modified commercial preparation.

1. **Ngari Chaphu:** Traditionally Ngari is produced by fermenting punti fish in earthen fermentation jar, locally known as Ngari chaphu (Figure 7.9). It is round-bottomed, narrow neck earthen pots specially designed for the purpose. In some localities of Manipur, this is also known as 'kharung.' The thickness of these pots is about 1.5–2.5 cm and can be of different capacities which can hold dried fish up to 15–75 kg. It is believed that the older re-used pots yield better quality Ngari as these become more airtight due to repeated oil processing and thus provide a better anaerobic condition for fish fermentation.

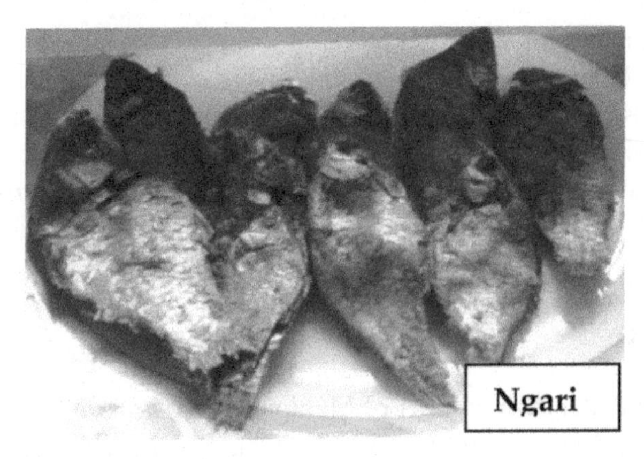

FIGURE 7.8 **(See color insert.)** Ngari.

FIGURE 7.9 **(See color insert.)** Ngari chaphu.

2. **Pre-Processing of Chaphus:** Pre-processing of chaphus is an essential step in Ngari production. During this process, the inner and outer walls of these chaphus are smeared with oil followed by drying them under the sun. Usually, vegetable oils like sunflower, mustard or palm oil, etc. are used for smearing purpose. The oil smearing and sun drying process are repeated for a week to two weeks time until the chaphus get fully saturated with the oil such that they are unable to absorb anymore oil especially on its inner wall. However, it is economic to use old and used chaphus because new chaphus absorb more oils and thus increase the cost of production. Oil smearing, in addition to its main role of blocking the micropores of the pots to happen fermentation in semi-aerobic to anaerobic condition, also

avoids adhering of fish on the inner surface of the pots. After proper oil smearing, chaphus are ready to be used for the production of Ngari. The earthen jars are tightly bound by wires in the outside to facilitate easy handling and to avoid leakage and breakage.

3. **Oils Used in Ngari Preparation:** Oil is chosen based on the availability and cost, and sometimes fish oil is also used, if available. Most of the commercial Ngari production centers in Manipur exclusively use vegetable oils like mustard oil and sunflower in order to reduce the cost of production. However, the use of fish oil has an additional advantage of adding essential n–3 fatty acids in the final product.
4. **Covering Paste:** The primary sealing of filled chaphus is done with covering paste. Usually, this is prepared from fish powder (FP) or even from some trash fishes as well as discarded materials during cleaning and sorting. They are ground well to make fine powders. This FP is moistened using a little amount of water (sometimes little oil is also added with water) to make a dough-like paste.
5. **Covering Leaf:** After sealing the chaphus with covering paste, it is wrapped with leaf. Any broad and thick leaf of creepers can be used as covering leaf. Leaves of *Musa paradacea* are most commonly used. But, leaves of bottle gourd and catesu leaves are also used. Sometimes, in case of unavailability of the above-mentioned leaves, the newspaper can be an ideal replacement.
6. **Final Sealing of Chaphus:** After putting covering paste over the filled chaphus, these are finally sealed airtight using a thick mud paste prepared from humus rich fine coarse soil. In some cases, old rug sacs are also used.

7.3.1 INDIGENOUS METHOD

The indigenous method uses raw material punti fish, which are usually collected from local fishponds or wetlands (*pats*). This process is more economical and usually done in the fishing seasons when punti fish are available in sufficient quantity as by-catch in many village ponds, low lying areas including paddy lands. Generally, whole punti fish is used for Ngari production. Fish are first washed thoroughly and dried under the open sun. When partially dried, these are rubbed with a little amount of mustard oil or fish oil and again dried for 3–4 days in sunlight. A little amount of salt is also added while rubbing oil, which is believed to facilitate absorption of oil deep into the fish body. This step is for getting a good texture of the final product.

The quantity of oil used for rubbing should be optimum as it decides the texture of the final product. Use of less oil makes the texture of the product very firm and not preferable. However, the optimum level of oiling is determined by experience only.

The oil smeared dried *Puntius* are put in the oil-processed chaphus. Filling of chaphus is compacted by pressing with legs after laying of each layer of fish. When the fish layer reaches to the neck region of chaphus, the mouths of the chaphus are then filled with cover paste and finally wrapped with the cover leaf. Before filling, the chaphus are installed over previously dug pits which are provided with moistened sacks. The chaphus are then removed from these pits and are kept in a sheltered place for 3–5 days. A thick layer of mud is put on the paste after removing the cover leaf, and thereafter these are kept undisturbed for about a week. The seals are observed carefully at the initial days, and as soon as any crack appears they are filled with mud paste immediately. The chaphus are kept in this condition for 4–6 months at room temperature. After an adequate period allowed for fermentation, the chaphus are ready to be opened which is locally known as 'chaphu kaiba' in Manipur. The final product, i.e., Ngari is taken out carefully layer-by-layer from the chaphus and then packed in smaller containers or pots. Special care is taken while handling the final product and 'chaphu kaiba' is usually done in a dry day so as to prevent the incidence of fungal infestation.

7.3.2 MODIFIED COMMERCIAL METHOD

For commercial production of Ngari, fish are imported from adjoining states like Assam and West Bengal, because locally available fishes are not sufficient for mass scale production. Mostly vegetable oil is used for processing to chaphus because of its low prices in the local market. The production starts with the drying of raw material fish (*Puntius* spp.) under the sun for 4–5 days in order to reduce the moisture content in the fishes to its minimum to prevent chasing out any maggot if present in them. Following sun drying, the dried fish are soaked in water for a period of 10–15 minutes. Usually, water-soaking step is carried out in the evening hours because the next step of partial drying under shed for 8–10 hrs is passed at nighttime in the absence of any disturbances from birds and insects. The partially dried fish are of soft textured with sticky surface. In the next step, partially dried fish are smeared with vegetable oil followed by another drying under the sun for about an hour. Following this, the fish are then stamped or rolled by rollers to make them soft and tender. The oil smeared, partially dried fish are taken in a big tumbler in bulk and continuously pierced with rod or bamboo stick in order to remove air bubbles present inside the heap. The fish now become ready for filling in the chaphus. The pre-processed chaphus are buried in the pits excavated earlier to avoid breaking during compaction while filling and layered with moist sacks with half or one-third of the belly underground. Initially, 5–6 kg of fish is put in each chaphus in proper layers followed by pressing with feet until wet liquid is released from the fish. The filling is continued in the same manner until the containers are filled with fish up to the neck. In most of the cases, compaction of fish is done by stamping with legs on the product layer by layer in order to make it airtight. The chaphus are then sealed first with covering paste and finally with wet mud keeping covering leaf in between. The sealed chaphus are then kept under shed for a period of 3–6 months for maturation. After adequate maturation, the top layer of fish in the chaphus below the covering paste (locally called as 'phumai') is removed, and the rest is Ngari. Generally, 'phumai' is considered as low-quality fermented fish products; however, many people consume 'phumai' after a mild heat treatment. The shelf-life of Ngari is about 12 months to 18 months, when produced properly using good quality raw material fish.

Ngari is considered as a gifted product by the people of the Manipur state since decades for its nutritional as well as therapeutic values. In every household, Ngari is valued as an essential ingredient in the daily diet, due mainly to its taste, which is strong appetizing in nature. Due to its ever-increasing popularity, its value as a food ingredient has attracted people of other states of the northeastern region and even in neighboring country Myanmar. Considering the importance of Ngari to the people of Manipur, recently sun-dried *Setipinna* species is introduced as raw material for reducing the cost of production, so as to make this affordable by all sections of the society.

Thapa et al., (2004) reported the presence of bacteria such as *Lactobacillus lactis*, *L. plantarum*, *Enterococcus faecium*, *Bacillus subtilis*, *B. pumilus*, Micrococcus, and yeasts species of *Candida* and *Saccharomycopsis*. The biochemical quality as reported by Thapa and Pal (2007) were moisture–33.5%, pH–6.2, ash–21.1%, protein–34.1%, fat–13.2%, carbohydrate–31.6% and food value of 381.6 kcal/100 g.

7.4 HENTAK

Hentak is a fermented fish paste, produced, and consumed in Manipur and other adjoining states. Unlike Shidal and Ngari, Hentak is a different kind of fermented fish product having plant ingredients such as '*Hongoo'* or '*Paangkhok'* (*Colocasi*a sp.). The petioles of aroid plant (*Alocasia macro-rhiza)* are mostly used as plant ingredients, which add taste and also help in the fermentation of dry fish. It is worthwhile to be mentioned that since time immemorial, Hentak technology has been traditionally practiced in Manipur. The homemade Hentak is very tasty, appetizing, and is usually offered to mother in confinement and patients in convalescence for its nutritional and therapeutic value (Figure 7.10). The primary aim behind this indigenous Hentak technology is to preserve the animal protein along with vegetable ingredients for a longer duration without any spoilage. Although, there is vast demand of this product, however, as such there is no commercialization of this product. Hentak is usually consumed as curry as well as a condiment with boiled rice.

FIGURE 7.10 (See color insert.) Hentak.

Any of the dried fish like *Puntius* spp. (*Ngakha*); Mola (*Amblypharyn-godon mola*) (*Mukaangaa*); Darkina (*Esomus danricus*) (*Agashaang*), etc. either alone or mixed are used for production of Hentak. The dried fish are washed thoroughly and allowed to dry completely by exposing them under sun continuously for few days till the fish become easily crushed. Following drying, fish are ground into fine powder with the help of a grinder. In some localities, 'Shoombal' (a wooden, manually operated grinder provided with one stable big block of wood having one concave hole and another wooden heavy rod with an iron ring at the tip for easily crushing the material to be grinded) is used for proper grinding. Simultaneously, the petioles of aroid plants is finely chopped, washed thoroughly, and dried in the sun for about an hour. The dried petioles and powder of dried fish are taken at 1:1 ratio and ground in a grinder to make a homogenous paste. The texture of the paste is considered as most essential for its quality and shelf-life. In this process, texture of the product should not be too soft or too hard. Optimum texture quality, therefore, can be achieved by maintaining an appropriate ratio of vegetable component and fish power and also their adjustment by judging the texture during processing. Generally, the texture becomes soft with more vegetable component and hard when FP is more. The paste thus obtained are made into small balls of size not more than an egg, kept in an earthen container in airtight condition and allowed to undergo fermentation for four days.

The balls are taken out after four days and kneaded uniformly and again transformed into small balls as earlier and kept again for another six days in the same container in airtight condition. In the next step, the balls are taken out of the container after six days and kneaded well with addition of little quantity of mustard oil and small balls are made out of this paste. These are finally stored in the same container in airtight condition for ten days. The texture of the product is examined in this step. The too soft or humid texture gives the average taste of the final product. In such a condition, a little quantity of finely ground FP is added to control the excess moisture or softness of the product. In case the texture is too hard, a little quantity of onion is added to it and again ground and compacted as a whole. In this way, the taste of the final product can be enhanced. After completion of this stage of fermentation, the product in the form of small balls are wrapped with clean banana leaf and put in the earthen container which is made airtight with a paste made from clay and raw cow dung at 1:1 ratio. The product is finally ready for human consumption after storage of six months in the same condition. Although, the product is consumable after three months of storage, but the taste will differ from the quality product. Hentak get hardened on keeping for few months. During culinary preparation, these balls are crushed and again made to paste with little water.

The biochemical quality was reported as moisture – 40.0%, pH – 6.5, ash – 15.0%, protein – 32.7%, fat – 13.6%, carbohydrate – 38.7% and food value of 408.0 kcal/100 g. (Thapa and Pal, 2007). Microbiological quality was found similar to Ngari.

7.5 TUNGTAP

Tungtap is a traditional fermented fish paste, commonly consumed by the 'Khasi' and 'Jaintia' communities of Jaintia Hills district of Meghalaya (Figure 7.11). It is not only popular for its taste but also believed to have medicinal value against various diseases, especially common cold. Although the process of its preparation differs slightly between 'Khasi' and 'Jaintia' communities, but the basic process is as follows.

FIGURE 7.11 (See color insert.) Tungtap.

Any trash fish like *Puntius*, *Danio* sp., etc. are most commonly used for production of Tungtap. Fish are first eviscerated and sundried followed by mixing with salt and other spices and put in earthen container in airtight condition for fermentation for a period of 7 to 10 days. During culinary preparation, the semi-fermented fish is taken out and roasted on a dry pan followed by grinding by hand to a fine paste with addition of chili, onion, mint leaves, turmeric powder, and salt. This paste serves as an important side dish in the local cuisine of the 'Khasi' and 'Jaintia' communities of the state.

The biochemical quality was reported as moisture – 35.4%, pH – 6.2, ash – 18.9%, protein – 32.0%, fat – 12.0%, carbohydrate – 37.1%, and food value – 384.4 kcal/100 g. (Thapa and Pal, 2007). Microbiological quality was found similar to Ngari.

7.6 LONAILISH

Lonailish, a salt-fermented fish product, prepared exclusively from Indian shad (*Tenualosa ilisha*). Hilsa is a high-fat fish, and fat content ranges from

14–25%. Lonailish is a very popular and widely consumed in northeast India and Bangladesh mainly due to its typical flavor, aroma, and texture. The major producer of lonailish amongst the NE states is Tripura, and the product is exported to other neighboring states. The product is diagonally cut hilsa steaks with thickness of about 1.50 to 2.00 cm. Typical lonailish has a uniform pink color and glossy appearance when taken out of the brine. The texture is firm, and the flesh does not easily separate from its bone. The product has a characteristic strong aroma mixed with some sweet, fruity, and acidic notes along with some saltiness. When exposed out of the brine, the strong odor spreads in the air and gives the area a characteristic smell of Lonailish (Figure 7.12). Until consumption, Lonailish is kept immersed in saturated brine to prevent deterioration of quality. Once the product is exposed to air, the autooxidation of lipids is very fast followed by gradual perception of rancid smell. The uniqueness of this product remains in the fact that despite the presence of highly prooxidative compounds like salt and metals, the rancidity of this highly fatty fish is kept under control, and is not manifested as long as the Lonailish are kept immersed in the fermenting brine.

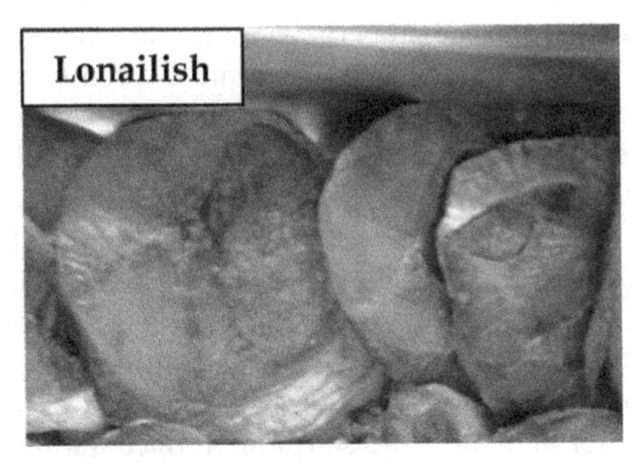

FIGURE 7.12 (See color insert.) Lonailish.

It is said that the Lonailish technology actually originated in the erstwhile undivided India (now Bangladesh) about 100 years ago on the bank of river 'Padma' and 'Meghna' under Noakhali district (Personal communication). It is also assumed that the technology evolved by compulsion when no such transportation, as well as preservation techniques except sun-drying and salting, were available to preserve hilsa during the glut period. Hilsa being a high-fat fish, sun-drying was not the suitable method of preservation due to rapid development of rancidity on being exposed to the sun. And salting-drying was also become difficult due to continuous spell of rain in July–August, that corresponds to the main glut period. All those factors might have compelled the fishermen to search for another alternative to quickly preserve large quantity of fish in an inexpensive way. Since the earlier days, the technology has not changed much, and the practice is still one of the major means of preservation of hilsa by the fishermen community of Bangladesh. Although, till today no other preservation techniques except salt-drying and to some extent canning are in practice. Recently, the large scale production is limited mostly in the Chandpur Subdivision in Noakhali district of Bangladesh. During partition of India in 1947, the migrated fishermen from erstwhile East Pakistan to northeast part of India started production of Lonailish in large scale to cater the demand of entire region. However, presently the consumption of Lonailish is restricted in Bangladesh and northeast part of India. The localized consumption of Lonailish in India may be due to the practical difficulties in its transport. However, there is a good demand for the product in the fish-eating communities in other areas, and also there is huge potential for its export adopting suitable packaging methods.

The Lonailish technology is simple and inexpensive. The fish (*Tenualosa ilisha*) are first washed properly with potable water to remove slimes and dirt. Then the fish are descaled with removal of tail and head, leaving gut inside. The fish are steaked extreme diagonally in such a way that the steak has more flesh exposed than that of the skin. The thickness of the steaks are usually 1.5 to 2.0 cm. This is followed by dry salting, when each steak is rolled thoroughly in salt (fish to salt ratio is 4:1) and kept in a bamboo made basket (locally called as 'tukri') layer after layer. Salt is put in the bottom of the basket before keeping fish and also sprinkled between each layer and above the top layer. The salted hilsa steak filled basket is covered with black polythene sheet so as to avoid entry of light, and stored in a dark place for 48 hours. The self-brine formed during this storage is allowed to drain. A substantial amount of moisture of fish is reduced during this process, and the color of the flesh becomes dull white, and the texture becomes somewhat tough.

Empty tin made containers (cap. 18L, used for cooking oil) are traditionally used for fermentation of hilsa. Although, recently some manufacturers are using LDPE barrel of 100 L capacity for bulk fermentation. Fermenting media is saturated brine. Saturated brine is prepared and boiled properly and cooled overnight. Following salt curing, the hilsa steaks are kept in the fermenting container layer after layer and compacted uniformly by hand after putting each layer till the fish layer reaches at least 5 cm below the top.

Salt-cured hilsa steaks are made free of adhering salt to some extent before storing in the container. After filling of the container with salt-cured hilsa steaks, cold saturated brine is pored slowly in the container over the fish to fill the voids between the steaks and maintain a level of brine about 2–3 cm above the top layer of fish. The container is closed with the lid. All the filled containers are stacked in one dark room and left undisturbed for 4 to 6 months for fermentation. The Lonailish tin (fermenting container) can be kept beyond one year, but the marketing starts after few months of maturation. Lonailish remains submerged in brine until sold. Once the container is opened, the materials are sold part by part. The product is fished out of the brine solution. Upon exposure of Lonailish to air and light for few hours, the pink-red color of the product gradually changes to grayish black, which is not acceptable.

There are few advantages observed in the traditional method of preparation of Lonailish. The period of Lonailish production, i.e., June to September which coincides with the peak fishing season of hilsa fish during their upward migration for breeding. Hilsa caught in this period contain a very high-fat content to the extent of 19–22% (Figure 7.13). An extreme diagonal cut while making steaks appears to be helpful in providing more surface area of fish flesh exposed to salt as well as halotolerant bacterial action. Loss of a significant amount of water from the fish as self-brine during salt curing prevents dilution of saturated brine during fermentation. Boiling of brine is also another advantage as this is done firstly for preparing a completely saturated solution and secondly to destroy undesirable microorganisms present in the salt that might contaminate the product. During fermentation or maturation, the texture of the fish meat changes and the final product obtain a texture of cooked meat. Fat is also enzymatically broken down to flavor bearing substances.

FIGURE 7.13 (See color insert.) Hilsa fermentation.

7.7 HUKOTI

Hukoti, an indigenous dry fish product, traditionally prepared by different tribal communities of upper Assam. Fatty fishes such as *Puntius* spp. are normally preferred for this product. During its preparation, fish are first washed and dressed to remove scales, fins, and gut. Salt and turmeric are mixed with the dressed fish and kept overnight in a container covered with a plastic sheet so as to prevent fly infestation. Although, there is no fixed ratio of the ingredients to fish and depends on the experience of the producer. In the next day, fish are washed again, and salt and turmeric are further added and spread uniformly on a perforated bamboo tray (locally called as 'chelani') and sundried for 3–4 days or dried over kitchen fire by placing it on a rack made up of wood or bamboo. This step is done to compose the texture tough enough to facilitate subsequent grinding operation. In the next phase of production, dried fish are ground well in wooden domestic grinder (locally called as 'dekhi'). The resultant powdered product is sieved using locally made bamboo sieve to remove extraneous matters like scales, bones, etc. in order to get homogenous matter. The powder so obtained is further grinded along with the stem of deskinned *Colocasia* spp. (cut into pieces). In usual practice, approximately 100 g of Colocasia stem is mixed with 1 kg of dry FP. Sometimes, leaves of a shrubby plant *Euphorbia ligularia* Roxb. (commonly known as 'leafy spurge' or 'milk hedge' and locally known as *Siju*) is also used @ 100 g per kg of dry FP along with Colocasia. The popular belief of the villagers regarding the purpose of incorporating such plants is to increase the adhesiveness of the prepared mixture. Also, some communities add a particular type of chili (*Capsicum chinense* and *C. frutescens*), ginger, garlic, etc. during grinding to suit their own taste and preferences. The end product after grinding is pasty in consistency and green in color. This green color paste is then stuffed into cylinders made from locally available matured bamboo (approx. 2.5 feet in length to accommodate nearly 2 kg of the paste). Filled bamboo cylinders are sealed primarily with dry banana leaves and finally with moistened clay soil to make it airtight. The sealed bamboo cylinders are then placed on a bamboo or wooden made rack hung over the kitchen for drying. Drying is done intermittently, and this is usually 4–5 times in a day for 10–15 min duration, and continued for a period of about 3 months for maturation. After the specified period, the final product, 'Hukoti' is ready for consumption and sold in the local markets. The mode of consumption of Hukoti is by frying in oil directly or with other ingredients/vegetables and usually consumed with rice. Consumption of Hukoti is believed by the tribal communities as a remedial measure to give relief from pain and prevent and cure malaria.

7.8 DANG PUI THU

Dang Pui Thu is a traditional fish product of Mizoram. Not much information could be collected about this product (Figure 7.14). Any variety of small fish or prawn can be used for this product. Small fish and prawn are dried over bamboo rack under the sun for 4–5 hours and ground to paste. The paste is stuffed in an earthen pot and fermented for 2–4 months.

FIGURE 7.14 (See color insert.) Dang pui thu.

7.9 NGAPTI

Ngapti or Napti is a fermented paste fish product of Tripura exclusively prepared by the Chakma tribes of South Tripura District (Figure 7.15). The technology actually originated in the Chittagong district of undivided India. Although not much information could be collected about this product, but small variety prawns are used as the raw material for this product. Dried prawns are mixed with salt and fermented in a cemented tank for a specified period. The product is brown colored and having a very pungent smell.

7.10 SMOKED FISH PRODUCTS OF NE

Northeast is famous for smoked and semi-smoked fish products (Figure 7.16). The state Manipur is especially famous for different variety of smoked fish. Smoked *Esomus danricus*, *Puntius sophore*, *Mystus bleekeri*, *Amblypharyngodon mola*, *Notopterus notopterus*, and *Glossogobius giuris* are some of the important smoked fish, indigenous of Manipur. Traditionally, the medium size fish are smoked by spreading the fish on a wire tray and then exposing to flame briefly to burn the skin, and the process is repeated after turning the fishes upside down. Sometimes, fish are dressed in butterfly style before smoking. The small size fish are simply spread on a wire tray, and these are exposed to smoke from burning paddy husk from a distance of about 30 cm below for about 2–3 h at 70–80°C. The smoking of fish in Manipur is done in a very crude method and restricted to a household activity only. So far no commercial plant for production of smoked fish in hygienic way has been developed in Manipur and other states of NE India.

FIGURE 7.15 (See color insert.) Ngapti.

FIGURE 7.16 (See color insert.) Traditional smoking kiln of Manipur.

7.11 NUMSING

Numsing is a traditional fish product developed by the 'Mising' community of Assam (Figure 7.17). It is a semi-dried and semi-smoked paste like product prepared from fish, petioles of arum (*Alocasia macrorrhiza*) and spices. Any small variety fish species are preferably used for this product. Fish are dressed to remove gills, scales, and intestines and washed with clean water. Dressed fish are dried on a bamboo-made rack fixed at a height of 3–4 feet over a fireplace. Heating is done by nearly a smokeless fire that is created by burning firewoods or dry bamboo. Drying and smoking is continued with occasional mild flame until the fish become moderately hard. Simultaneously, the edible part of arum is peeled and sliced and sundried for one day. Both the dried fish and arum slices are mixed together in the ratio of 4:1. Sometimes, other spices such as red pepper, green chili, ginger, and garlic are used for taste, but no salt is used. The mixture is ground in locally wooden made foot operated grinder (Dheki). The resultant pasty mixture is stuffed compactly in a bamboo cylinder (single internode of immature bamboo) keeping a headspace of about 10–15 cm for subsequent sealing. The filled bamboo cylinders are sealed with leaves of bladder fern (belongs to genus *Cystopteris* sp.) which are washed properly before use. Final sealing of the bamboo cylinders are done by moist clay soil to make it airtight and also to prevent entry of insects. The sealed bamboo cylinders are placed at a height of about 2–3 feet above the traditional mud-oven inside the kitchen or outdoor fireplace where the *Mising* people cook food. The bamboo cylinders with fish-plant mix are allowed to ferment in this condition, i.e., intermittent heating for about 30 days in the summer. After proper fermentation, the product becomes ready for consumption. The approximate shelf-life of the product is 2–3 years, and the quality deterioration is fast when the product is taken out of the bamboo cylinder and exposed to air and light. Usually, Numsing is consumed along with other vegetables.

7.12 SIDOL

Sidol is a semi-smoked fish product, traditional to the Rajbangshi communities in the lower Assam region of Assam State. This is a ready-to-eat type of product and usually consumed with main rice dish. The product has no offensive smell. In this process, drying of fish is accomplished by the thermal energy obtained from burning of wood, because of continuous rain spell during the months of monsoon in the NE region. Indigenous trash fishes such as *Puntius* and *Mystus* species are used for Sidol production. June to October is considered as the peak season of production. Raw material fish are collected from the markets or landing centers. Fish are first washed thoroughly followed by pre-processing by removing scales, viscera, etc. and placed over a bamboo made meshed tray for draining out of water. The dressed fish are spread over a bamboo made platform of 3–4 feet height with meshed top. A smokeless fire is provided at the bottom of the platform for heating the fish, and it is continued till the fish are dried completely.

FIGURE 7.17 (See color insert.) Numsing.

In the next step, dried fish are mixed with different spices and macerated along with petiole of aroid plants (*Colocassia*) to make a homogenous paste. The plant ingredients enhance the adhesiveness of the product. This paste is converted to small balls of similar sizes, wrapped tightly with leaves of aroid plants and put in an earthen container (pot) which is closed properly with a polythene sheet. The filled pot is kept in a dark place and left undisturbed for fermentation. After 4 to 5 days, the products become ready to consume. During culinary preparation, few balls of Sidol are taken out from the container, baked for some time, and little quantity of salt and mustard oil is added and made to paste again. This is usually consumed as 'chutney.' For long term storage, the wrapped balls of Sidol need to be dried periodically under direct sunlight, at least once in a month, and again put in the container. While doing this, the earthen containers are smeared with little quantity of mustard oil, which is believed to act as preservative. With this practice of periodical drying, the storage life of the product can be extended to about 2 to 3 years without any spoilage.

7.13 SAKUTI

Sakuti is smoked fish crushed into rough particles and very popular in the communities such as Hazarika, Bora, Kalita, Rajbangshi, Mohanta, etc. of Assam state. Fish soup is made from Sakuti by adding chili and garlic. It is strongly believed by the users that Sakuti helps in prevention of malaria. Any indigenous variety of trash fish can be used to prepare this product. This is an example of hurdle technology, where the combined preservative effect of salt and smoke generated from burning of wood is involved. The fresh trash fish are dressed by removing scales and viscera and washed thoroughly with clean water. The dressed fish are then immersed in brine (6–10%) for 10 to 15 minutes, followed by sun-drying for about 30 minutes. Partially dried fish are then placed over a bamboo made sieve (mesh size is according to the size of the fish), which is hung over a fire so that fish are exposed to smoke and heat produced by the burning wood. After about 4 to 5 hours (depends on size of the fish and previous experiences) the fish are turned upside down so that both sides of fish are exposed to smoke. When the fish are completely dried, they are ground with the help of an electrical grinder or 'Gaiyl-Chekai' (a traditional, wooden, and manually operated grinder) to a rough particulate form. The product is ready and is stored in airtight container and kept in a dry place. The product has a long shelf-life until exposed to air. Usually, Sakuti is consumed in pickles, soups, and curries, and is very common in the local markets.

7.14 GNUCHI

Gnuchi is a traditional smoked fish product of Sikkim, especially of the *Lepcha* community. The fish like *Schizothorax richardsonii, Labeo dero, Acrossocheilus* spp., *Channa* sp. are commonly used. Fishes are first washed and kept on a bamboo tray to drain off water. In the next step, fishes are degutted and rubbed well with salt and turmeric powder, and big size fishes are spread uniformly with upside down in the same bamboo tray. The tray with fish is kept above the earthen oven in kitchen. In case of small-sized fish, these are hung one after the other in a bamboo stripe above the earthen oven. The drying over the kitchen is continued for about 10–14 days. Usually, Gnuchi is consumed as curry and can be kept at room temperature for about 2–3 months.

7.15 SUKA KO MAACHA

This is a traditional smoked fish product of *Gorkha* community. Fishes of hilly river such as 'dothay asala' (*Schizothorax richardsonii* Gray) and 'chuchay asala' (*Schizothorax progastus* McClelland) are collected in a bamboo basket from the river or streams. The fish are degutted, washed, mixed with salt and turmeric powder. Degutted fishes are hooked in a bamboo-made string and are hung above the earthen oven in kitchen and kept for about 7–10 days. The product is very popular and usually consumed as curry. The shelf-life of the product is about 4–6 months in room temperature.

7.16 SIDRA

Sidra is a sun-dried fish product commonly consumed by the *Gorkha* community. Usually *Puntius sarana* is used, which is washed, and dried under the sun for 4–7 days. Properly dried fish are kept in airtight container and can be kept for 3–4 months at room temperature. Sidra pickle is popular cuisine of the community.

7.17 KHARANG AND KHA PYNDONG

This is very old and traditional technology of processing fish and is still practiced by the villagers of Umladkur and Thangbuli villages of Jaintia Hills district of Meghalaya. Kharang and Kha Pyndong are indigenous smoked fish products and several dishes like curry, chutney, etc. are prepared out of these products (Figure 7.18). Usually, the fish like Chocolate Mahseer (*Acrossocheilus hexagonalepis*), Rohu, Mrigal, Grass carp, etc. are preferable for production of Kharang and Khyrwong/Kha Pyndong. During the harvesting period of such fish, excess catch are processed and preserved in this method for their use in the lean period. Also during offseason, imported fish from other states are used for their production. A small hut or smokehouse is specially constructed for this process. The hut is roofed with leaves/grass (locally known as *Tynriew*), floor is muddy, and a special furnace is made in the center of the room. Wooden frames are fitted surrounding the furnace in such a manner that flame can flow through the center of the frame.

The fish are preprocessed before smoking. During this step, fish are first washed properly, descaled, and only viscera is removed by splitting the belly length-wise. When big sized fish (above 500g) is used, the product is known as Kharang, and in case of smaller size fish (less than 500 g), the product is called as Khyrwong/Kha Pyndong. After thorough washing of the dressed fish with plenty of water, these are pierced by bamboo sticks, single fish in case of Kharang and two fish at a time in case of Khyrwong/Kha Pyndong. Piercing is done by pushing thin bamboo stick through the mouth of the fish which pass through the belly up to the caudal peduncle region. Only a small portion (1–2 inch) of the stick remains protruded outside the mouth cavity to facilitate holding the fish in the wooden frame during smoking. As the fish are arranged in the wooden frame, smoking process starts by burning wood in the furnace. The door remains closed during smoking. Bigger fish (Kharang) are arranged nearer to the flame. Fish are placed with upside down so that the head and belly dries up first, and then gradually rolled-over/ turned in the frame by holding the protruded bamboo sticks for uniform smoking and drying. Smoking is continued till the fish become brown in color. Once the smoking is completed, they are removed from the wooden frame and allowed to cool and stout. The smoked fish is then kept in bamboo baskets, properly covered, and send to the market for sale. Different items such as soup, curry, chutney, fish balls, etc. are traditionally prepared from this smoked fish.

FIGURE 7.18 (See color insert.) Kharang.

7.18 FERMENTATION OF FISH BY NAGA TRIBES

Fish fermentation by Naga tribes is somewhat different from the others. The 'Lotha' tribe preserves fish through fermentation, and the product is a favorite food item of the localities. In this process, small sized fish are used as whole, whereas bigger ones are cut into smaller pieces. The washed fish are put inside a bamboo (commonly used bamboo is *Dendrocalamus hamiltonii*) and tightly plugged with leaves and kept over the fireplace for fermentation. Fermentation is completed within a few days and becomes ready for use as a tastemaker in vegetable curry. The shelf life of this fermented fish is only one month at room temperature.

A fermented product from crab has been reported as a traditional preparation of several tribes of Nagaland, such as 'Lotha,' 'Mao,' 'Angami' etc. The preferred raw material is the black species of crab with hard shell as it produces an aroma with good taste after fermentation. Crabs are first washed followed by removal of hard appendages, including entails. In the next step, the dressed crabs are ground and mixed properly with ground black til (*Sesamum orientale*) and wrapped in banana, *Macaranga indica* or *Phrynium pubinerve* leaf and allowed to ferment by placing these over the fireplace in the typical 'Naga' kitchen for about a week. Properly fermented product has a strong characteristic smell and is ready for use in cooking or chutney preparation (Figure 7.19).

FIGURE 7.19 (See color insert.) Smoking of fish in Naga kitchen.

The many different fermented fish products while noisome in their original state perform a beneficial role in human nutrition. They are rich in amino acids, nitrogen, and various trace elements, including sodium chloride, phosphorous, calcium, and fluoride. There is lacking of any complete study on the nutritional aspects of the fermented fish products of northeast India. However, the available data regarding nutritional facts indicates their superiority over fresh fish. Predominant bacteria in the fermented fish products of northeast India included coagulase-negative *Staphylococci* and lactic acid bacteria (LAB). Most important products, such as Shidal, Ngari, and Hentak, have also been reported to contain all the essential amino and fatty acids.

7.19 CONCLUSION

Traditionally processed fish products of northeast India have undoubtedly contributed to the survival and good health of entire generations that had only a small quantity of rice and vegetables to eat. Fermented fish products are mostly consumed as condiments; however, low salt fermented fish products can be consumed in large quantities. Flavor of fermented fish products is of first importance. Food fermentation involves mixed cultures of microorganisms that grow simultaneously or in succession. One can contemplate that indigenous people are using microorganisms unknowingly for varied purposes. The high degree of hydrophobicity by some LAB strains isolated by several workers from Ngari, Hentak, *and* Tungtap indicates the potential of adhesion to gut epithelial cells of human intestine, advocating their 'probiotic' character.

Although, there is a good market of smoked and fermented fish products in the entire northeast India, but still the products except Shidal and Lonailish are not manufactured at commercial scale. The age-old traditional techniques of production are still followed without much improvement. Other fermented fish and all smoked products are made at a very small scale in different communities. Practically no sanitation and hygiene is maintained in the production unit, and also there is no proper disposal of wastes. Smoked fish are produced without any control over smoke and quality. Similar is the condition of the dry fish markets, and no packaging is practiced during their retailing. A genuine study on the traditional fermentation process may reveal the intellectual richness of the indigenous people of our country in terms of their ability in preparing microbial products for varied purposes in addition to food and beverages. Fermentation is not only a method of preservation; in addition, the fermented foods can also have the added benefits of enhancing flavor, increased digestibility, improving nutritional value, and provide pharmaceuticals. Fermented foods are associated with a unique group of microflora, which increases the levels of proteins, vitamins, essential amino acids, and fatty acids. Malnutrition associated with protein and vitamin deficiencies are the major problems of the third world countries.

KEYWORDS

- **ethnic groups**
- **fish preservation**
- **Northeast India**
- **shidal**
- **traditional technology**

REFERENCES

Deepayan, R., Ranendra, K. M., Satyendra, K. M., & Hemant, H. T., (2015). Biochemical and microbial changes associated with fermentation of *Setipinna phasa*. *Fishery Technology, 52*, 26–33.

Devananda, U., Deepayan, R., Ranendra, K. M., & Prabir, D., (2015). Diversified traditional cured fish products of certain indigenous tribes of Tripura, India. *Indian Journal of Traditional Knowledge, 14*(3), 440–446.

Majumdar, R. K., & Basu, S., (2010). Changes in the nitrogenous compounds during fermentation of hilsa steaks. *Fishery Technology, 47*, 167–172.

Majumdar, R. K., & Basu, S., (2010). Characterization of a traditional fermented fish product *Lona ilish* of Northeast India. *Indian Journal of Traditional Knowledge, 9*(3), 453–458.

Majumdar, R. K. Basu, S., & Nayak, B. B., (2005). Studies on the biochemical changes during fermentation of salt-fermented Indian shad (*Tenualosa ilisha*). *Journal of Aquatic Food Products Technology, 15*(1), 53–69.

Majumdar, R. K., Basu, S., & Nayak, B. B., (2009). Assessment of nutritional quality of 'shidal'–a fermented fish product of Northeast India. *Journal of Indian Fisheries Association, 36*, 25–34.

Majumdar, R. K., Nayak, B. B., & Basu, S., (2008). Involvement of *Bacillus licheniformis* and *Micrococcus kristinae* during ripening of salt fermented Indian Shad (*Tenualosa ilisha*). *Journal of Aquatic Food Products Technology, 17*(4), 423–440.

Ranendra, K. M., (2016). *Final Report of Research Project "Fermented Fish Products of North East India: Characterization of Microbes and Their Healthful Metabolites for Application and Scale Up."* DBT, New Delhi.

Ranendra, K. M., Deepayan, R., Sandeep, B., & Bhaskar, N., (2016). An overview of some ethnic fermented fish products of the Eastern Himalayan region of India. *Journal of Ethnic Foods, 3*(4), 276–283.

Ranendra, K. M., Deepayan, R., Sandeep, B., & Bhaskar, N., (2015). Chemical and microbial properties of shidal, a traditional fermented fish of Northeast India. *Journal of Food Science and Technology, 53*(1), 401–410.

Ranendra, K M., Deepayan, R., Snehal, S., & Bhaskar, N., (2017). Scientific evaluation of Shidal technology–An age-old traditional practice of fish preservation of Northeast India. *Journal of Traditional and Folk Practices, 5*(1), 16–24.

Ranendra, K. M., Sandeep, K. B., Deepayan, R., Snehal, S., Apurba, S., & Bhaskar, N., (2015). Biochemical and microbial characterization of *Ngari* and *Hentaak*–traditional fermented fish products of India. *Journal of Food Science and Technology, 52*(12), 8284–8291.

Sarojnalini, C., & Singh, W. V., (1988). Composition and digestibility of fermented fish foods of Manipur. *Journal of Food Science and Technology, 25*, 349–351.

Sarojnalini, C., & Vishwanath, W., (1995). Nutritional characteristics of the two fermented fish products Hentak and Ngari of Manipur. *Journal of Indian Fisheries Association, 25*, 75–81.

Sarojnalini, C., & Vishwanath, W., (1987). Biochemical composition and fungal flora of fermented fish paste 'Hentak' of Manipur. *International Journal of Academy of Ichthyology, 8*(1), 9–12.

Sarojnalini, C., & Vishwanath, W., (1994). Composition and nutritive value of sun-dried *Puntius sophore, Journal of Food Science and Technology, 30*(6), 480–483.

Tamang, J. P., & Nikkuni, S., (1996). Selection of starter culture for production of kinema, a fermented soybean food of the Himalaya. *World Journal of Microbiology and Biotechnology, 12*, 629–635.

Tamang, J. P., (2001). Food culture in the Eastern Himalayas. *Journal of Himalayan Research and Cultural Foundation, 5*, 107–118.

Thapa, N., & Pal, J., (2007). Proximate composition of traditionally processed fish products of the Eastern Himalayas. *Journal of Hill Research, 20*(2), 75–77.

Thapa, N., Pal, J., & Tamang, J. P., (2006). Microbial diversity in ngari, hentak and tungtap, fermented fish products of Northeast India. *World Journal of Microbiology and Biotechnology, 26*(6), 599–607.

Thapa, N., Pal, J., & Tamang, J. P., (2006). Phenotypic identification and technological properties of lactic acid bacteria isolated from traditionally processed fish products of the Eastern Himalayas. *International Journal of Food Microbiology, 107*(1), 33–38.

CHAPTER 8

Advances in Fish Processing Technology

R. K. MAJUMDAR

College of Fisheries, Central Agricultural University (I), Lembucherra, Tripura–799210, India

ABSTRACT

Fisheries and aquaculture play a pivotal role in the sector of food production in order to support nutritional security to the country like India. At present, fish and fish products have emerged as the largest group in agricultural exports of India and contribute about 10.51 lakh tonnes in terms of quantity and Rs. 33,442 crores in value. The global food industry is mainly driven by the major factors like health, nutrition, and convenience. As fish is highly perishable, proper processing and packaging help in maintaining the quality of fish. Worldwide, an array of preservation techniques are followed, ranging from a simple chilled or ice storage to most recent and advanced high pressure (HP) and electromagnetic field application. Recently, a trend in the consumers' demand, which encompasses high quality processed foods with minimal changes in nutritional and sensory properties, has been observed. To satisfy the consumers' requirement, recently fish processing technologies are targeted towards technology up-gradation, value addition, product diversification, and quality assurance. In fish processing industry particularly, the word 'value addition' is most popular and appropriate, mainly because of the increased opportunities and demand, the activity is attractable for earning foreign exchange. India appears to have good potential to increase its share in international fish trade by exporting value-added and diversified fish products. The age-old fish processing technologies such as drying, salting, smoking, etc., aimed mainly for their preservation, and not compatible for the present day markets and marketing systems. Some advanced but very common traditional preservation techniques such as chilling and super chilling are generally insufficient to maintain quality and also for offering long shelf life and safety of the product. In case of freezing, usually adopted for long term preservation, is not at all free from quality deterioration with the progress of the storage life. Some newly developed technologies are coming into prominence, which targets inhibition or complete elimination of spoilage and pathogenic microorganisms from the product to prevent quality loss. Few of the emerging technologies that find application in fish processing are High-Pressure Processing (HPP), Irradiation, Pulsed light (PL) technology, Pulsed Electric Field (PEF), Microwave Processing (MW), Radiofrequency (RF), Ultrasound, etc. Besides these, packaging technologies like Modified Atmosphere, Active, and Intelligent packaging also play an important role in fish preservation. The emerging technologies in the field of preservation of fish and fishery products have proven to be successful, not only in ensuring the safety, but also in extending the shelf life, because of their high effectiveness in inactivating spoilage and pathogenic organisms with-out deteriorating product quality.

8.1 INTRODUCTION

Indian fisheries and aquaculture play a key role in the sector of food production in order to support nutritional security to the country. Since independence, India has shown continuous and sustained increments in fish production. The fisheries sector contributes to a great extent to agricultural exports, and about 14 million people are engaged in different activities. Fisheries resources of India are wide and diverse and composed of a number of oceans, vast river systems, both fresh and brackish water lakes in addition to huge man-made water bodies in plains and in the mountains. The phenomenal annual growth rate of over 6% achieved by the aquaculture sector in the last three decades has been instrumental in bringing the 'Blue Revolution' in the country. In terms of biodiversity of fish and shellfish species are concerned, the country's contribution is more than 10% globally. India contributes about 6.3% of the global fish production and shares 1.1% of the GDP and 5.15% of the agricultural GDP of the country. Out of the present total fish production of 10.07 million metric tonnes, nearly 65% is contributed by the inland sector and nearly the same from the culture fisheries. Shifting of the trend towards increasing contributions from the inland sector and further from aquaculture is being observed significant over the years. For India's development, the fish processing industry is of paramount significance because of the vital linkages and synergies, and it promotes between the two pillars of our economy, industry, and agriculture. Rapid growth in the fish processing sector along with concurrent improvement in the expansion of value chain is also of great importance to achieve favorable terms of trade for Indian agriculture both in the domestic and the international markets.

The last decade witnessed India's shift from a situation of insufficiency to surplus in food. Now, the fish processing industry in India has become an assured track of growth and profitability in the arena of trade of food commodities. In agricultural exports of India, fish, and fish products have presently emerged as the largest group with a significant contribution of about 10.51 lakh tonnes in terms of quantity and Rs. 33,442 crores in value. Further, the fisheries sector contributes about 10% of the total exports of the country and nearly 20% of the agricultural exports. Presently, more than 50 different types of fish and shellfish products are exported to 75 countries around the world. The sector, however, has to go a long way to contribute in nutritional as well as food security of the country, besides earning of foreign exchange.

The definition given by the United States Federal Food, Drug, and Cosmetic Act, Section 201, Chapter II for the processed food as "any food other than a raw agricultural commodity and includes any raw agricultural commodity that has been subject to processing, such as canning, cooking, freezing, dehydration, or milling." This definition ascertains parameters for the Food and Drug Administration (FDA) to regulate quality and safety in the fish processing industry. Fish processing operation means a procedure in which fish is prepared for consumption and product development, and sometimes, with added value. This sector employs a large number of people, many of whom are unskilled laborers and traders. However, processing fish is also usually intended to make fish which is nutritious, and can include activities such as food fortification, in which vitamins and minerals are added to fish during processing to increase the nutritional value. Fish are a highly perishable commodity and generally spoil faster than other muscle foods. Quality changes of freshly caught fish towards deterioration are the overall results of autolysis and bacterial activity. Although, the extent of these changes with time determines the shelf-life of the product. Therefore, proper storage conditions are essential to prevent/defer spoilage activities in the fish and fishery products. Moreover, safety is considered a major concern in food processing, especially industrial processing, to create packaged foods which are sold commercially. Making fish safe for consumption includes processing to the extent to remove any potential risks, such as the presence of pathogens, in addition to maintaining strict safety procedures to reduce the risk of re-entering of harmful organisms during processing.

The global food industry is mainly driven by the major factors like health, nutrition, and convenience. Presently, people have become more selective in their food choices and are ready to spend more on convenience food such as assembled meals, rather than preparing from scratch. Such preferences are the result of changes in work cultures and lifestyles, availability of disposable income, and lack of leisure for household works. Health-oriented food and nutraceuticals are preferences of today's consumer. The conventional products are disappearing from the market, and their place is being taken up by the hygienically prepared and attractively packed products in the form of as 'ready-to-cook,' 'ready-to-serve' convenience foods. This is possible due to modern processing facilities available for value addition to fish and such hygienically, and attractively packed value-added fish products have potential as well as a good prospect to catch domestic as well as international markets. Subsequently, today's fish industries are exposed to a market that constantly raises its demands for fresh, high quality, and new products. Newer technologies are being fast introduced in the market, and our consumers are also accustomed to the different newer high-quality fish products. These are made possible by using the latest state-of-the-art technology that is also giving the industry a competitive edge. These will help the people to enjoy processed fish items that are out-of-season or not native to the region. As fish is highly perishable, proper processing and packaging help in maintaining the quality of fish. Worldwide, an array of preservation techniques are followed, ranging from a simple chilled or ice storage to most recent and advanced high pressure (HP) and electromagnetic field application.

8.2 NUTRITIVE VALUE OF FISH

Fish plays an important role to meet the nutritional requirement of a human being in respect of animal protein. From a nutritional point of view, it is the balanced diet one can easily think of, when consumed along with cereals. These are the richest source of essential amino acids which are required for the overall growth of a person. Besides amino acids, fish contains a number of fatty acids of immense nutritional value which could help to reduce the risk from cardiovascular diseases. That is why fish is also known as 'heart food.' Further, some of the fatty acids help in the development of the brain, and for this reason, fish is known as 'brain food.' People are now more health conscious. Fish is such a diet which is low in fat and cholesterol with high vitamins and minerals are often preferred by people over meat. In the

21st century, there is an increasing trend of fish becoming a 'health food' due comparatively to its high nutraceutical value, compared to meat products, which are viewed as unsafe due to incidences of diseases like mad cow disease; good quality fish is extremely safe.

The global food industry is mainly driven by the major factors like health, nutrition, and convenience. There is a constantly growing demand for fresh, high quality, and new products from fish. Presently, fish has attracted considerable attention due to its high-quality protein, lipids, vitamins, and minerals.

Fish contains protein of high biological value (BV), essential minerals, vitamins, and fats. It has been reported that 85% to 95% of the fish protein is assimilable. The chemical score or amino acid score of any food is determined by comparing its pattern to that of whole egg protein. The proteins efficiency ratio (PER), which measures protein quality is around 3.5 for fish, and it is much higher than beef (2.3), milk (2.5), and close to that of an egg (3.92). Fish is also rich in taurine, a non-protein amino acid with multiple functions like neurotransmission in the brain, stabilization of cell membrane and in the transport of ions such as sodium, potassium, calcium, and magnesium. Apart from this, fish protein has some unique nutritional characteristics. It is now established that fish protein behaves as a hypocholesterolemic agent, probably because of the characteristic amino acid composition. Another recent finding is that fish protein is a strong cardioprotective agent and can control the symptoms of experimentally induced myocardial infarction in albino rats. It protects the antioxidant system and thus minimizes the chances of tissue damage caused by oxidative stress. It has also been reported recently that fish protein has a clear protective effect in diabetic renal diseases.

Lots of systematic research has been conducted throughout the world on the nutritional value of fish, and as a result, researchers all over the world have repeatedly emphasized the beneficial effect of eating fish. The fact that coronary heart disease, while being one of the biggest killers in the world, is practically unknown among the Eskimos. Several investigations reported that their diet is mostly fish based and is rich in long-chain ω–3 (*omega*-3) polyunsaturated fatty acids. Additionally, they also have a reduced tendency to blood clotting and longer bleeding times compared to other people. The result of all such researches showed that men who ate fish once or twice per week were protected against coronary heart disease, and, it was concluded that intake of fish oils in the diet results in a marked reduction in blood cholesterol and triglyceride levels and also thrombosis problem (Tables 8.1–8.3).

TABLE 8.1 Proximate Composition of Some Indian Freshwater Food Fish and Shell Fish

Common name	Scientific name	Moisture	Protein	Fat	Ash
Rohu	*Labeo rohita*	76.90	19.10	0.20	0.90
Mrigal	*Cirrhinus mrigala*	77.10	19.00	0.10	1.40
Catla	*Catla catla*	76.30	19.60	1.30	0.90
Carpio	*Cyprinus carpio*	74.84	20.84	3.15	1.17
Calbasu	*Labeo calbasu*	76.00	20.50	0.60	1.70
Bhagna	*Cirrhinus reba*	78.56	19.46	0.22	1.33
Tilapia	*Tilapia mossambica*	77.30	20.47	0.58	0.90
Magur	*Clarius batrachus*	78.70	18.20	1.42	0.97
Channa	*Channa striatus*	77.50	20.47	0.58	1.45
Wallago	*Wallago attu*	79.10	15.80	3.16	1.30
Puntius	*Puntius sarana*	74.84	20.84	3.15	1.17
Freshwater prawn	*Macrobrachium rosenbergii*	78.29	21.17	0.27	0.37
Small prawn	*Acetes* sp.	83.50	11.00	0.60	2.50

TABLE 8.2 Clinical Disorders of Human Reported to be Reduced/Prevented by *n*–3 PUFAs

• High cholesterol	• Diabetes
• High blood pressure	• Osteoporosis
• Heart diseases	• Bipolar disorder
• Attention-deficit/hyperactivity disorder (ADHD)	• Inflammatory bowel disease (IBD)
• Asthma	• Menstrual pain
• Colon cancer	• Breast cancer
• Rheumatoid arthritis	• Skin disorders
• Depression	• Schizophrenia
• Prostate cancer	• Macular Degeneration

TABLE 8.3 Recommendations of the American Heart Association (AHA) for Omega-3 Fatty Acid Intake

Population	Recommendation
Patients without documented coronary heart disease (CHD)	Eat a variety of (preferably fatty) fish at least twice a week. Include oils and foods rich in alpha-linolenic acid (flaxseed, canola, and soybean oils; flaxseed and walnuts).
Patients with documented CHD	Consume about 1 g of EPA+DHA per day, preferably from fatty fish. EPA+DHA in capsule form could be considered in consultation with the physician.
Patients who need to lower triglycerides.	2 to 4 grams of EPA+DHA per day provided as capsules under a physician's care.

Fish and shellfish contain high levels of polyunsaturated fatty acids (PUFA) and arachidonic acid, eicosapentaenoic acid (EPA) and docosahexaenoic acid (DHA) constitute the major components. PUFAs of fish oil are mostly belong to the *omega*-3 family, which tend to lower blood cholesterol by depressing low-density lipoprotein (LDL) concentration. There are reports that *Omega*-3 fatty acids are involved in reducing the levels of plasma triglyceride, especially very low-density lipoprotein (VLDL). EPA is believed to play important roles in maintaining the health of heart and circulatory system, whereas DHA is involved in the functioning of the brain, nerves, etc. Since these acids cannot be synthesized in the human body to the required extent, it is essential that adequate quantities are supplied through food. Fish and other marine products are the best sources for these acids.

Further, researchers believe that fish oil suppresses the synthesis of pro-inflammatory eicosanoids from arachidonic acid and thus produce an overall anti-inflammatory effect. Fish lipid positively affects gene expression or the activities of signal transduction molecules involved in the control of cell growth, differentiation apoptosis, angiogenesis, and metastasis. Fish oil suppresses excessive production of nitrogen oxide (NO) during chronic inflammation and thereby helps prevent DNA damage and impaired DNA repair. The role of fish oil in decreasing estrogen production and thus reducing the estrogen-stimulated growth of hormone-dependent cancer cells has been reported. Other roles of fish lipids include improvement of insulin sensitivity and cell membrane fluidity and thereby probably help prevent metastasis through these effects.

Apart from protein and lipid, fish is also a good source of almost all the minerals and B group vitamins. The mineral contents of fish range from 0.4 to 1.5%, and these include iron, calcium, zinc, iodine (from marine fish), phosphorus, selenium, and fluorine. These minerals are highly 'bioavailable,' i.e., they are easily absorbed by the body.

Present day consumers are very much sensitive in the nutritional and sensory properties of fish, and that created a demand for high quality processed fish, which have the minimum impact of processing on the change of essential features. The novel food processing technologies are being implemented to cater the need of the consumers, i.e., to provide safe, fresher-tasting, nutritive foods without the use of heat or chemical preservatives. Therefore, the development and implementation of new technologies enhance food quality and safety, resulting in the development of some innovative food products with unique product attributes. Many emerging technologies have the potential to extend shelf life. Few of the emerging technologies that have application in fish processing are high-pressure processing (HPP), irradiation, pulsed light (PL) technology, pulsed electric field (PEF), microwave processing (MW), radio frequency (RF), ultrasound, etc. Packaging technologies like Modified atmosphere, active, and intelligent packaging also play an important role in fish preservation.

8.3 HIGH-PRESSURE PROCESSING (HPP)

HPP is a promising 'non-thermal' food preservation technology that has been developed with the aim of producing microbiologically safe food products while avoiding undesirable changes in the sensory, physicochemical, and nutritional properties of foods. This is an innovative food processing technology being its major advantage remains in the maintenance of fresh quality attributes. The HPP reportedly provides long shelf-life, retains a fresh taste and minimum quality loss since it does not have many of the undesirable changes that are associated with thermal processing (Swami Hulle and Rao, 2016). Although, the HPP processed food has been reported to show enhanced flavor with no indication of oxidation; however, structure, texture, and color can be negatively affected. The mechanism of processing food by this non-thermal technology includes placing of food product to be treated in a pressure vessel capable of withstanding the required pressure, and the product is submerged in a liquid, which acts as the pressure transmitting medium. Usually, water, castor oil, silicone oil, sodium benzoate, ethanol, or glycol is used as the pressure transmitting medium. The pressure transmitting fluid is selected based on the factors such as its protection capacity of the inner vessel surface from corrosion, the specific HP system being used, the process temperature range and the viscosity of the fluid under pressure, etc.

Basically, two fundamental laws govern HPP of food. Le Chatelier's Principle, which addresses changes to equilibrium as a result of pressure application. This principle states that any phenomenon (phase transition, change in molecular configuration, chemical reaction) accompanied by a decrease in volume is enhanced by pressure. When pressure, i.e., extensive variable changes, the equilibrium shifts in a direction that tends to reduce the change in the corresponding intensive variable, i.e., volume. Therefore, the pressure shifts the system to that of the lowest volume. Another law is the Isostatic Rule, which states that pressure is instantaneously and uniformly transmitted throughout a sample under pressure, whether the sample is in direct contact with the pressure medium or hermetically sealed in a flexible package that transmits pressure. Throughout the sample, the pressure is transmitted in a uniform (isostatic) manner and the time required for pressure processing is therefore independent of the sample size, and that is in contrast to thermal processing. Hydrostatic pressure alters inter-atomic distances and for which non-covalent interactions are mostly affected. Several studies have revealed that covalent bonds that constitute the primary structure of proteins can withstand the pressure up to 1,500 MPa and remain unaffected. The capability of hydrostatic pressure to maintain covalent bonds intact (unaffected) has been the central hypothesis for the preservation of biological activity of functional biomolecules in foods (Figure 8.1).

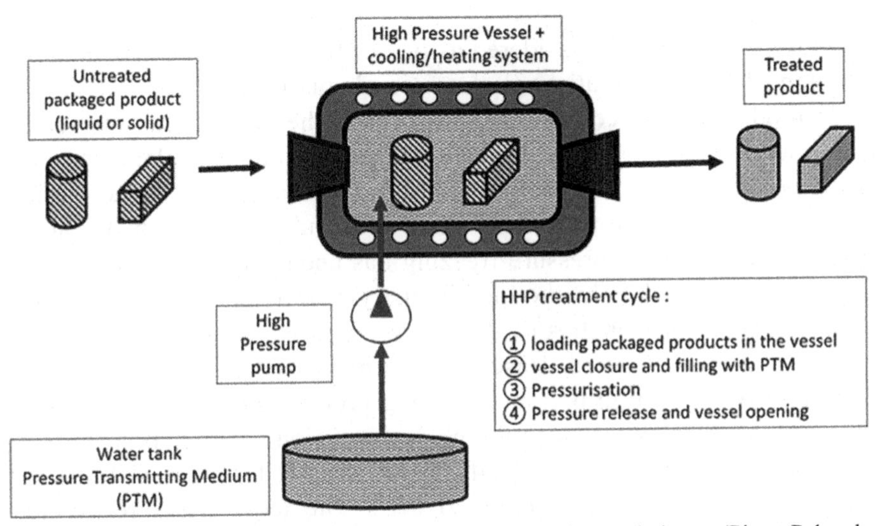

FIGURE 8.1 Schematics of high-pressure food processing techniques (Picart-Palmade et al., 2018).

Preservation of foods by high-pressure processing is the combined effect of elevated pressures (up to 900 MPa or approximately 9000 atmospheres) and moderate temperatures (up to 120°C) over a short period. Additionally, this technology also has advantages like uniform pressure application, negligible heat damage to food and thereby the potential for altering functional properties of foods. High-pressure processing, thus, increase shelf-life of food with minimal heat processing, and in this way satisfy the consumers' demand for fresher and higher quality heat-sensitive foods, which appears to be difficult to accomplish using conventional food preservation methods. Besides fish preservation, gelation of different kinds of surimi from pollack, sardine, skipjack tuna and squid could also be induced using hydrostatic pressure, as it is reported from Japan. A pressure of 300 MPa to 400 MPa was found to be effective to destroy microorganisms in surimi paste. Some workers reported that a pressure level of 220 MPa and a 30 min holding time were optimal and most effective in prolonging the storage period of tuna muscle, as well as in reducing the proteolysis activity, texture degradation, total volatile basic nitrogen (TVBN) and histamine formation.

8.4 HIGH-PRESSURE FREEZING PROCESSES

Although, freezing is considered as one of the most successful methods for long-term preservation of the natural quality attributes of perishable foods, but the ice crystals that form in the product can also cause considerable drip loss resulting textural, color, and organoleptic modifications after thawing. Therefore, the size and shape of these ice crystals play a critical role for the final quality of the frozen product. During the last decade, high-pressure freezing has been emphasized as a promising freezing method by the food industry, mainly due to its potentiality for improving the kinetics of the process and the characteristics of the ice crystals thus formed. Based on the phase diagram of water, three different types of high-pressure freezing processes can be distinguished in terms of the way in which the phase transition occurs: High-Pressure Assisted Freezing (HPAF), High-Pressure Shift Freezing (HPSF) and High-Pressure Induced Freezing (HPIF). The term 'pressure-assisted' means phase transition under constant pressure, 'pressure-shift' means phase transition due to a pressure release, and 'pressure induced' means phase transition initiated by a pressure increase and continued at constant pressure.

Usually, HPAF is conducted under constant pressure while the temperature is lowered to below the corresponding freezing point. The only difference with the traditional freezing process is that the HPAF is done under HP rather than atmospheric conditions. It has been reported that the ice nucleation, an integral part of this technique, occurs only in the outermost part of the product which is in direct contact with the cooling medium, and therefore, cooling of the sample proceeds from the surface to the center. The ice crystals are reported to be large, needle-shaped, radially oriented and show a marked size gradient from the surface of the sample to the center. Once the sample reaches the final temperature, and the freezing plateau is complete, the pressure is released.

In HPSF, cooling of the sample to less than 0°C is done under pressure and kept in a non-frozen state. As soon as the products acquire the desired temperature, the pressure is released, which induces uniform supercooling throughout the sample due to the isostatic nature of pressure. This supercooling induces the uniform formation of nuclei throughout the sample, so that latent heat is released, raising the sample temperature to the corresponding freezing point. Since the PSF creates smaller and uniformly distributed ice crystals, problems like tissue deformation and shrinkage (mostly due to large ice crystals) are not encountered, resulting in a significant improvement in the product quality. Therefore, the PSF has been generally accepted as a more advantageous high-pressure freezing process over glycol/water bath freezing and conventional air freezing. In a comparative study, Atlantic salmon's fillet samples were frozen by pressure shift freezing at 100 MPa (–8.4°C), 150 MPa (–14°C) and 200 MPa (–20°C), whereas, by conventional air freezing at –30°C and glycol/water bath freezing at –20°C.

8.5 PRESSURE ASSISTED THAWING

Pressure-assisted thawing is a technique where a food material is thawed under constant pressure. This process helps in reducing thawing time as well as drip loss. Normal freezing preservation depends on careful pre-freezing preparation, control of the freezing rate, storage conditions, and thawing conditions. Not many investigations have been reported at improving the thawing rate. Since thawing is done in minimal ambient temperature, therefore, this needs to be ensured. The use of HP for thawing offers a new alternative to this conventional process. Especially for thawing biological substances, HP offers a unique alternative to conventional processes in that it decreases the phase change temperature of pure water (down to –22°C at 220MPa). When the melting point of ice is lowered under HP, it allows the temperature gap between the heat source and the phase change front to increase, and thus enhances the heat flux rate in the case of pressure assisted thawing. The phase change temperature increases with pressure above 220MPa, and thus 220MPa appears as the maximum pressure of interest in thawing with respect to the heat transfer criterion.

8.6 PULSE LIGHT TECHNOLOGY

PL technology, one of the novel technologies which have gained increasing interest in recent years. The technique of PL finds its application in food processing due mainly to its non-thermal nature, which involves the discharge of high voltage electric pulses (up to 70 Kilovolt/cm) into the food product placed between two electrodes for few seconds. The potential applications of PL can be categorized as sterilization/pasteurization of food, packaging materials, and food contact surfaces; food enhancement in molecules of interest, such as vitamins; and the improvement of functional properties of some molecules. One of the major applications of this process is for decontamination of food so as to reduce pests, spoilage microorganisms, and pathogens without much effect on its quality. Due to the absence of any harmful residues, chemical, and toxic by-products in pulsed treated foods, this technology is proven to be safe (Forney and Moraru, 2009).

This technique involves the use of light energy in concentrated form, which is exposed on the substrate as intense short bursts of light (pulses). The term light mean radiations having wavelength ranging from 180nm to 1100 nm, which is further classified as ultraviolet rays (UV 180 nm to 400 nm, roughly subdivided into UV-A, 315 nm to 400 nm, UV-B, 280 nm to 315 nm, UV-C, 180 nm to 280 nm), visible light (400 nm to 700 nm) and infrared (IR) rays (IR 700 nm to 1100 nm). In a typical food processing operation, about one to twenty flashes per second are applied. PL is produced with the help of Xenon lamps that can produce several flashes per second. The ultraviolet (UV) part of the light spectrum is mostly responsible for destroying microorganisms in the food. The photochemical action of UV on the microbes causes thymine dimerization in the DNA chain, preventing replication and ultimately leading to cell death. This technology of PL treatment of foods has been approved by the FDA in 1996 under the code 21CFR179.41. Therefore, application of PL could lead to a reduction in the need for preservatives or chemical sterilizing agents. Compared to the conventional methods of inactivation of microbial load, the PL is a fast process, as well as a green technology because it consumes very less energy during its application. Several research findings have shown that this can be used for the reduction of microbial load of *Listeria monocytogenes*, *Salmonella enteritidis*, *Pseudomonas aeruginosa*, *Bacillus cereus*, and *Staphylococcus aureus*, *Escherichia coli*, etc. Many works have also shown extended shelf life in shrimp treated with PL (Figure 8.2).

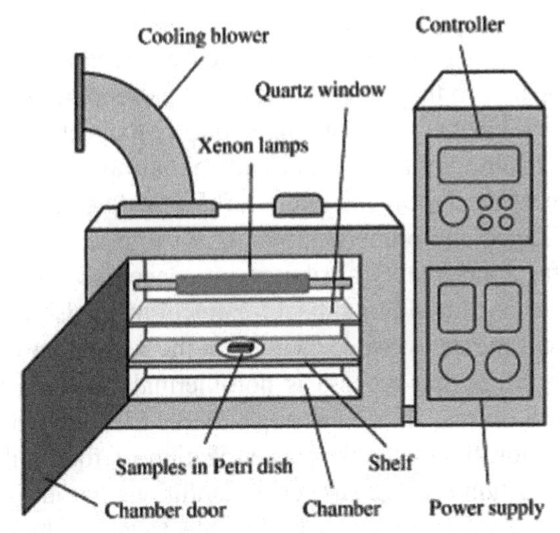

FIGURE 8.2 Meat processing unit with pulse light technology (Cheigh et al., 2013).

8.7 PULSED ELECTRIC FIELD (PEF) PROCESSING

The PEF processing is a novel, non-thermal cell disintegration or preservation process, basically governed by the principle of application of short pulses of high electric fields with a duration of microseconds and intensity of 10–80 kV/cm. The PEF has the potential to produce foods with excellent sensory and nutritional quality and shelf-life. The process is a continuous type and suitable for preserving liquid and semi-liquid foods, but not suitable for solid food products that cannot be pumped. The advantage of this technique is that a rapid rupture is possible in microbial cells and that too in limited areas without affecting the entire cell membrane, but the extent of effect depends on the external electric field applied. The causative factor for inactivation of organisms is the creation of an electric field with high voltage. When any liquid food is placed in an electric field, electric current flows into the liquid food and is transferred to each point in the liquid because of charged molecules present. The primary effect of PEF on biological cells may be attributed to the local structural changes and breakdown of the cell membrane, which is a highly important component of the biological cell because of its semipermeability nature regulates mass transfer and plays a key role in the synthesis of nucleic acids, protein, and cell wall components as well as many other complex metabolic activities. The processing time is calculated by multiplying the number of pulses with an effective pulse duration. This process provides a short-time (milliseconds) and low energy treatment, compared to the traditional thermal pasteurization of liquid food products. PEF has been accepted as an alternative to mechanical, thermal, or enzymatic cell disintegration of plant and animal raw materials. However, the crucial part of the process is that after the treatment, the food must be aseptically packaged and properly stored under refrigeration to have a long shelf life. So far, food quality attributes are concerned, PEF technology is considered superior to traditional heat treatment of foods because it avoids or greatly reduces the unfavorable changes of the sensory and physical properties of foods. Since the process is non-thermal, foods treated this way retain their fresh aroma, taste, and appearance. PEF technology has been successfully demonstrated for the pasteurization of foods like fish soups, tomato juice, and liquid eggs. However, conditions like the absence of air bubbles in liquid food and foods with low electrical conductivity are to be ensured before its application.

8.8 MODIFIED ATMOSPHERIC PACKAGING

The shelf-life of food is critical in determining both its quality and profitability. Fish and fish products spoil rapidly in air due to moisture loss or uptake, reaction with oxygen and the growth of aerobic microorganisms, i.e., bacteria, and molds, results in unfavorable changes in texture, color, flavor as well as the nutritional value of the food. Modified atmosphere packaging (MAP) is a process where the perishable food products are packaged in an atmosphere which has been modified so that its composition is different from the air. In this packaging system, ambient air is replaced either totally or partially by a gas, usually an inorganic gas. MAP is a technologically viable method, and its application has been restricted as a supplement to ice or mechanical refrigeration to reduce the losses and extend the storage life of fresh fish/seafood products. In this process, retardation of microbial activity and/or products' discoloration is accomplished by replacing the atmospheric air inside the pack with different gas mixtures. The proportion of each component gas is fixed when the mixture is introduced into the package; however, no control is exercised during storage. The initial gas composition of fresh MAP foods changes along with the progress of the storage due to chemical, enzymatic, and microbial activity of the product. The gaseous atmosphere in the storage atmosphere is primarily enriched with carbon dioxide, which controls microbial growth and results in the extension of shelf-life of products. Carbon dioxide effects lowering of the intra- and extra-cellular pH of tissues and possibly that of microorganisms. This pH shifting may possibly affect the membrane potential of microorganisms and influence on the equilibrium of decarboxylating enzymes of microorganisms. The gaseous mixture normally employed are carbon dioxide, mixtures of carbon dioxide and nitrogen, carbon dioxide and oxygen and carbon dioxide, oxygen, and nitrogen with the primary objective to extend the shelf-life of the product beyond that obtained in conventional refrigerated storages. Amongst all gases, inhibitory activity by carbon dioxide was found to be more effective when the product was stored at the lowest range of refrigerated temperatures.

In MAP, packaging materials of choice are flexible films of nylon/surlyn laminates, PVC molded trays laminated with polythene, polyester/low-density polythene film, etc. The gaseous composition used for MAP of fresh fish varies, depending upon whether the fish is lean or oily fish. For lean fish, a ratio of 30% oxygen, 40% carbon dioxide, 30% nitrogen is recommended. In practice, carbon dioxide fraction is kept higher for fatty and oily fish with a comparable reduction in the level of oxygen in the mixture leading to 40% to 60% nitrogen. The development of oxidative rancidity in fatty fish can be retarded by excluding oxygen in the mixture of gases. Although there is a very wide variation in the sensitivity of anaerobes to oxygen, but oxygen can inhibit the growth of strictly anaerobic bacteria like *C. botulinum*. However, sometimes, the inclusion of only some oxygen with nitrogen or carbon dioxide will not prevent botulism with absolute certainty (Figure 8.3).

FIGURE 8.3 MAP processed prawn.

8.9 VACUUM PACKAGING

Vacuum packaging may be defined as the "packaging of a food product with complete removal of air from inside the pack followed by a hermetic seal." Exclusion of air from the package creates a vacuum inside the packs, and lack of oxygen inside the packages may lower the oxidative deteriorative reactions as well as aerobic bacterial growth. The residual oxygen of package is absorbed due to chemical reaction and microbial activity. This is considered as a hypobaric (low pressure) storage and considerably extend the effective shelf-life, especially of cooked products by reducing the lipid peroxidation reactions. Application of vacuum packaging, in gas impermeable and heat stable materials, has many advantages, which include; no or negligible risks of post-pasteurization contamination, ease of handling, inhibition of growth of aerobic spoilage organisms and inhibition or slowing of deteriorative oxidative reactions in the food during storage due to oxygen barrier properties of the packaging material. The packaging requirements for vacuum packaging include, high durability, i.e., ability to withstand considerable mechanical stresses during packaging, handling, and transport, retention of flexibility even at low temperatures (–2°C to 4°C) to enable satisfactory handling in the packaging and refrigeration rooms, ability to withstand heating to at least 150°C without structural damage and leaching of potentially toxic plastics or plasticizers to the food products inside. Other requirements are that it must be impermeable to liquids, including oils and fats, macromolecules, and gases, in particular, oxygen, so that oxidative deterioration of the packaged foodstuffs is minimized or inhibited. The packaging materials must be manufactured from non-toxic, food acceptable, odorless materials and must be able to create airtight durable heat seals to close packs. Many of these criteria have been met by a range of materials, mostly multi-laminated plastics.

In vacuum packed condition, foods maintain their freshness and flavor 3 to 5 times longer than with conventional storage methods, and this is exhibited mainly because of lack of oxygen. Microorganisms such as bacteria mold and yeast cannot grow in a vacuum, and thus foods maintain their texture and appearance. Since the foods no longer become dehydrated from contact with cold, dry air, and this way freezer burn is also avoided. Most foods do not get dried due to lack of air to absorb the moisture from the food. On the other hand, dry, and solid foods also maintain their original texture, because they are not exposed to the air and, therefore, cannot absorb moisture from the air. Moreover, foods rich in fats and oils are also prevented from becoming rancid, because of lack of oxygen, which is considered to be responsible for this (Figure 8.4).

FIGURE 8.4 Vacuum packed fish steaks (O'sullivan, 2016).

8.10 ADVANCES IN PACKAGING

Freshness quality of fish deteriorates fast, and potential quality shelf-life is reduced if they are not packaged, handled, and stored properly. Therefore, quality and type of packaging materials and mode of packaging are of great importance for preserving the quality of fish. Recently a great deal of improvements has taken place in conventional packaging practices, which includes active and intelligent packaging technologies.

8.10.1 ACTIVE PACKAGING

Active packaging can be defined as 'a type of packaging method that alters the condition of the packaging and maintains these conditions throughout the period of storage to extend shelf-life or to improve safety or sensory properties while maintaining the quality of packaged food.' It is an innovative concept in which certain additives are incorporated into the packaging film or within the packaged containers with the aim to maintain and extend products' quality shelf-life. Basically, this technique is either scavenging or emitting systems added to emit (e.g., N_2, CO_2, ethanol, antimicrobials, antioxidants) and/or to remove (e.g., O_2, CO_2, odor, ethylene) gases during packaging, storage, and distribution. The additives, which are added in active packaging systems include O_2 scavengers; CO_2 emitters; moisture regulators; antimicrobial packaging; antioxidant release; release or absorption of flavors and odors etc. These packaging systems are having dual functionality consisting of a combination of oxygen scavengers with carbon dioxide and/or antimicrobial/antioxidant substances) etc. Although, several researches are going on in this field to develop some easily adaptable technologies so as to maintain the edible state of the food inside for a longer duration. In this direction, other active packaging systems that are expected to find increased attention in the future include color-containing films, light absorbing or regulating the system, susceptors for microwave heating, gas permeable/breathable films, anti-fogging films, and insect repellent package, etc. (Figure 8.5).

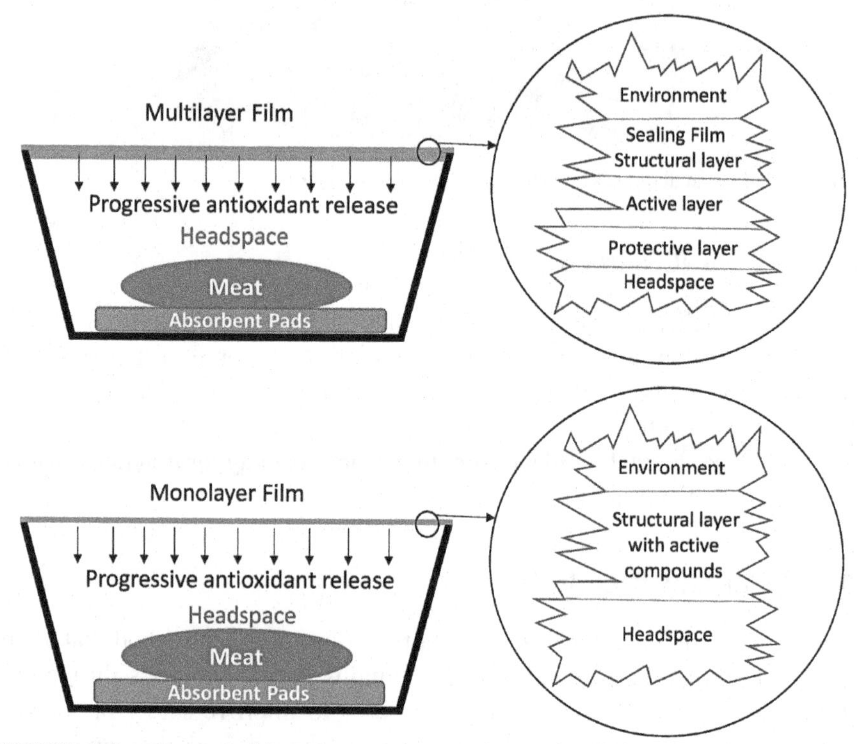

FIGURE 8.5 Meat packed in active packaging (Domínguez et al., 2018).

8.10.2 *INTELLIGENT PACKAGING*

Intelligent packaging may be defined as the 'inclusion of some sort of sensors in the packaging materials that monitor the condition of packaged foods and informs the manufacturer, retailer, and consumer about the condition/state of the food well ahead before its consumption.' This packaging system is also sometimes referred to as smart packaging, because it can sense some properties of the food it encloses or the environment in which it is kept. These include time-temperature indicators, leakage indicator, freshness indicator, etc. Active and intelligent packaging systems contribute to the improvement of food safety and extend the shelf-life of packaged foods. However, these are evolving technologies in the seafood area, and many of these systems are in the developmental stage.

8.11 VACUUM COOLING

Vacuum cooling is a method which differs from the conventional refrigeration methods in that, the cooling effect in vacuum cooling is achieved by evaporating some water from a product directly, rather than by blowing cold air or other cold medium over the product. The process involves rapid cooling by evaporative method, especially for porous and moisture foods to control the growth of microorganisms and preserving the quality of the products. Vacuum cooling is based on the principle, in that, once the pressure inside the chamber is reduced to the saturation pressure corresponding to the initial temperature of the product, evaporation of water takes place as the latent heat required for the evaporation is supplied by the product itself. The product which has free water and removal of such water does not damage the structure of the product, can only be processed by this method. The efficiency of the vacuum cooling is dependent on the ratio of evaporation surface area of food to the mass of the food. Vacuum cooling has been felt like a promising technology in respect of enhancing quality shelf-life of cooked meats, fishery products, and ready meals. In practice, tuna are caught at sea and frozen in brine immediately followed by transportation to canning plants where they are thawed and then steam cooked to 65°C. When the cooked tuna is cooled fast to 35–40°C by vacuum cooling, a 3–4% weight loss is incurred.

8.12 IRRADIATION

Irradiation is a process where food is exposed to the ionizing radiation, such as from gamma rays, x-rays or electron beams. This is a process where foods are treated with electromagnetic rays, and the energy is transmitted to the food without direct contact to the source of energy, i.e., the radiation resulting ionization in the targeted food. Food irradiation primarily extends the shelf-life of irradiated foods by effectively reducing the risk of food-borne illness, inhibiting spoilage microorganisms, delaying or eliminating natural biological processes such as ripening, germination, or sprouting in fresh food, and controlling insects and invasive pests. The approved sources of gamma rays for food irradiation are the radionuclides cobalt-60 (most common) and cesium-137. Although irradiation is lethal to food-borne bacteria, but produces some chemical changes in the food resulting production of small amounts of radiolytic products, but do not affect the nutritional quality of the food. Application of gamma irradiation to fish and fish products has been reported to considerable reduce the microbial population both in quality and quantity and thus preserve the fish for a longer duration. The lethal dose required for irradiation of fish was determined to be of 2 kGy to 7 kGy, capable of reducing important food pathogens such as *Salmonella, Listeria*, and *Vibrio* spp., as well as many fish-specific spoilers such as *Pseudomonadaceae* and *Enterobacteriaceae* are also significantly decreased in number. Although there is more negative consumer perception of foods treated with irradiation as compared to other methods of preservation because of people's imagination that the irradiated food is radioactive or mutated. On the contrary, all independent researches, such as the U.S. FDA, the World Health Organization (WHO), the Center for Disease Control and Prevention (CDC), and U.S. Department of Agriculture (USDA) have confirmed irradiation to be safe.

8.12 ULTRASOUND PROCESSING

Ultrasound refers to the sound that is just above the range of human hearing, i.e., above a frequency of 20 MHz. An ultrasound, when propagated through a biological structure, induces compressions and depressions of the medium particles imparting a high amount of energy to the material. Ultrasonic waves are capable of causing physical, mechanical, or chemical changes in the material leading to upsetting the physical integrity, hastening of certain chemical reactions through the generation of massive pressure, shear, and temperature gradient in the medium. Ultrasonics has been successfully used to inactivate *Salmonella* spp., *E. coli, L. monocytogenes, S. aureus*, and other pathogens. Ultrasound is probably the most simple and most versatile method for the disruption of cells and for the production of extracts. The process is efficient, safe, and reliable.

8.13 MICROWAVE PROCESSING (MW)

Conventional food heating methods employ heat energy generated externally and then transferred to the food material by conduction, convection, or radiation. In case of products containing particulates, especially when particulates are very large, conventional heating methods require such excessive heat processing to degrade the outer portion of the particulates. Accordingly, there is a considerable need for alternative technologies that perform rapid and uniform heating throughout the food that results in desired microbial lethality without altering or degrading overall food quality. Therefore, microwave heating, ohmic heating (OH), use of IR and RF, etc., generate heat in the product itself and find application in various food processing operations.

Microwave heating finds vast applications in the field of food processing, including fish and fish products such as cooking, drying, pasteurization, sterilization, thawing, tempering, baking, etc. Theoretically, microwaves are electromagnetic waves having frequency varies within 300 MHz to 300 GHz. Microwave heating takes place due to the ability of the materials to absorb microwave energy and convert it into heat. The energy required in MW is generated mainly due to dipolar and ionic mechanisms. Water is dipolar in nature, and its presence in food causes dielectric heating. When water molecules are exposed to an oscillating electric field, the permanently polarized dipolar molecules try to realign in the direction of the electric field. Because of the high-frequency electric field, this realignment occurs at a very fast rate (million times per second) and causes internal friction of molecules resulting in the volumetric heating of the material. Also, oscillatory migration of ions in the food in the presence of high frequency oscillating electric field can generate heat in the food. Application of MW in blanching operation is carried out for color retention and enzyme inactivation of fish which is conventionally carried out by immersing food materials in hot water, steam or boiling solutions containing acids or salts. This technique can also be applied in drying operation to remove moisture from fish and fishery products, because of its advantages of fast drying rates and improving the quality of the product. Because in microwave drying, vapors are generated inside the product and an internal pressure gradient is developed resulting removal of water. Unlike conventional sun drying or mechanical drying, shrinkage of food materials is prevented in microwave drying. Nevertheless, one of the disadvantages of microwave drying is that excessive temperature along the edges of food products results in roasting and production of off-flavors, especially during the final stage of drying.

8.14 OHMIC HEATING (OH)

OH is somewhat similar to microwave heating but with different frequencies. This is an emerging technology with a large number of actual and future applications in blanching, evaporation, dehydration, fermentation, extraction, sterilization, pasteurization, and heating of foods. OH is one of those alternative processing techniques, and its advantage is that it uniformly heats the food in contrast with the non-uniform distribution of microwave heating. OH or may be called as direct resistance heating is a process in which food liquids and solids are heated simultaneously by passing an electric current through them. It is a direct resistance heating by applying an electrical current through foods, so that heating takes place by heat generation through the conversion of electric energy to thermal energy within the material, where the food acts as an electrical resistor. The internal thermal energy generation technology enables the material to heat at extremely rapid rates from a few seconds to a few minutes. For foodstuff applications, OH provides a considerable advantage by not only avoiding the degradation of heat labile compounds due to over-heating but also by reducing the fouling of treated food surfaces during processing. Additionally, improvement of food quality, saving cost, and energy to processors are the special advantages of OH process. OH, therefore, have the potential to produce safe, high-quality food, and to validate any commercial process by experimentally demonstrating its application. Moreover, the shelf-life of foods processed ohmically is comparable to that of canned and sterile, aseptically processed products. However, further researches are essential to understand all effects produced by OH to food products, lethal effects of applied electric field and electric frequency during OH on different microorganisms and foods, cold spot determination, etc.

8.15 INFRARED (IR) AND RADIO FREQUENCY (RF) PROCESSING TECHNOLOGIES

Electromagnetic radiation (EMR) is a form of energy that is transmitted through space at an enormous velocity (speed of light). The heat generation in the material exposed to EMR could be due to the vibrational movement (as in case IR) or rotational movement (as in case of RF and MW) of molecules. Application of EMR heating in food processing has definite advantages over the conventional processes that include faster and efficient heat transfer, low processing cost, uniform product heating and better retention of organoleptic and nutritional value in the processed material, etc. In a conventional heating system like hot air heating, the heat is applied at the surface which is carried inwards through conduction mode of heating. In the case of EMR/dielectric heating, the waves can penetrate the material to be absorbed by inner layers. Quick energy absorption causes rapid heat and mass transfer, leading to reduced processing time and better product quality. However, the main advantage of electromagnetic heating over conventional electric and gas oven based heating is its high thermal efficiency in converting the electrical energy to heat and its dissipation in the food. On the other hand, in ordinary ovens, a major portion of the energy is lost in heating the air that surrounds the food, and fairly a good amount escapes through the vent, besides being lost through the conduction to the outside air. On the contrary, almost all the heat generated by EMR, which reaches the interior of the oven, is produced inside the food material itself.

8.16 EXTRUSION

In the modern cereal based industry, extrusion technology plays a central role, especially for the production of ready-to-eat breakfast cereals, expanded snacks, pasta, flat-bread, soup, and drink bases. Presently, the extruded products are shifting from baby foods to the adult markets, which needs to improve the eating quality with respect to texture, color, flavor, and most importantly, the nutritional facts. Basically, an extruder is a pump, heat exchanger and bio-reactor that simultaneously transfer, mixes, heats, shears, stretches, shapes, and transforms chemically and physically at elevated pressure and temperature in a short time. The operations take place during raw-material processing by extrusion include gelation, extrusion cooking, molecular disintegration, sterilization, mixing, shaping, and expansional drying. Consequences of cooking of the food ingredients during the extrusion process are the gelatinization of starch, denaturation of protein, inactivation of many raw food enzymes, which can cause food deterioration during storage, the destruction of anti-nutritional factors (ANF) such as trypsin inhibitors in soybeans and the reduction of microbial counts in the final product. In extrusion cooking, gelatinization of starch and denaturation of protein ingredient is achieved by the combined effect of temperature and mechanical shear. The extrusion cooking process is also referred to as High-Temperature Short-Time process (HTST). Through extrusion technology, underutilized fisheries resources could be efficiently utilized for the development of extruded snack like products. Extrusion cooking of fish mince with cereals and other ingredients has a good prospect to develop a number of fish protein-rich products shelf-stable at ambient temperature (Figures 8.6 and 8.7).

FIGURE 8.6 Extrusion cooking.

FIGURE 8.7 *Sous-vide* cooking.

8.17 *SOUS-VIDE* COOKING

Sous vide is a French term, which means "under vacuum" and refers to the process of vacuum-sealing of food in a bag followed by cooking it under controlled conditions of temperature and time. This may be defined as the "process where raw materials or complete foodstuff are cooked under controlled conditions of temperature and time inside heat stable vacuum pouches." In the traditional method of cooking, hardly there is any control over heat and temperature; moreover, the process is time-consuming to cook great food consistently. Sometimes, foods are overcooked on the outside and only a small portion in the center that is cooked to the temperature as desired resulting loss of characteristic flavor and texture of the food concerned. Importance of vacuum sealing in this process has several benefits, as it allows heat to be efficiently transferred to the food, increases the food's shelf-life by eliminating the risk of re-contamination during storage, inhibits off-flavors due to lipid oxidation and prevents evaporative losses of flavor, volatiles, and moisture during cooking and also reduces aerobic bacterial growth. *Sous vide* technology comes under low-temperature long time (LTLT) process where products are cooked at 65°C to 95°C for a long period. The factors which determine the microbial safety and shelf-life of *sous vide* products are the intensity of heat used for cooking, cooling time and temperature, control of chilled storage temperature. In the case of fish and fish products, the *sous vide* process helps to retain more health beneficial omega-3 fatty acids and nutrients compared to traditionally cooked fish.

8.18 BIO-PRESERVATION

Bio-preservation is accomplished by using natural preservatives of both plant and animal origin. Some natural compounds such as essential oils of spices, chitosan, nisin, lysozyme, and bacteriocins have been investigated and found to contain certain antioxidative and antimicrobial compounds, which could preserve food without using any chemicals. Thus, the potential of biological compounds as a replacement of chemical preservatives facilitate to obtain green label products. Bacteriocins are a heterogeneous group of antibacterial proteins (synthesized by bacteria to defend itself), vary in the spectrum of activity, mode of action, molecular weight, genetic origin, and biochemical properties. Recently, interests are growing to preserve food without chemicals, and various spices and essential oils have been found potential antioxidative and antimicrobial, and are being used to extend the storage life of fish and fishery products.

8.19 APPLICATION OF ENZYMES

Since a long time, enzymes have been used in various food-related industries as processing aids. Presently, enzyme technology has evolved as an integral part of the food industry, due mainly to their application in several steps in food processing make the things easier and safe, besides avoidance of chemicals which consumers do not prefer in the food. In the early stages, one major application of enzyme in the food industry was on the production of fermented food products, where different substrate-specific endogenous proteases were used under appropriate conditions.

Presently, modern food industries have adopted enzymatic methods as an important and essential part of food processing to produce a large and diversified range of products for human consumption. This could be possible only due to high specificity in the activity of enzymes, their high activity at very low concentrations and under mild conditions of pH and temperature, all of which seems to be responsible for developing fewer unwanted side-effects in foods and by-products in the production process. Due to growing interest in this field, demand for food and application-specific enzymes with specific properties is high, and for this, various enzyme sources are being investigated. Application of enzyme technology in food processing has also been found to be important in improving the production capacity and operating characteristics of various unit operations.

Cold-active enzymes including elastase, collagenase, chymotrypsin extracted from Atlantic cod was used in various food processing applications. The other applications of cold-active enzymes include caviar production, extraction of carotenoprotein, etc. It has been reported that recovery of the proteins from the fish frame or shrimp shell waste can be easily accomplished by treatment with proteases for a few hours under mild operating conditions. The role of transglutaminase in surimi production is well established. The gel strength of surimi can be improved by the application of extracellular microbial transglutaminase. Lipase extracted from *Pseudomonas* spp. can be used to produce PUFA enriched cod liver oil. In developed countries, enzymatic de-skinning of fish fillets is done by partial denaturation of collagen protein of skin using a gentle heat treatment followed by immersion in enzyme solution for several hours at low temperature (0–10C).

8.20 CONCLUSION

Considering the present trend the world's population will rise to over 9 billion in the coming years, and such population explosion will be accompanied by increased stress on our food supply and resources, and therefore, would invite a challenging task for the food industry. Aquatic foods play a vital role in addressing the issue of food and nutritional security. The four dimensions of food security are food availability, access to food, food utilization and use, and food stability. The concern related to an expected disproportion of these four dimensions of food security with the growing population in the coming years could easily be addressed by all these emerging and novel technologies, and many of such are yet to be commercialized. Product diversification, value addition, increasing exports, and reducing post-harvest losses can contribute to economic growth and reduce hunger in the world. Advances in food processing science enable us to preserve foods longer, improving food safety as well as to enrich the food with neutraceuticals beyond making food only for taste. A lot of food, i.e., approximately 1.3 billion tons (nearly one-third of world's food) is lost, wasted or spoiled each year at the different steps, such as production, harvest, post-harvest, and processing due to lack of adequate storage and processing facilities. Therefore, the application of novel food processing technologies in place of conventional methods might be the appropriate answer for reducing food loss, especially in the developing and underdeveloped nations. Moreover, present-day consumers demand high quality processed products with minimal changes in nutritional and sensory properties. Emerging technologies in processing and product development of fish will not only help in extending shelf life, maintain or improve sensory and nutritive properties, but also will ensure safety and reduction of processing wastes. Further, product diversification through advanced technologies will facilitate exports/cross-boundary trade, resulting in increase of economic value.

KEYWORDS

- **emerging technologies**
- **fishery product**
- **packaging**
- **processing**
- **quality**
- **shelf life**

REFERENCES

Balasubramaniam, V. M., Sergio, I., Martınez, M., & Gupta, R., (2015). Principles and application of high pressure-based technologies in the food industry. *Annual Review of Food Science and Technology*, *6*, 435–462.

Balasubramaniam, V. M., Farkas, D., & Turek, E. J., (2008). Preserving foods through high-pressure processing. *Food Technol. Chicago*, *62*(11), 32–38.

Cheigh, C. I., Hwang, H. J., & Chung, M. S. (2013). Intense pulsed light (IPL) and UV-C treatments for inactivating Listeria monocytogenes on solid medium and seafoods. *Food Research International, 54*(1), 745–752.

Da-Wen, S., (2005). *Emerging Technologies for Food Processing*. Elsevier Academic Press. The UK.

Domínguez, R., Barba, F. J., Gómez, B., Putnik, P., Kovačević, D. B., Pateiro, M., ... & Lorenzo, J. M. (2018). Active packaging films with natural antioxidants to be used in meat industry: A review. *Food Research International, 113*, 93–101.

Fellows, P., (2000). *Food Processing Technology: Principles and Practice* (2nd edn.). CRC, New York.

Forney, L. J., & Moraru, C. I. (2009). Ultraviolet Light in Food Technology: Principles and Applications. CRC Press.

Hogan, E., Kelly, A. L., & Sun, D. W., (2005). High-pressure processing of foods: An overview. In: Sun, D. W., (ed.), *"Emerging Technologies for Food Processing."* Elsevier Academic Press. 525 B Street, California, USA.

Hugas, M., Garriga, M., & Monfort, J. M., (2002). New mild technologies in meat processing: High pressure as a model technology. *Meat Science, 62*, 359–371.

Kerry, J. P., O'Grady, M. N., & Hogan, S. A., (2006). Past, current and potential utilization of active and intelligent packaging systems for meat and muscle-based products: A review. *Meat Science, 74*, 113–130.

Li, B., & Sun, D. W., (2002). Novel methods for rapid freezing and thawing of foods—A review. *Journal of Food Engineering, 54*(3), 175–182.

O'sullivan, M. (2016). A Handbook For Sensory and Consumer-Driven New Product Development: Innovative Technologies For the Food and Beverage Industry. Woodhead Publishing.

Oms-Oliu, G., & Martín-Belloso, O., (2010). Pulsed light treatments for food preservation: A review. *Food Bioprocess Technology, 3*, 13–23.

Picart-Palmade, L., Cunault, C., Chevalier-Lucia, D., Belleville, M. P., & Marchesseau, S. (2018). Potentialities and limits of some non-thermal technologies to improve sustainability of food processing. *Frontiers in Nutrition, 5*.

Ravishankar, C. N., (2016). Recent advances in processing and packaging of fishery products: A review. *Aquatic Procedia, 7*, 201–213.

Swami Hulle, N. R. & Rao, P. S. (2016). Effect of high pressure and thermal processing on quality changes of aloe vera-litchi mixed beverage (ALMB) during storage. *Journal of Food Science and Technology, 53*(1), 359–369.

Venugopal, V. V., (2006). *Seafood Processing.* Taylor and Francis, Boca Raton, London, New York.

FIGURE 7.1 Punti shidal.

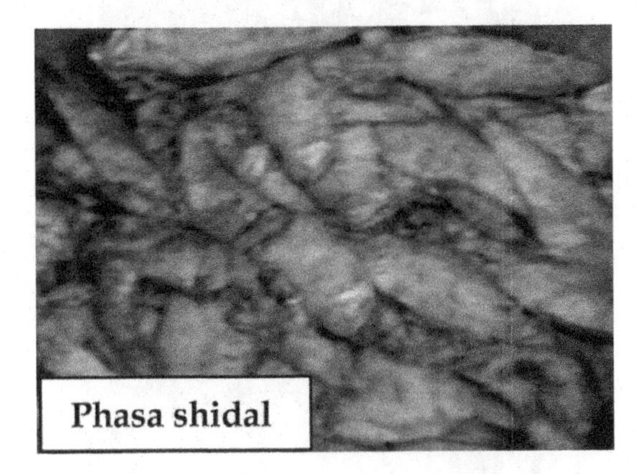

FIGURE 7.2 Phasa shidal.

FIGURE 7.3 Oil-smeared matkas.

FIGURE 7.4 Water soaking.

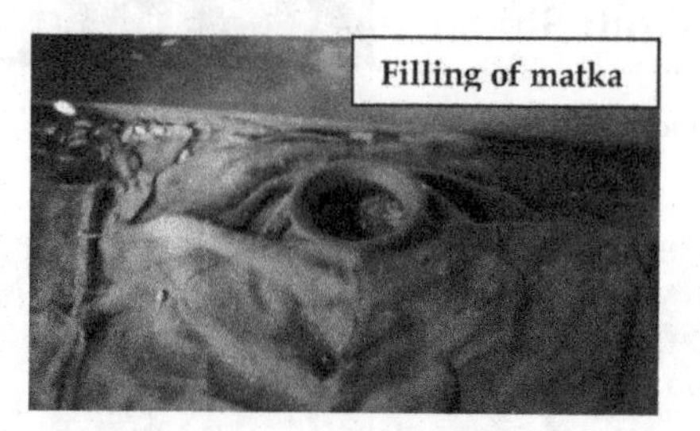

FIGURE 7.5 Filling of matka.

FIGURE 7.6 Filled matkas.

FIGURE 7.7 Fermentation shed.

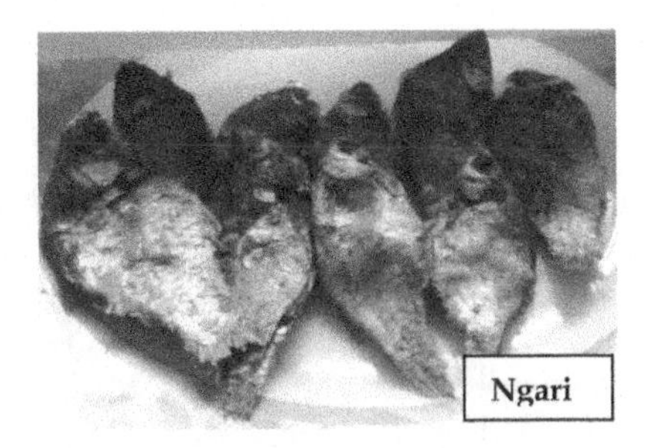

FIGURE 7.8 Ngari.

FIGURE 7.9 Ngari chaphu.

FIGURE 7.10 Hentak.

FIGURE 7.11 Tungtap.

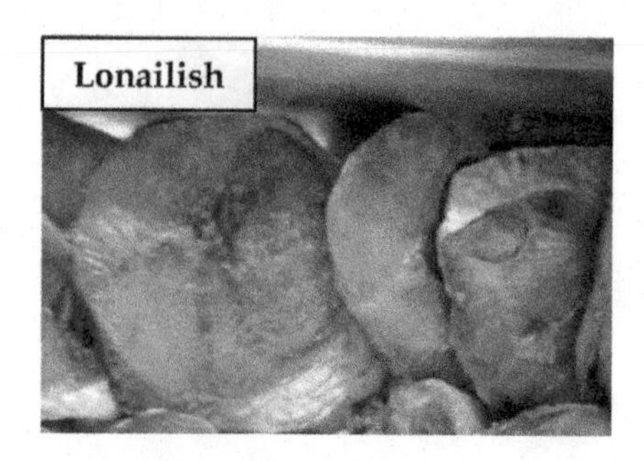

FIGURE 7.12 Lonailish.

FIGURE 7.13 Hilsa fermentation.

FIGURE 7.14 Dang pui thu.

FIGURE 7.15 Ngapti.

FIGURE 7.16 Traditional smoking kiln of Manipur.

FIGURE 7.17 Numsing.

FIGURE 7.18 Kharang.

FIGURE 7.19 Smoking of fish in Naga kitchen.

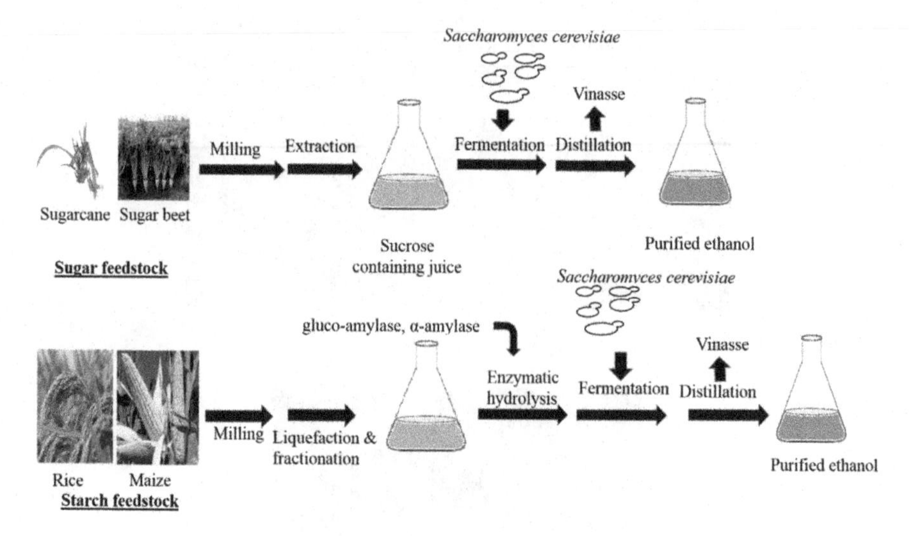

FIGURE 10.1 Bioethanol production from first generation feedstock.

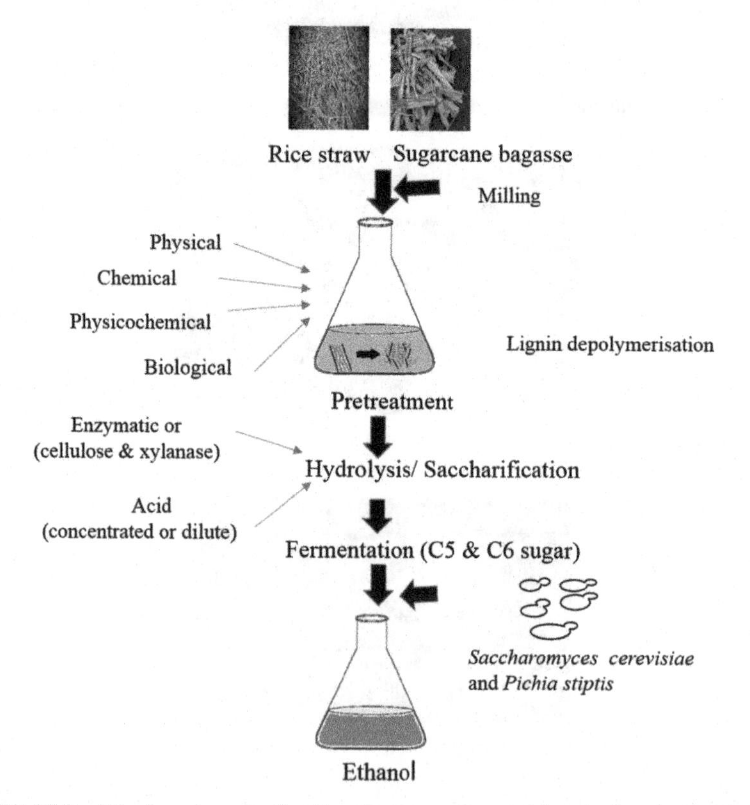

FIGURE 10.2 Bioethanol production from the second-generation feedstock.

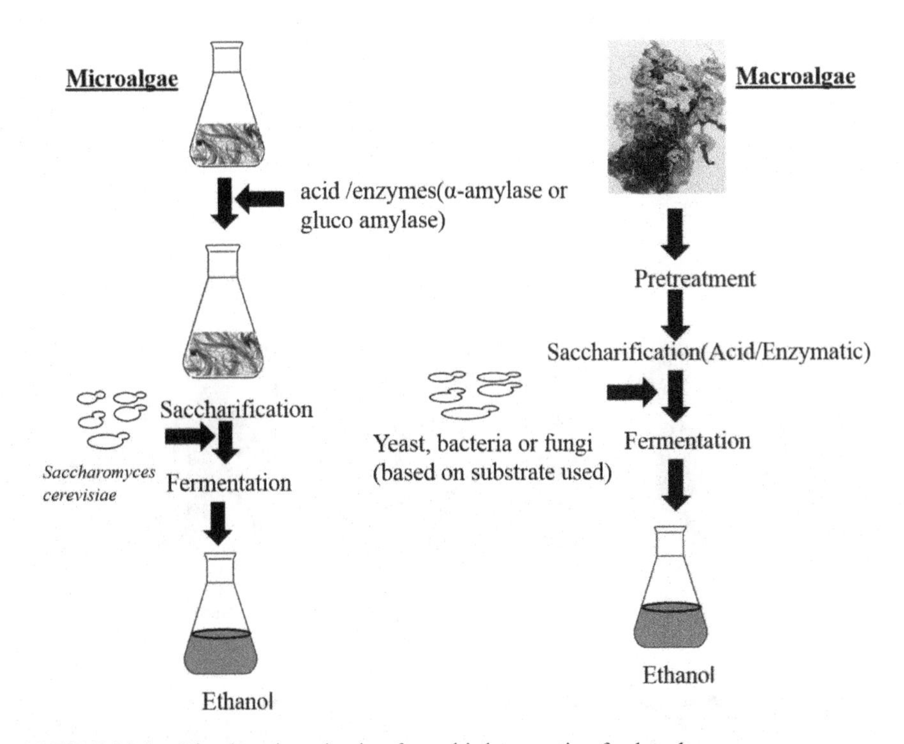

FIGURE 10.3 Bioethanol production from third generation feedstock.

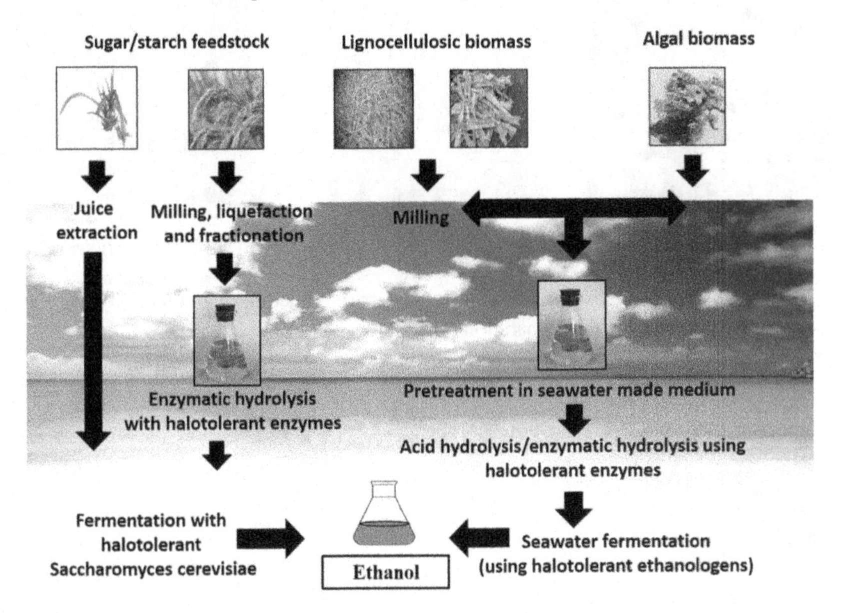

FIGURE 10.4 Bioethanol production from 1st generation, 2nd generation, and 3rd generation feedstock using seawater as the reaction medium.

(A) Fermentative cake of Rabha Community (B) *Saccharum officinarum* (C) *Ananascomosus (L.) merr*

(D) *Capsicum frutescens* (E) *Artocarpus heterophyllus* (F) *Cynodon dactylon*

FIGURE 11.4 Fermentative cake and the plant parts added as ingredients by the Rabha-Hasong community during the preparation of alcoholic starter cake.

CHAPTER 9

Nutritional Values and Processing of Tropical Tuber Crops

B. K. SAKHALE and NAMRATA A. GIRI

University Department of Chemical Technology (UDCT), Dr. Babasaheb Ambedkar Marathwada University, Aurangabad-431004, Maharashtra, India, E-mail: bksakhale@gmail.com, namrata_cft@yahoo.in

ABSTRACT

Tropical tuber crops having a significant role in food security, nutrition also climate change adaptation. These crops are essential diet components in many countries. Tropical tuber crops comprise of major and minor (under-exploited) plant species. The edible species include major crops like cassava, sweet potato, taro, yams, Chinese potato, canna, arrowroot and starchy curcuma and aroids. Non-edible medicinal crops include *Coleus forskohlii*, wild *Dioscorea*, wild *amorphophallus* species etc. These nutritionally important tuber crops possess health-promoting components, antioxidants, and pharmacologically active principles. The exploration of these tuber crops is a new area of food processing, that address hitherto unmet nutritional needs and the tropical tuber crops have been bounteous in providing hundreds of diverse natural products. Tropical tubers crops may be used for the preparation of functional products in order to meet the nutritional requirements of the community. The main aim of this chapter is to focus on the potential health benefits of tropical tuber crops and their utility for the development of functional food products.

9.1 INTRODUCTION

Tuber crops are important sources of starch after cereals, besides being used as a staple or supplementary food. Sweet potato and cassava tubers widely used as food preparations, animal feed and industrial purpose. These are considering as main tuber crops whereas other tubers consumed as a vegetable in the homestead or semi-commercial level. Tuber crops, cassava, in particular, being perishable deteriorate rapidly after harvest and are often unfit for food or feed within a few days. They are generally consumed as a vegetable after boiling it in water and eaten along with some spicy curry or by extracting starch. Besides serving as an insurance crop during times of food scarcity, it can also function as a source for the development of products having industrial value. These crops have short shelf life due to perishable nature leads to problems in storage, transportation and marketing for farmers. In order to overcome these problems, processing, value addition, and product diversification of the tubers near the farm site are recommended. The produce will also ensure the promotion of cottage and small-scale industries besides ensuring food security.

Carbohydrate is the main component of tuber crops and staple food for tropical and subtropical countries due to the high source of energy (Liu et al., 2006). Moreover, tubers also contain health beneficial nutrients such as resistant starch (RS) and mucilage. RS offers the slow digestion, which helps to release glucose slowly, also beneficial in the management of obesity, diabetes and other related diseases (Liu et al., 2006). Tubers having angiotensin converting enzyme inhibitory and antioxidative activities because of mucilage content. The specialty of tuber crops, do not contain any gluten, could be used as an alternative to wheat for the development of gluten-free foods for celiac patients.

The tropical tuber crops flour is well suited for the preparation of food for celiac patients. Gluten-free products are designed especially for celiac patients who are strict to the gluten-free diet. Celiac disease (CD) is a chronic autoimmune disorder, approximately 1–2% peoples were suffering from CD (Reilly and Green, 2012). Ingestion of gluten by celiac patients leads to damage of microvillus of the small intestine and leads to cramping, bloating, diarrhea, weight loss, vitamin, and mineral deficiencies, etc. (Green and Cellier, 2007). Consumption of gluten-free diet as a strict life-long strategy is the only known treatment for CD, as even trace amounts of gluten could trigger an immune response in the patients (Rubio-Tapia and Murray, 2010; Sciarini et al., 2008). Gluten-forming proteins are necessary for proper dough development and texture of most of the baked food products. Gluten present in the proteins of wheat, rye and barley comprises two functions such as glutenins and gliadins (Hill et al., 2005). Glutenin has a rough rubbery texture on hydration while gliadin on hydration yields a fluid mass. These researchers found that during dough development, glutenin crosslinks with itself and then associates with gliadins to form gluten strands. Imparting the necessary dough consistency is a major challenge in the development of gluten-free foods.

The present chapter highlights on the nutritional benefits and processing of underutilized tropical tuber crops which need attention by the researchers for further processing tubers into value-added functional food products which provides health benefits to the consumers.

9.2 TROPICAL TUBER CROPS

Many tropical tuber crops like cassava, sweet potato, taro, yam species, elephant foot yam, costus, coleus, typhonium species, canna species, tacca species, giant taro and aroids are grown in many tropical regions of the world. In addition to the food value, the edible tuber crops posses several health benefits and medicinal properties. The tuber crops are not only a store house of carbohydrates, but also rich sources of various antioxidant compounds like carotenoids, polyphenols, phenolic acids, flavonoids and triterpenoids. Apart from the edible tuber crop species, which are important for their food value, a large number of tuber crops possess physiological and pharmacologically active principles and are important for their medicinal value are listed in Table 9.1, these crops are well studied extensively for their therapeutic potency. The important medicinal tuber crop species comprise of *Acorus calamus*, *Safed musli, Asparagus, Alocasia, Alpinia, Amorphophallus* (wild species), *Coleus forskohlii, canna, ceropegia, costus, Ipomea.*

9.3 SWEET POTATO (*Ipomoea batatas* L.)

Sweet potato (*Ipomoea batatas* L.) is the sixth important crop rich in carbohydrate and staple food for Africa, Indonesia, Central and South America, Japan, the Caribbean, Polynesia, Hawaii and Papua New Guinea. Alone, North Carolina produces 40% of the national supply in the USA (NCDA and CS, 2012). It was reported that sweet potato is an important crop of developing countries but fewer vital in some developed countries (FAO, 2011). China is one of the main producer of sweet potato, approximately 80 to 85% of total production in the world followed by rest of the countries in Asia and then by Africa and Latin America (Centro International de la Papa, 2009). It was reported that sweet potato is un-exploited tuber though having nutritional benefits (Grant, 2003). In the past few years, it is also known for the "food security" or "famine relief" crop, particularly in developing countries (Grant, 2003).

TABLE 9.1 Tropical Tuber Crops Potential Sources of Antioxidants

S. No	Botanical name and family	Parts used	Main Chemical Constituents	Biological activities	References
1.	Cassava *Manihot esculenta* Family: Euphorbiaceae	a. Cassava starch b. Rind c. Stem	a. Phenolic compounds b. Proanthocyanidin and phenolics content c. Phenolic compounds	a. Antioxidant and free radical scavenging activity b. Antioxidant activity c. Antioxidant activity	a. Amornpitak et al., 2012 b. Siyumbwa et al., 2014 c. Yi et al., 2010
2.	Taro *Colacasia esculenta* Family: Araceae	a. Tubers b. tubers, stem, and leaves	a. Phenolics compounds, peptides b. Octadecadienoic acid, Hexadecanoic acid	a. Anticancerous activity (antimetasta-tic activity) b. Anticancerous activity	a. Kundu et al., 2012 b. Wei et al., 2011
3.	Elephant foot yam *Amorphophallus paeonifolius* Family: Araceae	Tubers	Flavonoids	Antioxidant activity	Jagatheesh et al., 2010
4.	Sweet potato Ipomoea batatas Family: Convolvulaceae	a. Leaves b. Leaves c. Tubers d . Purple sweet potato tubers e. Storage roots	a. Carotenoid b. Phenolic and flavonoids content c. Anthocyanins and phenolics content d. Anthocyanins e. Phenolic compounds	a. Anticancerous b. Antioxidant activity c. Radical-scavenging activity d. Radical scavenging effects e. Antioxidant activity	a. Fidrianny et al., 2013 b. Hue et al., 2012 c. Boliavar et al., 2003 d. Jiao et al., 2012 e. Teow et al., 2007
5.	Giant swamp Taro, Cyrtosperma merkusii Family: Araceae	Tubers	Mucilage content	Antioxidant activity	Nguimbou et al., 2014

TABLE 9.1 *(Continued)*

S. No	Botanical name and family	Parts used	Main Chemical Constituents	Biological activities	References
6.	West Indian Arrowroot *Maranta arundinacea*.L Family: Marantaceae	Whole plants (tubers, leaves, and flowers)	Phenols and flavonoids	Anticancerous activity	Nishaa et al., 2012
7.	*Curcuma zedoaria, Curcuma angustifolia* and *Curcuma caesia.* Family: Zingiberaceae	Rhizomes	Phenolics compounds	Antioxidant activity and Free radical quenching ability	Dhal et al., 2012

Sweet potato contains natural health-promoting component having a functional value such as β-carotene, phenolic acids, anthocyanins, carbohydrates, fibers, potassium, zinc, calcium, iron, vitamin B complex, vitamins A and C and high-quality protein (Grace et al., 2014). It is having medicinal properties like anti-cancer, antidiabetic, and anti-inflammatory. Sweet potato has been considered as diabetic food due to the low glycaemic index (<55) (Björck et al., 2000). It is predicted that the diabetic population to be increased from 4% in 1995 to 5.4% by 2025, approximately 170% in developing countries, particularly India ranks first, followed by China. It was recommended by FAO-WHO Expert Consultation that there is increased in consumption of low GI food having RS, non-starch polysaccharides and oligosaccharides (Goni and Valentin-Gamazo, 2003). It is reported that food having low glycemic response helps in the management of type 2 diabetes mellitus and in weight management (Gelencsér et al., 2008). Sweet potato is reported having low GI use as diabetic friendly, and the component responsible for this effect have been isolated and studied from white-skinned sweet potatoes (Kusano and Abe, 2000).

There are different varieties of sweet potato such as pale cream-colored, orange-fleshed and purple colored rich in phenols, beta carotene, anthocyanins etc. The high concentration of stable anthocyanins present in the purple-fleshed sweet potato tubers and leaves could be an excellent substitute for artificial color (Truong et al., 2010). This variety having anthocyanin, phenolics with antioxidant and anti-inflammatory activities (Grace et al., 2014). The anthocyanin from purple-fleshed sweet potato is a highly stable and free radical-scavenging activity (Oki et al., 2002), memory-enhancing effects (Shan et al., 2009; Lu et al., 2012) and hepatoprotective activity (Zhang et al., 2013). Moreover, orange-fleshed sweet potato is having β-carotene (Grace et al., 2014), responsible for preventing vitamin A deficiencies and night blindness.

The consumption pattern of sweet potato tubers is found different in the world. It can be processed in diverse food products such as primary, intermediate and secondary products. The primary processed products include steamed, boiled and grilled or processed to be a chip or 'kolak' (traditional food) whereas intermediate products of sweet potato such as flour, paste, puree and mash (Herawati and Widowati, 2009) and secondary products like pasta, noodles, spaghetti etc.

In worldwide, only 1% of the total production of sweet potato enters world trade with Canada, the United Kingdom, France and the Netherlands being the major importing countries (Katan and De Roos, 2004). The USA exports 35% sweet potato of world trade followed by China (12%), Israel (9%), France (7%), Indonesia (6%) and Netherlands/France (5%). The majority of the tubers consumed directly with a little portion of produce utilize for industrial purpose and animal feed.

9.3.1 NUTRITIONAL COMPOSITION

Sweet potato is starchy and sweet in taste used as a root vegetable, belongs to the family Convolvulaceae. The growing duration of this root is 90 to 120 days only. It is a concentrated source of calories with phytochemical used for human nutrition and animal feeding. The nutritional benefits helps to meet the requirements of human includes carbohydrates, fiber, carotene, thiamine, niacin, potassium, zinc, calcium, iron, vitamin A and C etc. The nutritional composition presented in Table 9.2 whereas total starch, α-amylase activity, and trypsin inhibitor activity of sweet potato roots are presented in Table 9.3.

TABLE 9.2 Nutritive Value of Sweet Potato*

Nutrient	Unit	Value per 100 g
Water	g	77.28
Energy	kJ	359.00
Protein	g	1.57
Total lipid (fat)	g	0.05
Ash	g	0.99
Carbohydrate	g	20.12
Fiber, total dietary	g	3.00
Calcium, Ca	g	30.00
Iron, Fe	mg	0.61
Magnesium, Mg	mg	25.00
Phosphorus, P	mg	47.00
Potassium, K	mg	337.00
Sodium, Na	mg	55.00
Vitamin C	mg	2.40
Pantothenic acid	mg	0.80
Vitamin B-6	mg	0.21
Vitamin A	IU	14187

Source: USDA, (2009).

TABLE 9.3 Total Starch, α-Amylase Activity and Trypsin Inhibitor Activity in Sweet Potato

Genotype	Dry matter (%, dry basis)	Total starch (%, dry basis)	α-amylase activity (Ceralpha unit/g, dry basis)	Trypsin inhibitor activity (U/mg, dry basis)
Hi-dry	33.5±0.9	73.6±0.5	0.41±0.01	16.5±1.84
Yan1	29.3±1.6	55.3±0.1	0.81±0.01	18.6±2.56
Chao1	22.6±0.6	46.8±2.0	1.73±0.06	3.90±0.18
Yubeibai	27.9±0.1	52.6±1.1	1.25±0.18	4.99±0.17
Guang7	26.9±1.2	57.6±3.4	1.14±0.04	8.74±0.89
Guang16	24.3±0.4	49.6±1.1	1.44±0.04	21.8±1.74
Mean	27.4	55.9	1.13	12.41
LSD (0.05)	2.59	4.8	0.20	3.50

Source: Zhang et al., (2013); Zhang, et al., (2002).

The whole parts of sweet potato plants are useful and have medicinal value. The iron and protein content of leaves more as compared to roots. The leaves are used to treat type 2 diabetes, inflammatory by Ghana and Brazil, respectively (Pochapski et al., 2011). Sweet potato has been eaten raw to treat anemia, hypertension and diabetes in regions of Kagawa, Japan (Namsa et al., 2011).

Sweet potato is rich in carbohydrates containing starch, sugar particularly sucrose, glucose, and fructose and little amount of pectin, cellulose, hemicelluloses (Onabanjo and Ighere, 2014). Apart from carbohydrate, it also contains dietary fiber, β-carotene, vitamins and minerals. Sweet potato tubers play an important role in monitoring the blood glucose level in diabetics as it is having a low glycemic index (GI), based on studies done by researchers it was found it's ability to manage to stabilize blood sugar level and lower insulin resistance (Preedy et al., 2011). Orange-fleshed sweet potatoes have anti-carcinogenic properties whereas purple-fleshed sweet potatoes have better cancer-fighting abilities due to cyanidins and peonidins, which prevents the growth of cancer cell. It also contains antioxidant enzymes such as copper/zinc superoxide dismutase and catalase.

Sweet potato also considered as an "insurance crop" due to it's ability to be grown under different agro-climatic conditions. This is the crop of Southeast Asia, Oceania and Latin America regions and China. It is also considered as space food. Sweet potato has special features and nutritive value, National Aeronautics, and Space Administration (NASA) has selected it as a candidate crop to be grown and incorporated into the menus for astronauts on space missions (Bovell-Benjamin, 2007).

9.3.2 SWEET POTATO FLOUR

Sweet potato tubers are consumed freshly due to perishable nature and having a short shelf life. Recently it was suggested by China that fresh perishable sweet potato tubers could be converted to the processed products having a longer shelf life and fulfill the demand of consumer in offseason. The numerous types of products could be processed from sweet potato, but as per new analytical assessment approved in developing countries, dried chips, starch and flour were recognized as among the most useful. In order to make sweet potato tubers available throughout the year and to maximize the utilization, it should be processed into flour. Sweet potato tubers harvested, at three and ten months maturity. The tubers were washed free of dirt and manually peeled and sliced to round discs of approximately 5 mm thickness. The slices were sundried for 36 h till the moisture content was brought down to <10% and powdered in a hammer mill into fine flour and pass through 85 mesh sieve of BSS standard (particle size 0.177 mm). The prepared flour was packed in airtight containers and stored at room temperature (30±1°C) for further use. The flow chart for the preparation of sweet potato flour presented in Figure 9.1.

9.3.3 COMPOSITE FLOUR

Sweet potato flour is able to simply use as an alternative for wheat flour in different baked products such as biscuits, cookies, cake, muffins, bread etc. and due to rich in carotene content. Nevertheless, the cost of sweet potato flour has to be competitive with wheat flour and have excellence quality (Van Hal, 2000).

It was reported that sweet potato flour was suitable to alternate wheat flour in bread preparation. Sweet potato flour from peeled and unpeeled tubers was not affecting to yield and color but unpeeled flour found high in ash and crude fiber content. Different researchers found the making of sweet potato powder under various conditions. The moisture content of slices up to 6.6% would be achieved when slices were dried in cabinet tray drier at 60–70°C.

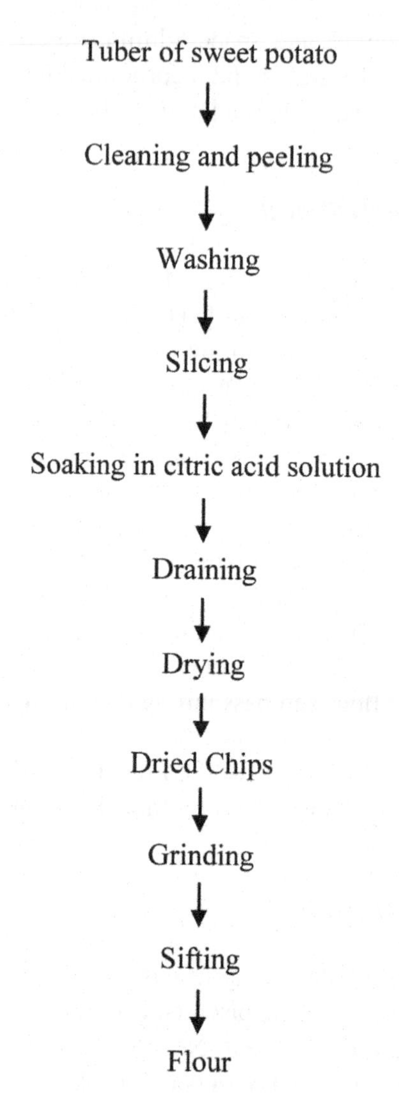

FIGURE 9.1 Scheme of sweet potato flour processing.
(Source: Elisabeth, 2015)

Hagenimana et al., (1992) investigated that the incorporation of orange-fleshed sweet potato in food products such as buns, chapattis and mandazis significantly increased the amount of total carotenoids. However, the inclusion of sweet potato flour to wheat flour at different level improves the nutritional value with respect to fiber and carotenoids. It also helps to lower the gluten content and prevent from CD (Tilman et al., 2003). Replacement of wheat flour with soy meal affects on decreasing the water absorption capacity and bulk density of flour blends (Iwe and Onuh, 1992).

The functional properties of gluten-free composite flour blend of sweet potato and maize flour was studied. The maize flour was substituted with sweet potato flour at 10–100%. It was observed that as the level of sweet potato flour increased in flour blend, water binding capacity increased and starch swelling power, bulk density and dispensability decreased (Adeyeye and Akingbala, 2016).

9.3.4 NUTRITIONAL AND FUNCTIONAL PROPERTIES OF COMPOSITE FLOUR BLEND

Tropical countries mainly depend on root tuber crops, where wheat does not flourish fine. These countries completely depend on imported wheat, which leads to the increased cost of wheat, put financial pressure on poor people. So, there is a need to examine the opportunity of providing wheat flour with flour from comparatively low-priced and easy-to-produce root tuber, such as cassava, potato and sweet potato. These composite flour blends could reduce the demand of import wheat and could meet the demand of production of pasta, noodles, bread, cookies etc. From various studies, it was investigated that wheat flour could be substituted up to 20% with root tuber flour without changing the nutritive and sensory value of the products (Kwame et al., 2017).

The nutritional and functional properties of composite flour blend were investigated by Ehimen et al., (2017). The composite flour includes unripe cooking banana, pigeon pea and sweet potato flour. The functional properties of flour blend significantly affected as the level of ingredients varied in the blend. It was investigated that as the level of sweet potato flour and unripe cooking banana flour increased, dispersibility, bulk density, water, and oil absorption capacities of the blends increased. This formulation of flour found suitable for preventing malnutrition in Nigeria and developing new food formulations.

The blending of wheat flour with fermented sweet potato flour improved the functional properties of the blend. Composite flour bends containing wheat flour with unfermented and fermented sweet potato flour in the ratios WF:USPF:FSPF:100:0:0,0:100:0, 0:0:100,90:10:0,90:0:10. The foaming capacity, emulsion capacity and least gelation were improved due to fermentation whereas unfermented sweet potato flour (100%) had highest bulk density and oil absorption capacity (Adams et al., 2017).

9.3.5 SWEET POTATO PUREE

Sweet potato could be converted into purees and dehydrated forms which are suitable for food products as functional ingredients. The processing steps include for preparation purees, peeling, cutting/grinding and pre-cooking/finish-cooking with a temperature-time program suitable for starch conversion by endogenous amylolytic enzymes to obtain the products with targeted maltose levels and viscosities. The shelf life of sweet potato purees could be increased by refrigeration, frozen storage, canning or aseptic packaging. The aseptic packaging was found as the best method to store purees. The dehydrated sweet potato chips can be produced by solar drying and cabinets and tunnels drying which converted into flours. Sweet potato powder for specific functionality can be produced by extrusion technology and chemical treatments. Sweet potato purees and dehydrated forms high in β-carotene and anthocyanin are incorporated in food as functional ingredients to improve textural properties and nutritional value in processed food products (Van-Den and Avula, 2010).

9.3.6 COOKIES

Cookie-type biscuits are widely consumed due to their long shelf life and crisp texture (Mareti et al., 2010). Cookies are made traditionally from wheat flour using a number of additives such as sugar, chocolate chips, peanut butter etc. Functional cookies have been attempted by various researchers using several types of alternative non-wheat flours such as buckwheat flour, cassava flour, quinoa flour etc. (Mishra et al., 2015). Gluten-free cookies have been developed from rice, corn, buckwheat and potato flours (Sarabhai et al., 2015).

Sweet potato flour-based gluten-free cookies suitable for the celiac patient were developed using sweet potato flour (40–60%), rice flour (RF) (20–25%), sorghum flour (15–20%) and cassava flour (5–15%). Minerals content in sweet potato flour based gluten-free cookies were higher than Maida based cookie. Gluten-free cookies prepared with sweet potato flour showed the maximum crude fiber content as compared to *maida*-based cookie. Results indicated that sweet potato flour based gluten-free cookies could be prepared using sweet potato flour 60% with RF 20%, sorghum flour 15% and cassava flour 0.5% (Giri et al., 2016).

The study was conducted by Adeyeye and Akingbala (2016) to develop gluten-free cookies for celiac patients using sweet potato and maize composite flour blends. Replacement of maize flour with sweet potato flour at the level of 10–100%. As the level of substitution of sweet potato flour increased, there were significantly increased in ash and sugar contents, whereas the total calorie content of cookies decreased. The panel members accepted the optimum substitution level up to 40% of sweet potato flour.

Sweet potato rich in β-carotene helps to fight against vitamin A deficiency. Biscuits fortified with orange-fleshed sweet potato (OFSP) developed to make maximum utilization of this variety (Andualem et al., 2016). The biscuits were developed by incorporating 30% OFSP flour to wheat flour. The nutritional properties were evaluated. It provides 9.88–11.06% protein, 0.54–6.01 μg/g β-carotene and 360.7–379 kcal/100 g calories.

Gluten-free crackers for celiac patients were developed by Deborah et al., (2011) using *Moringa oleifera* and *Ipomoea batatas.* Butter and cream crackers were made from 100% cassava flour and 100% sweet potato flour using 100% wheat flour as the control. It was found that the β-carotene levels of the leaves ranged from 4.76 mg/100 g to 11.54 mg/100 g for the sweet potato leaves with *Moringa oleifera* having the highest β-carotene content of 23.43 mg/100 g. The sweet potato cracker provides 4.87% moisture, 3.31% ash, 1.79% fiber, 4.23% protein, 13.29% fat, 77.38% carbohydrates and 446.09% calories. Similarly, gluten-free crackers were developed for celiac patients enriched with prawn as a source of protein. An attempt was made to prepare prawn crackers using cassava starch, pink and orange-fleshed sweet potato at the ratio of 100%, 90:10% and 80:20% (Mbaeyi-Nwaoha and Itoje, 2016).

9.3.7 BREAD

Composite flour mixture of sweet potato, maize starch and soybean flour in different proportions were utilized for the manufacturing of bread. Xanthan gum at the rate of 0.5% added in each formulation. The bread prepared from composite flour containing sweet potato flour 40%, maize starch 40%, soybean flour 19.5% and xanthan gum 0.5% was found most acceptable with respect to physical and sensory properties (Elisa Julianti et al., 2017).

β-carotene enriched flatbread for children of Ethiopia using an orange-fleshed sweet potato to combat the vitamin A deficiency was developed with different ratio of sweet potato to maize flour: 25%:75%, 30%:70%, 35%:65% and 0%:100% (control). The vitamin A (μg RAE) content of the control bread was observed to be 0 whereas 269.63 μg RAE for flatbread containing 35% orange-fleshed sweet potato flour. It was reported that flatbread enriched with vitamin A source helps to fulfill 61.63 to 86% of the recommended dietary allowance of pre-school children (Menen Zegeye et al., 2015).

9.3.8 PASTA

Sweet potato is considered as a low glycemic food having GI less than 55, which is suitable for consumption by diabetic people (Bjorck et al., 2000). Various workers have attempted the use of sweet potato for making noodles and pasta (Limroongreungrat and Huang, 2007). Sweet potato flour pasta was developed with soy flour, water, Arabic gum and carboxymethylcellulose (CMC) and effect of these ingredients on quality responses (sensory, solids loss and hardness) of the pasta product were investigated. Responses were more significantly affected by the level of soy flour and gum and to a lesser extent by sweet potato flour and water levels. It was concluded that, the maximum sensory score (33.8), minimum solids loss (16.6%) and maximum texture hardness (5616 g) were identified at 674 g/ kg sweet potato flour, 195 g /kg water, 110 g /kg soy flour, 10.6g /kg Arabic gum and 10.1 g /kg CMC levels (Singh et al., 2004). The flow chart for the preparation of pasta from sweet potato flour presented in Figure 9.2.

Sweet potato is known as low GI food and various workers developed pasta and spaghetti from sweet potato flour (Jyothi et al., 2011, 2012). Low GI spaghetti was developed from sweet potato with banana or legume starches and fiber sources (Renjusha et al., 2015). But starch noodles have very low nutritive value as only starch used as an ingredient (Jyothi et al., 2012). There is a need to fortify starch noodles or pasta with resistance starch (RS) due to it's health benefits to control obesity and diabetics. Pasta prepared from wheat, banana and sweet potato having high RS (Jyothi et al., 2012).

Today there are rapid increases in coronary heart disease, cancer, obesity, and diabetes due to change in lifestyle and false diet habits. The consumption of food containing dietary fiber provides health benefits to reduce the risk of lifestyle diseases (Buttriss and Stokes, 2008). NUTRIOSE is a source of resistance starch used in the formulation of functional foods having a low GI. NUTRIOSE is used as a source of fiber in food preparation, which is a hydrolyzed form of starch of wheat or corn (Guerin-Deremaux et al., 2011). Dietary fiber-rich functional sweet potato pasta was prepared using dietary fiber sources like oat bran, wheat bran and rice bran. Sweet potato pasta fortified with these fiber sources improved the crude protein content to 5–10% in the pasta. Two varieties of sweet potato roots were used by Jyothi et al., (2012) such as Sree Arun and Sree Kanaka. The starch digestibility was studied to develop pasta, slow digestible and diabetic friendly. The level of fortification of fiber into pasta was 10 and 20% and all fiber fortified pastas show slow starch digestibility as compare to control pastas. The retention of RS in pastas from Sree Arun was 38–49% and Sree Kanaka was 39–55% whereas control pasta was only 14–17%. The high fiber content with high RS made sweet potato pastas ideal for diabetic and obese people.

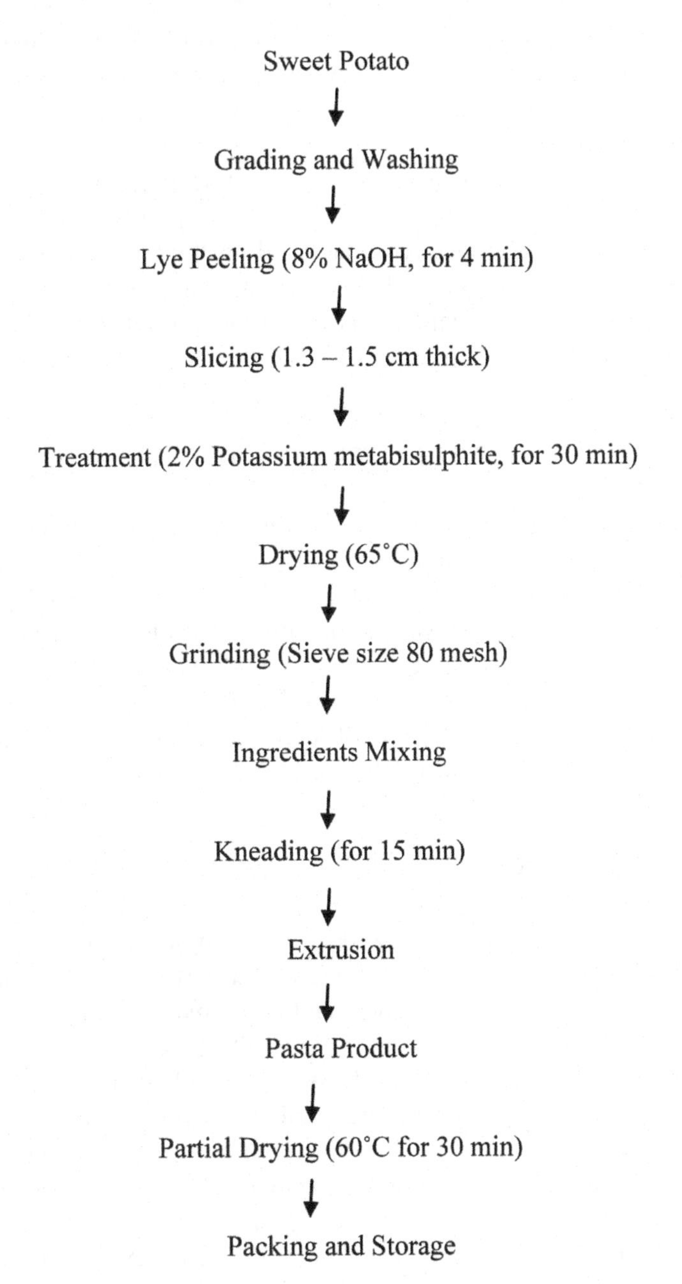

FIGURE 9.2 Flow sheet for the pasta preparation from sweet potato flour.

Pasta fortified with protein-rich sources such as whey protein concentrate (WPC), defatted soy flour (DSF) and fish powder (FP) was attempted by Krishnan et al., (2011) using cream and orange-fleshed sweet potato. It was found that WPC provides tough starch-protein network and starch digestibility also get reduced. Among all protein fortified pasta, WPC fortified pasta had higher protein quality, essential amino acid index and calculated protein efficiency ratio.

The ready to eat extrudate snacks for fasting purpose were prepared using vari rice, sweet potato and unripe banana powder. The composite flour mixture of vari rice, sweet potato and unripe banana powder in the ratio of 40:40:20 extruded at 180°C die temperature, screw speed of 170 rpm with 16% moisture content was found most acceptable (Rathod and Annapure, 2016).

9.3.9 NUTRITIVE VALUE OF SWEET POTATO LEAVES

Sweet potato leaves have not been much used as human food, though it has a rich source of vitamins, minerals and protein. These leaves are used as a vegetable in some part of the world. Various workers studied the nutritive value of sweet potato leaves. It contains high antioxidants, particularly phenolic compounds (Yoshimoto et al., 2005). This phenolic compound includes caffeoylquinic acid derivatives having an antimutagenic effect. Leaves also attributed high content of lutein to the extent of 29.5mg/100g, which has eye protection effect (Ishiguro and Yoshimoto, 2005).

9.4 CASSAVA (*Manihot esculenta* Crantz)

Cassava (*Manihot esculenta* Crantz), known as tapioca, is a dicotyledonous, perennial shrub belonging to the Euphorbiaceae. It is an excellent source of energy and staple food of tropical and subtropical region of the world. Cassavas are having the wide applications like food for direct consumption, animal feed and industrial products. It is also known as food security crop due to it offers source of income, particularly producer. Cassava offers a low-cost vegetative propagation with flexibility in harvesting time and seasons, it has drought-resistant root (Haggblade et al., 2012). It was reported that cassava roots could be stored in the ground for over two years, so it is boon for the farmer in West African Countries such as Nigeria (Falade and Akingbala, 2010). Cassava tubers can be processed into various value-added products having an industrial value such as flour, alcohol, starches for sizing paper and textiles, sweeteners, biodegradable products.

In olden days, cassava was not considered as important crop due to low protein value and high cyanide content (Charles et al., 2005), but now it's gaining importance because it is a concentrated source of energy (Falade and Akingbala, 2010) whereas, the high protein content of cassava leaves and some macronutrients are presents. These tubers are consumed as a vegetable in tropical and subtropical countries (Burns et al., 2012). The nutritional composition of cassava roots and leaves mentioned in Table 9.4.

TABLE 9.4 Nutritional Composition of Cassava Roots and Leaves

Proximate composition	Raw cassava (100 g)	Cassava roots	Cassava leaves
Food energy (kcal)	160	110–149	91
Moisture (g)	59.68	45.9 to 85.3	64.8 to 88.6
Dry weight (g)	40.32	29.8 to 39.3	19 to 28.3
Protein (g)	1.36	0.3 to 3.5	1.0 to 10.0
Lipid (g)	0.28	0.03 to 0.5	0.2 to 2.9
Carbohydrate, total (g)	38.06	25.3 to 35.7	7 to 18.3
Dietary fiber (g)	1.8	0.1 to 3.7	0.5 to 10.0
Ash (g)	0.62	0.4 to 1.7	0.7 to 4.5
Vitamins			
Thiamin (mg)	0.087	0.03 to 0.28	0.06 to 0.31
Riboflavin (mg)	0.048	0.03 to 0.06	0.21 to 0.74
Niacin (mg)	0.854	0.6 to 1.09	1.3 to 2.8
Ascorbic acid (mg)	20.6	14.9 to 50	60 to 370
Vitamin A (µg)	---	5.0 to 35.0	8300 to 11800
Minerals			
Calcium (mg)	16	19 to 176	34 to 708
Phosphorus, total (mg)	27	6 to 152	27 to 211
Ca/P	0.6	1.6 to 5.4	8 2.5
Iron (mg)	0.27	0.3 to 14.0	0.4 to 8.3
Potassium (%)	---	0.25 (0.72)	0.35 (1.23)
Magnesium (%)	---	0.03 (0.08)	0.12 (0.42)
Copper (ppm)	---	2.00 (6.00)	3.00 (12.0)
Zinc (ppm)	---	14.00 (41.00)	71.0 (249.0)
Sodium (ppm)	---	76.00 (213.00)	51.0 (177.0)
Manganese (ppm)	---	3.00 (10.00)	72.0 (252.0)

Source: United States Department of Agriculture (USDA) (2009).

The cassava roots could be processed into various products having value for the industry, food and animal feed, etc. The schematic presentation of different processed products of cassava mentioned in Figure 9.3.

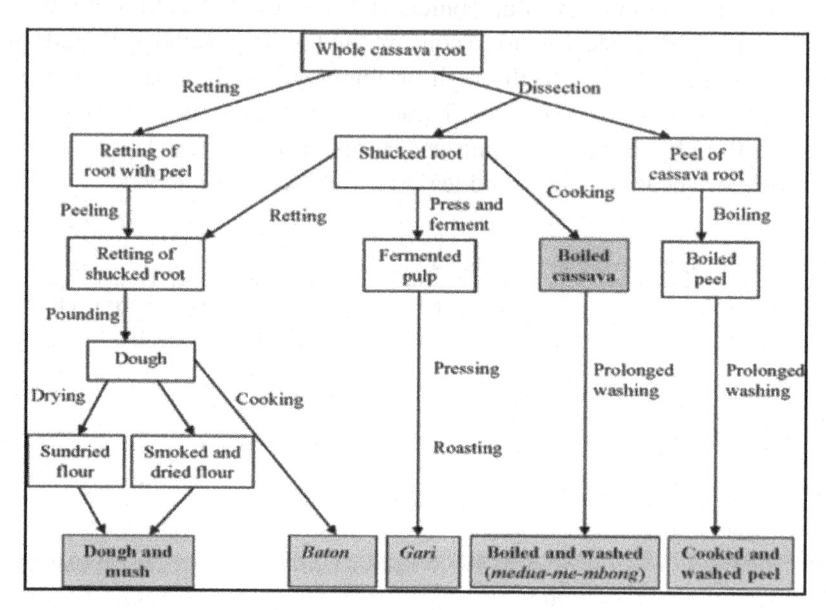

FIGURE 9.3 Different processing techniques for whole cassava root. The edible forms of cassava root are shaded in gray (Montagnac et al., 2009).

9.4.1 UNFERMENTED CASSAVA PRODUCTS

9.4.1.1 TAPIOCA

Tapioca grit is a partly gelatinized flake normally consumed in various countries, especially in Africa as a convenience food (Adebowale et al., 2007).

9.4.1.2 CASSAVA CHIPS AND PELLETS

Cassava chips and pellets are minimally processed (MP) form of cassava means for exporting purpose to fulfill the consumer demand (Adamade and Azogu, 2013). Dried cassava chips having a size between 2 and 5 cm packed in a jute bag or paper bag for exporting (Adebowale et al., 2007). Pellet is prepared similar to the chips only difference in the lesser moisture content of about 9% compared to chips. Therefore, pellets having long shelf life than chips (Falade and Akingbala, 2010). As they both have lower moisture content helps to avoid both quality and quantity postharvest loss (Adamade and Azogu, 2013).

9.4.1.3 CASSAVA STARCH

Tapioca starch is widely utilized in the food industry as a basic ingredient for various purposes. It is processed by peeling, washing, grating of roots followed by sieving to separate fiber and then the mixture is kept undisturbed to sediment the starch followed by decanting to collect the starch (Raji et al., 2008). The cassava starch has low gelatinization temperature, high viscosity, high water binding capacity and it is used directly in food preparation as a thickener and binder (Taiwo, 2006).

9.4.1.4 UNFERMENTED CASSAVA FLOUR

Unfermented cassava flour is also commonly known as high-quality cassava flour (HQCF). The production of HQCF will promote the maximum utilization of cassava incorporated composite flour and also decrease the requirement of imported cereals and grains in developing countries (Taiwo, 2006). The processing involves, harvesting, sorting, peeling, washing, grating, dewatering, grinding, drying, fine milling and sifting the milled flour with a motorized flour sifter 250 μm followed by packaging (Jekayinfa and Olajide, 2007). The processing of cassava roots to HQCF presented in Figure 9.4.

9.4.2 FERMENTED CASSAVA PRODUCTS

Fermentation helps to process cassava into products with improved flavor, taste, and extend the shelf life (Falade and Akingbala, 2010). Fermentation also helps to reduce the cyanide level and detoxification of the root.

9.4.2.1 CASSAVA BREAD

Cassava bread is a fermented product was attempted by Shittu et al., (2008), from the combination of wheat flour and cassava flour in the ratio of 5:1. This proportion was most acceptable by panel members. The cassava flour as composite flour could be used in bakery industries.

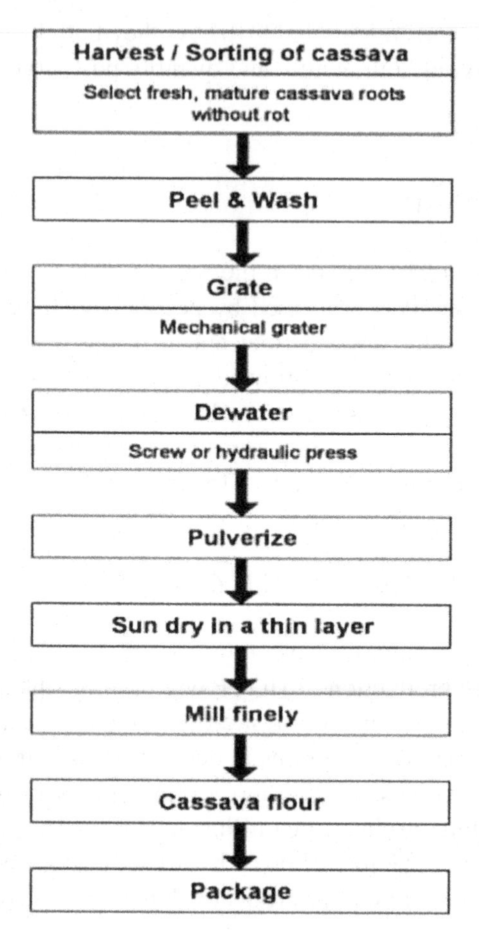

FIGURE 9.4 Process flow chart for high-quality cassava flour (Cassava Master Plan, 2006).

9.4.2.2 FERMENTED CASSAVA STARCH

It is an adapted form of starch produced by fermentation of cassava root. It can be particularly utilized for frying and baking of cheese bread in Brazil (Srinivas, 2007). The cassava roots processed to fermented starch by steeping grated roots in a tank of water for a period of 20 to 70 days to carry fermentation. After fermentation, starch is dried and converted to powder.

9.4.2.3 CASSAVA FUFU

Fufu is a famous traditional food for the West African countries, which is produced by an acid-fermentation of cassava roots (Oyewole and Sanni, 1995). It is produced by submerged fermentation of roots without peels in water for 2 to 3 days and then it is screened, kept undisturbed condition, water removed with cloth bags, cooked, and finally grinded to fufu. This product having high deterioration rate due to the high moisture content of about 50% and will not be utilized for commercial purposes.

9.4.2.4 GARRI

It is cream-white in color, pregelatinized granular and high energy source with a sour taste. It is prepared by peeling, grating, mashing of cassava roots following fermentation of grated pulp in a sack which are placed under a heavy stone for 3 to 4 days to dewater the pulp (Falade and Akingbala, 2010). Garri is recognized as precooked convenient food eaten as a snack, which could be roasted to improve the shelf life (Fadeyibi, 2012).

9.4.3 BIOENERGY (ETHANOL FROM CASSAVA)

The worthy use of cassava, which becomes the scientific interest globally is it's use as a renewable energy source (Adelekan, 2010). Cassava considers as an energy crop because it is a cheaper alternative to conventional energy sources. The development of ethanol from cassava also helps to reduce the post-harvest losses and enhance the value addition of cassava root (Ubalua, 2007). Ethanol fuel (ethyl alcohol) is like the alcohol present in alcoholic drinks and utilized as a biofuel option for gasoline in the USA (Adelekan, 2010).

9.4.4 CASSAVA BASED ADHESIVES

i) **Liquid Starch Adhesives:** These are used particularly in plastic or lined metal drums, jerry cans and bottles.
ii) **Pre-Gel Starch Adhesives:** These are exported by packing in paper bag or sacks which are water proofed.
iii) **Dextrin Based Adhesives:** These are of two types of liquid and dry forms. The liquid dextrin adhesives are packed as the liquid starch

adhesives, while the dry dextrin adhesives are packed as the milled pre-gel adhesives.

9.4.5 SWEETENERS

Sweeteners in the form of high fructose syrup, glucose and sorbitol could be produced from cassava starch and HQCF. The hydrolysis of starch and flour takes place to convert into glucose first, which further get purified to different sweeteners.

9.5 ARROWROOT (*Maranta arundinacea L.)*

Arrowroot (*Maranta arundinacea L.)* plant is a tall herb with short-leafy branched stem from Marantaceae family which yields thick, fleshy, creeping, spindle-shaped, long (15 to 20 cm) and dull white to creamy colored root-stock rhizomes which contain fine easily digestible starch. It is indigenous to tropical America. It is highly under-utilized and the rhizomes are processed as food products and starch. The crop comes up well under shaded conditions and no serious pests and diseases are noted in the crop. It is tolerant to shade conditions and can be cultivated under coconut and areca nut plantations and can yield 20–30 tonnes fresh rhizomes per hectare.

Extraction of starch from arrowroot rhizomes can be done even in households by adopting a simple procedure. It can be economically important crop as it serves as a raw material for cottage industry by unemployed women and rural youth. The starch possesses demulcent and anti-diarrheal properties and is used in the treatment of intestinal disorders, which adds medicinal value to the crop. It is widely distributed throughout the tropical countries like West Indies particularly St. Vincent, which produces about 95% of the world's commercial supply; India, Sri Lanka, Indonesia, Philippines and Australia. In India, arrowroot is mainly cultivated in Uttar Pradesh, Bihar, Orissa, West Bengal, Assam and Kerala. It's powder is one of nature's finest carbohydrates. The nutritional facts of raw arrowroot rhizomes given in Table 9.5.

9.5.1 STARCH EXTRACTION PROCESS

Fresh arrowroot rhizomes contain about 25–30% starch on a fresh weight basis. The rhizomes are washed thoroughly, cleaned and scales should be separated from rhizomes followed by washing, then converted into pulp using mortars which lead to the extraction of milky liquid due to continuous beating action. The liquids are allowed to pass through a cloth and pure starch will be separated, which is insoluble then it is kept in undisturbed condition to allow the starch to settle down in container at the bottom. The wet starch is dried in the sun or in a drying house. The result is a starch powder, which is quickly packed for the market in airtight cans, packages, or cases (Spennemann, 1992). It is the best source of easily absorbable starch. It is nourishing food for children, chronically ill and convalescents. Arrowroot starch has many medicinal uses as well. It is mostly given to baby because of easy digestibility. Due to urinary antiseptic properties, arrowroot helps in urinary problems. When taken daily, it helps to lower cholesterol because it has functional constituents like alkaloids, cardiac glycosides, phenolic compounds, terpenoids, saponins, flavones and gum. Arrowroot starch has extensive application in medicinal, nutraceutical, industrial and bakery products. Applications of arrowroot starch have been given in detail below. Arrowroot starch extraction is outlined in Figure 9.5.

TABLE 9.5 Nutritional Composition of Arrowroot Rhizomes*

Principle	Nutrient value	% RDA	Principle Vitamins	Nutrient value	% RDA
Energy	65 Kcal	3	Folates	338 µg	84%
Carbohydrates	13.39 g	10	Niacin	1.693 mg	10.5%
Protein	4.24 g	7.5	Pyridoxine	0.266 mg	20%
Total Fat	0.20 g	1	Riboflavin	0.059 mg	4.5%
Cholesterol	0 mg	0	Thiamine	0.143 mg	12%
Dietary Fiber	1.3 g	3.5	Vitamin A	19 IU	<1%
			Vitamin C	1.9 mg	3%
Electrolytes			**Minerals**		
Sodium	26 mg	1.5	Calcium	6 mg	0.6%
Potassium	454 mg	10	Copper	0.121 mg	13.5%
			Iron	2.22 mg	28%
Phytonutrients			Magnesium	25 mg	6%
Carotene-ß	11 µg	—	Manganese	0.174 mg	7.5%
Carotene-α	0 µg	—	Phosphorus	98 mg	14%
			Selenium	0.7 µg	1%

Source: USDA, (2009).

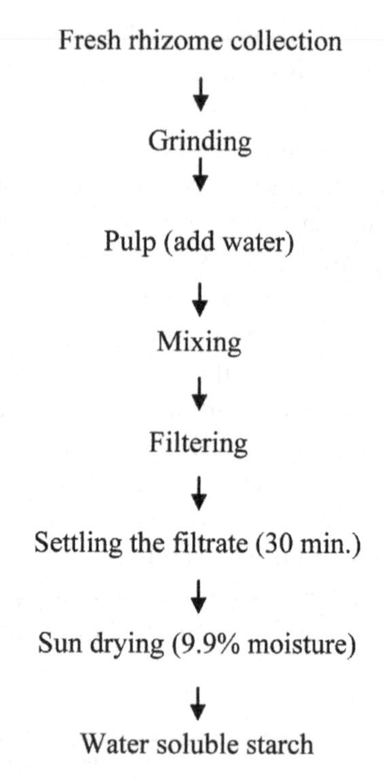

FIGURE 9.5 Process of arrowroot starch extraction.

9.5.2 USES OF ARROWROOT STARCH

9.5.2.1 BAKERY AND CONFECTIONARY PRODUCTS

Arrowroot commonly utilized in biscuits, pasta, bread, cakes, puddings, jellies and hot sauces. Arrowroot starch addition in ice cream prevents the crystal formation whereas makes fruit gel clear and transparent. This starch also utilized as thickner in different food preparations such as puddings and soups. Nowadays, it is used as the best substitute due to the ability to thicken and transparent nature. It is also used in gluten-free products (The Wealth of India, 1962; Spennemann, 1992).

9.5.2.2 INDUSTRIAL AND COSMETIC PRODUCTS

Arrowroot is extensively used in many cosmetic products. Arrowroot starch is used as a replacement of talcum powder or chemically laced moisturizing creams. It is considered as ingredients for the herbal treatment for skin and oil absorbent. It is also best substitute for baby powder due to the fine micro-structure of arrowroot starch. Due to its excellent thickening property, it is also used in hair dye. In Britain, Canada, Europe, and the U.S.A, it also has non-food uses such as in carbonless paper manufacturing for computers.

9.5.2.3 MEDICINAL AND PHARMACEUTICAL USES

Arrowroot flour has been known to be used in many medicines and health-related products. Arrowroot having nutritional benefits such as vitamins and minerals. Widely used in primitive cultures, arrowroot has had such traditional uses as healing wounds from poison arrows, scorpion bites and gangrene. This starch helps to prevent birth defects, support proper growth and development, and improve your circulation, blood pressure, weight, and digestion. It is also remedies for preventing the bloating and stomach pain prevalent in this condition. Arrowroot has also been found to be effective for treating diarrhea in people with irritable bowel syndrome. Arrowroot promotes bile production, which increases cholesterol uptake by your gall-bladder for necessary bile synthesis and helps to optimize cholesterol levels. It acts as anti-inflammatory and antiseptic. Arrowroot is predominantly known for it's soothing properties. A natural healer for small irritations, it assists with the drying out of wounds, rashes, or blemishes (Mathew, 2007).

9.6 TIKHUR *(Curcuma angustifolia)*

Curcuma species is a minor tuber crop found widely all over south India and many other tropical areas with adequate rainfall. *Curcuma angustifolia* is most commonly cultivated in India, especially in the northeast and western coastal plains and hills, particularly, Maharashtra, Madhya Pradesh, Andhra Pradesh, Himachal Pradesh, Orissa, Chhattisgarh, Tamil Nadu and Kerala. This species can also be found in Burma, Laos, Nepal and Pakistan (Paramapojn et al., 2009). Although highly nutritious and high in medicinal properties, the usage of Curcuma species is limited to tribal hamlets and traditional medicinal preparations (Chaveerach et al., 2008). Curcuma starch is similar to arrowroot starch, having high digestibility. It could be used for the preparation of value-added products at commercial scale. The rhizome of *Curcuma angustifolia* contain starch and is reported to be used in tribal of India for the preparation of milk puddings. Since tikhur starch is of little attention, despite their potential application in the food and textile industries (Jothy et al., 2003). The chemical composition of tikhur rhizomes and starch mentioned in Table 9.6.

TABLE 9.6 Proximate Composition of *Curcuma Angustifolia**

Sl. no.	Parameters	*Curcuma angustifolia* rhizome (%)
1	Moisture	71–76
2	Ash	0.88–0.92
3	Crude Fiber	42
4	Fat	0.17–0.19
5	Protein	1.5–1.7
6	Carbohydrate	21.75–23.44

(*Source:* Lisona et al., 2011; Nayak et al., 2013).

The interesting ingredients present in Curcuma species is starch due to it's high viscosity as compare to traditional other sources of starch. Based on the characteristics of Curcuma starch, it was estimated that it is having high thickening and gelling properties and high stability when agitated (Rani et al., 2012).

Curcuma angustifolia produces a digestible starch. Starch from rhizome was extracted by washing, peeling followed by immersing in water for overnight and ground to get a soft paste. The paste is kept in a vessel and is allowed to settle. Then the excess water is separated and settled particles of starch washed repeatedly till to obtain clean starch. Then it is allowed to dry completely to convert into powder. The completely dried starch is stored in bottles for future use. Traditional processing techniques such as fermentation, soaking and cooking help to remove antinutritional factors (Preetham Sarkar, 2007).

It was found that *Curcuma angustifolia* has compounds with medicinal value such as hydroxyhexanoic acid, α–thujene, β-pinene, Tolune, Caryophyllene, Napthalenediol, Camphor, α-amorphene etc. The investigation through GC-MS analysis of the rhizome proves the presence of these compounds which have a role in the pharmaceutical industry.

9.6.1 *Curcuma angustifolia* STARCH

Curcuma angustifolia starch can be used as a substrate for single cell production. The main outlets for this starch in India are pharmaceutical uses, food industry and the textile industry. *Curcuma angustifolia* starch is utilized as a fitter in the pharmaceutical and food industry. In addition to major industries which use starch, there a number of applications of starch which are unexplored in many fields (Villarreal et al., 2013). It was found that tikhur starch extracted by using 1% ammonium oxalate and 0.03% M ammonium solution was obtained 38.46 and 37.64%, respectively. The starch grain is rounded, oval to elliptical, spherical and elongated and 3.32 to 32.55 in length and 2.29 to 8.47 in width.

In some part of India, *Curcuma angustifolia* is used as weaning food for babies. The starch obtained from the rhizomes is given to babies three-four times a day after boiling in water or milk. This preparation is given as a traditional food for babies. Sometimes it is used as a replacement of mother's milk. This may be due to the high content of total soluble protein, reducing sugar and total free amino acids in this species. Tikhur is found as a primary ingredient in cakes, fruit preserves, biscuits and puddings (Shukla et al., 2007).

It was reported that tikhur rhizomes are remedies on burning sensations and stomach pains, removal of stone from kidney. It also provides a cooling effect for the treatment of headache (Nag et al., 2006). The rhizome of the Tikhur is lightly bitter, demulcent, non-irritating, nutritive and fragrant.

9.6.2 MEDICINAL USES

The rhizome of *Curcuma angustifolia* is used to soothe coughs and to treat bronchitis. Essential oils from Curcuma having antifungal properties, whereas the Curcuma leaves having antibacterial properties. Curcuma starch properties make it's replacement with corn starch and wide use in tablet preparation (Doble et al., 2011). Presence of phenols plays a vital role against oxidative stress in the cell and thereby acts as an antioxidant and protects the human system from various ailments including cancer, cardiovascular diseases. Flavonoids have a significant role in pharmacological activities such as antimicrobial, anti-inflammatory, anticancer, antioxidant activities (Sushama et al., 2011).

9.6.3 FOOD USES

Curcuma rhizomes could be converted into powder to mix with milk or water to form a nutritious drink used as a replacement of breast milk for babies. Curcuma starch also utilized widely as weaning food. Curcuma flour found as a suitable ingredient in the preparation of bakery products and puddings (Moorthy et al., 1994). *Curcuma angustifolia* starch provides ample scope for diversification and value addition. There lies a vast opportunity for non-traditional uses of these crops in the form of value-added products. The demand for quick cooking of instant food production is expanding at a phenomenal reaction in developed and developing countries.

Preetham et al., (2007) attempted to develop *burfi, halwa* and *jalebi* using Curcuma starch as a major ingredient. *Tikhur* starch has cooling effects, so *Tikhur sarbat* is prepared in the Chhattisgarh region of India during summer (Preetham et al., 2007).

9.6.3.1 TIKHUR BARFI

The preparation of *Tikhur barfi* required *Tikhur* starch, sugar and water in the ratio of 1:1.5:2, respectively. The preparation method of tikhur barfi given in Figure 9.6.

9.7 TARO (*Colocasia esculenta)*

Taro (*Colocasia esculenta*) is an extensively cultivated root crop of the tropical areas like South East Asia, the Pacific Islands, Africa, the United States of America and the Mediterranean. It is also widely cultivated and utilized in the northeast states of India. Taro corms contain very less amount of fat and protein but it is excellent source of carbohydrates (90.8–95.5%) and minerals such as magnesium (32.9–382 mg/100 g), calcium (25.4–192 mg/100 g) and potassium (3.5–59.7 mg/100 g) (Kaur et al., 2011; Njintang et al., 2011). Besides it's nutritional importance, it also contains a high level of gum, which helps to reduce blood pressure, hypercholesterolemia and diabetes (Njintang et al., 2011). Although taro constitutes up to 16% of the total calorie intake in Polynesia, it's value addition is meager. Taro is reported to be rich in mucilage, which has hypocholesterolemic and hypoglycemic effects (Njintang et al., 2011). Njintang et al., (2008) examined the potential of taro flour for bread and biscuits making and found that up to 10% level, it did not alter the wheat dough alveography. Himeda et al., (2014) reported that use of wheat-taro composite mixes in biscuits making and reported that up to 10–15% substitution level was acceptable. The nutritional composition of taro corm presented in Table 9.7.

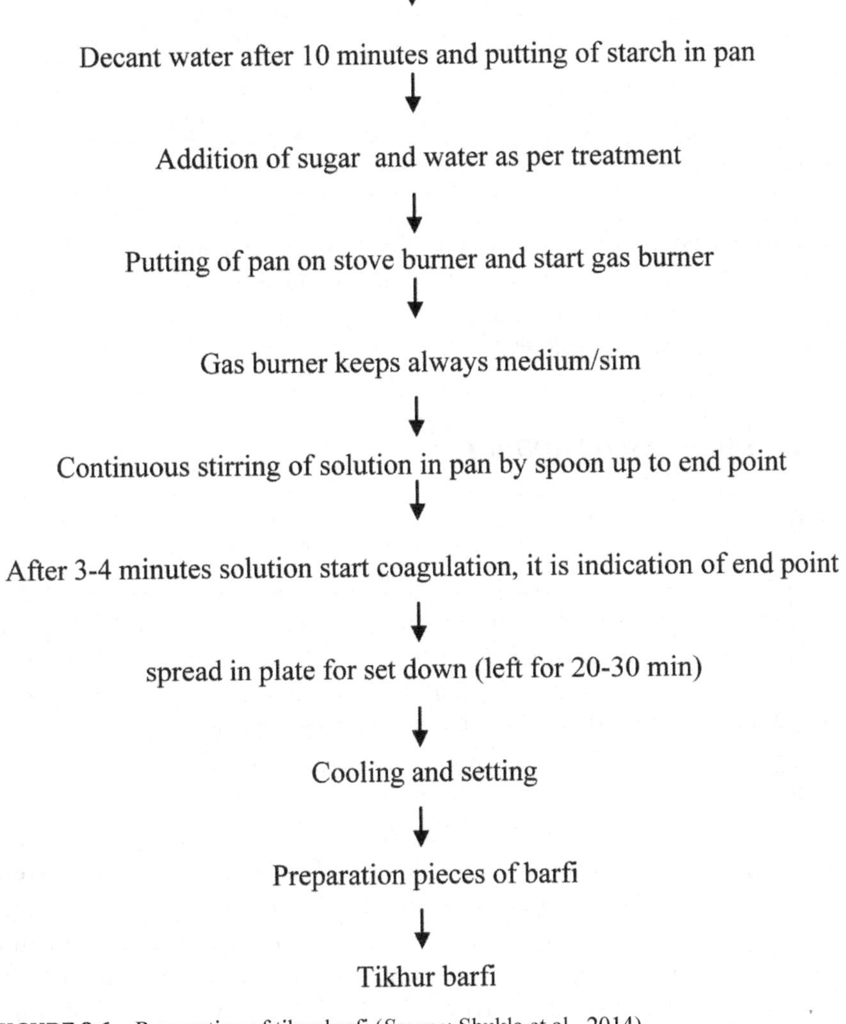

FIGURE 9.6 Preparation of tikur barfi (*Source*: Shukla et al., 2014).

TABLE 9.7 Nutritional Value of the Taro Corm on a Fresh Weight Basis*

Component	Content
Moisture	63–85%
Carbohydrate (mostly starch)	13–29%
Protein	1.4–3.0%
Fat	0.16–0.36%
Crude fiber	0.60–1.18%
Ash	0.60–1.3%
Vitamin C	7–9 mg/100 g
Thiamine	0.18 mg/100 g
Riboflavin	0.04 mg/100 g
Niacin	0.9 mg/100 g

(*Source*: Onwueme, 1994).

Taro is easy to digest due to the micro size of the starch grain and it's hypoallergenic behavior responsible for it's uniqueness. Taro corms are used as an ingredient for the preparation of weaning food and food for elder ones due to it's digestibility (98.5%).

9.7.1 TARO-BASED PRODUCTS

9.7.1.1 FLOUR

Taro corms could be converted to flour for extension of shelf life and to use for value addition. The utilization of taro flour as a component for product preparation is at present of new importance (Njintang et al., 2006). Taro powder was manufactured by washing, slicing, blanching and drying at 50°C in a cabinet drier for 4 h and powdered in a mixer. The prepared powder was passed through sieve no. 72 (British Sieve Standards) to get flour (Kaur et al., 2011).

Ikpeme-Emmanuel et al., (2010) prepared taro powder in a different way by giving chemical treatments to taro slices before drying. Taro corms were peeled, sliced (1 mm thick) and washed followed by soaking in slices water (1:2 volume of water) for 7–8 h, followed by immersing in 0.25% sodium bisulfate solution for 3 h. Taro chips were dried in cabinet drier at 60°C for 12 h. Then dried chips were converted into powder, sieved to pass 60 mesh screen sieve (British standard) to obtain sieve size of 0.250 mm.

Blanching of taro slices before drying helps to prevent the action of enzymes which could not be killed by the sun-drying process. Njintang et al., (2007) designed the production of taro powder utilized as starting material for the manufacturing of various food products which relatively helps to manage the food insecurity problem and prevent the post-harvest losses. The use of taro flour in combination with other flour, it is essential to differentiate their nutritional composition, physical and functional properties (Aboubakar et al., 2008). It was reported that taro flour had low protein, fat and starch whereas more sugar and fiber content than wheat flour.

Nurtama and Lin, (2010) reported that the moisture sorption isotherm of taro powder had sigmoid shape at 18, 25 and 35°C. Srikaeo et al., (2011) investigated that the blend of unripe banana, edible canna and taro powder had significant amounts of fiber and RS (1 to 26 g per 100 g), and GI varied from 67% to 99%. These flours could successfully be used for the preparation of low GI food having slow digestibility. This blend of flour can alternate commercial modified starches to reduce GIs of noodles and similar types of foods.

Taro flour showed the higher WAC and lowest foaming capacity, it means that taro flour provides good body to food and can be used as a thickener or gelling agent in various food products (Kaur et al., 2011). The nutritive value, physicochemical and functional properties of taro flour was studied by Amon et al., (2011). These properties changed as variation in boiling time. It was concluded that as the boiling time changes, there is significantly reduction in the iodine affinity of starch, total carbohydrate, total phenolic compound, reducing, and total sugars contents, whereas the moisture content, water absorption capacity, water solubility index, paste clarity and foam capacity increased significantly.

9.7.1.2 PASTE

Taro paste is popular in Asia, prepared by cooking and mashing taro into a soft paste to which sugar, fat and seasoning are commonly incorporated. Taro paste having particular unique texture and aroma could be applied as filler for bakery products or desserts. The staling of taro paste takes place when it was allowed to store for a longer time due to the retrogradation of starch (Hsi-Mei et al., 1998). The paste of taro prepared by boiling, peeling and grinding in mortar to a soft texture known as *Achu*, which is a traditional product of taro. The texture of *Achu* could be soft due to starch damage. It was reported that reconstituted *Achu* had a high degree of starch gelatinization.

Boiling the whole corms before drying could improve the texture of the reconstituted *Achu*. The texture of reconstituted *Achu* could be better if taro corms were boiled before drying. Precooking time affects on the functional properties of the finished product (Njintang and Mbofung, 2006).

9.7.1.3 STARCH

Taro starch having good digestibility because of the smaller size of starch granules. The starch in taro having four parts of amylopectin and one part of amylose. The ratio of amylose to amylopectin of starch affects the properties of starch. Amylopectin, in particular, is responsible for the properties of starch. Taro starch has high swelling power, good clarity and forms a hard coating layer (Adebayo and Itiola, 1998). Taro starch has been found highly digestible due to micro granule size (0.5–5 μm) of starch, which is responsible for the formation of soft gel (Jirarat et al., 2006). Taro starch found not much desirable as an ingredient for industrial purpose due to the smallness of starch granules and well suitable for infant food preparation. Taro starch also utilized as a filling agent for the biodegradable polyethylene film and as a fat replacer (Alam and Hasnain, 2009).

9.7.1.4 COOKIES

Taro flour is suitable for the preparation of cookies. It was reported that taro flour might be utilized as a substitute for the *maida* in cookies preparation. Snap type cookies were formulated using taro flour with shortening, sugar and less amount of wheat flour by Nip et al., (1994). A drop-type chocolate chip cookie was prepared by substituting 51% of the maida with taro powder and found acceptable by panel members.

9.7.1.5 NOODLES

Rosarlo et al., (1999) developed the noodles using a mixture of wheat and taro flour. It was investigated that, as the level of taro flour increases, mixograph peak times also increased along with peak viscosities and final setback values. Also increased in cooked weight of noodles. In order to improve the protein content, green gram flour or soy proteins were added. Taro flour also utilized for the development of gluten-free noodles with rice and pigeonpea flour. The different proportions of taro flour at the rate 20%–60% blended with rice and pigeonpea flour. The noodles found most sensory acceptable containing 50% taro flour with rice and pigeonpea flour.

9.8 ELEPHANT FOOT YAM (*Amorphophallus paeoniifolius)*

Elephant foot yam belongs to the Araceae family and an important major crop of tropical and subtropical countries due to it's nutritive value (Ravindran and George, 2008). Elephant foot yam is broadly cultivated and consumed in southeastern countries like India, Philippines, Malaysia, Indonesia. It is considered as an economic crop, particularly in India because of it's high production potential, market acceptability and lucrative economic returns (Misra et al., 2002). In India, it is a major crop of Andhra Pradesh, West Bengal, Gujarat, Kerala, Tamil Nadu, Maharashtra, Uttar Pradesh and Jharkhand.

9.8.1 MAJOR NUTRIENT AND CHEMICAL CONSTITUENTS OF Amorphophallus CORM

The presence of anti-oxidant vitamins such as Vitamin C and Vitamin E enhances the health value of Elephant foot yam tubers. The tubers are especially rich in Zinc containing 2.0–2.3 mg/100 g, as against a recommended dietary intake of 0.015 mg. As per the reports, Zinc is an important component of more than 300 enzymes in the human body that participate in the synthesis and catabolism of carbohydrates, lipids, proteins and nucleic acids. Zinc also plays a vital role in the immune system and is associated with cellular and humoral immunity. Elephant foot yam tubers contain flavonoids and tannins contributing to health effects. Quercetin was a major flavonoid isolated and characterized from Elephant foot yam, which has reported antioxidant activity. The nutritional facts of elephant foot yam given in Table 9.8.

9.8.2 TRADITIONAL USES OF CORMS

The elephant foot yam tubers are popular as a vegetable in food preparations. It is having blood purifier properties and have been used traditionally for the treatment of piles, abdominal disorders, tumors, enlargement of spleen, asthma and rheumatism (Misra and Sriram, 2001). These tubers also used traditionally in the treatments of arthralgia, elephantiasis, tumors, inflammations, hemorrhoids, hemorrhages, vomiting, cough, bronchitis, asthma, anorexia, dyspepsia, flatulence, colic, constipation, helminthiasis, hepatopathy, splenopathy, amenorrhea, dysmenorrhoea, seminal weakness, fatigue, anemia and general debility.

TABLE 9.8 Proximate Composition of Elephant Foot Yam Tuber

Principles	Elephant foot yam
Major nutrients (g/100g dwb)	
Carbohydrates	70.75
Protein	11.53
Fat	3.52
Crude fiber	14.32
Ash	6.90
Minor nutrients (mg/100g dwb)	
Potassium	3.81
Zinc	2.31
Iron	34.02

9.8.3 VALUE ADDITION

Elephant foot yam is not only used as a vegetable, but also different value-added products like chips, pickles and dried cubes are prepared. It is also used as a thickening agent in food preparation. It was reported that elephant foot yam was well suited for the preparation of osmodehydrated slices, bread (Singh and Wadhwa, 2012). It is reported that *Amorphophallus* corm flour (20%) could substitute wheat flour in bread preparation. Elephant foot yam peel contains enzymes and phytochemicals, which are having antioxidant activity. Thus Jimikand (*Amorphophallus*) peel wastes have high-value products and important medicinal constituents.

9.9 CONCLUSION

The present chapter concluded that tropical tuber crops are underutilized though having nutritional health benefits such as the good source of starch, vitamins, minerals and also having a low GI, high content of β-carotene, anthocyanine, phenolic compounds, antioxidants etc. These health benefits properties of tropical tuber crops could be a very good vehicle to address the health-related problems like diabetics, CD, infant food etc. It also serves as a food security crop. The natural colorant and antioxidant present in orange and purple-fleshed sweet potatoes can be used for developing functional foods. There is further research needed to develop the functional food products from tropical tuber crops flour, which are gluten-free, low GI, and diabetic free. There is no adequate research on the development of low calorie, low GI and gluten-free functional food products using these crops. There is a need to explore the nutritional benefits of tropical tuber crops for development of industrial based food products which could address the specific health-related problems. Available evidence suggested that the nutritional benefits of these tuber crops are not recognized potentially from the researchers and consumers, which also need further efforts.

KEYWORDS

- **arrowroot**
- **bioavailability**
- **cassava**
- ***Curcuma angustifolia***
- **elephant foot yam**
- **nutrition**
- **processing**
- **sweet potato**
- **taro**
- **tuber crops**

REFERENCES

Aboubakar, N. Y. N., Scher, J., & Mbofung, C. M. F., (2008). Physicochemical, thermal properties and microstructure of six varieties of taro (*Colocasia esculenta* L. Schott) flours and starches. *J. Food Eng.*, *86*, 294–305.

Adamade, C., & Azogu, I., (2013). Comparison of proximate composition, physio-mechanical properties and economics of production of cassava pellets derived from cassava chips and mash. *J. Agril. Eng. Tech.*, *21*, 18–26.

Adams, O. K., Adams, I. M., & Orungbemi, O. O., (2017). The effect of fermentation on functional properties of sweet potato and wheat flour. *Afri. J. Food Sci. Tech.*, *8*(2), 014–018.

Adebayo, A. S., & Itiola, O. A., (1998). Properties of starches obtained from *Colocasia esculenta* and *Artocarpus communis*. *J. Nat. Prod. and Med., 02*, 29–33.

Adebowale, A. R., Sanni, L., Awonorin, S., Daniel, I., & Kuye, A., (2007). Effect of cassava varieties on the sorption isotherm of tapioca grits. *Int. J. Food Sci. Tech., 42*, 448–452.

Adelekan, B., (2010). Investigation of ethanol productivity of cassava crop as a sustainable source of biofuel in tropical countries. *Afri. J. Biotech., 9*, 5643–5650.

Adeyeye, S. A. O., & Akingbala, J. O., (2016). Quality, functional, and sensory properties of cookies from sweet potato–maize flour blends. *J. Culinary Sci. Tech., 14*(4), 363–376, doi: 10.1080/15428052.2016.1160016.

Alam, F., & Hasnain, A., (2009). Studies on swelling and solubility of modified starch from taro (*Colocasia esculenta*): Effect of pH and temperature. *Agric. Conspec. Sci., 74*, 45–50.

Amon, A. S., Yade, S. R., Bony, K. P. K., Ahipo, D. E., & Lucien, K. P., (2011). Biochemical characteristics of flours from Ivorian Taro (Colocasia esculenta cv Yatan) corm as affected by boiling time. *Adv. J. Food Sci. Technol., 3*(6), 424–435.

Amornpitak, T. Y., Koguchi, M., & Teramoto, Y., (2012). Antioxidant activity of herbal wine made from cassava starch. *World Applied Sciences Journal, 16*(6), 874–878.

Andualem, A., Kebede, A., & Abadi, G. M., (2016). Development of pro-vitamin a and energy-rich biscuits: Blending of orange-fleshed sweet potato (*Ipomea batatas* L.) with wheat (*Triticum vulgare*) flour and altering baking temperature and time. *Afr. J. Food Sci., 10*(6), 728–773.

Björck, I., Liljeberg, H., & Ostman, R., (2000). "Low glycaemic index foods." *British J. Nutr., 83*(1), 149–155.

Boliavar, A., Cevallos, C., & Zevallos, L. C., (2003). Stoichiometric and kinetic studies of phenolic antioxidants from Andean purple corn and red-fleshed sweet potato. *J. Agri. Food Chem., 51*, 3313–3319.

Bovell-Benjamin, A. C., (2007). "Sweet potato: A review of its past, present, and future role in human nutrition." *Adv. Food and Nutr. Res., 52*, 1–59.

Burns, A. E., Bradbury, J. H., Cavagnaro, T. R., & Gleadow, R. M., (2012). Total cyanide content of cassava food products in Australia. *J. Food Compo Analy., 25*, 79–82.

Buttriss, J. L., & Stokes, C. S., (2008). Dietary fiber and health. *Nutrition Bulletin, 33*, 186–200.

Cassava Master Plan, (2006). *A Strategic Action Plan for the Development of the Nigeria Cassava Industry* (pp. 42–50). UNIDO.

Centro International de la Papa, (2009). *Sweet Potato*. Retrieved from: http://www.cipotato.org/sweet potato/ (Accessed on 8 June 2019).

Charles, A., Sriroth, K., & Huang, T., (2005). Proximate composition, mineral contents, hydrogen cyanide and phytic acid of 5 cassava genotypes. *Food Chem., 92*, 615–620.

Chaveerach, A., Sudmoo, R., Tanee, T., Mokkamel, P., & Sathyasai, N., (2008). Two new species of Curcuma used as cobra bite antidotes, *J. Systematics and Evolution, 46*(1), 80–88.

Deborah, O., Ellis, W. O., Oduro, I. (2011). Development of crackers from cassava and sweet potato flours using *Moringa oleifera* and *Ipomoea batatas* leaves as fortificant. *Am. J. Food. Nutr., 1*(3), 114–122.

Dhal, Y., Deo, B., & Sahu, R. K., (2012). Comparative antioxidant activity of non-enzymatic and enzymatic extracts of *Curcuma zedoaria*, *Curcuma angustifolia* and *Curcuma caesia*. *Int. J. Plant, Animal Environ. Sci., 2*(4), 232–239.

Doble, B., Sumeet, D., Kushagra, D., & Joshi, H., (2011). "Pharmacognostical and antimicrobial activity of leaf of Curcuma angustifolia Roxb." *Inter. J. Drug Disco. Herbal Res., 1*(2), 46–49.

Ehimen, R., Ohizua, A. A., Adeola, M. A., Idowu, O. P., Sobukola, T., Adeniyi, A., Raphael, O. I., Simeon, O. A., Tolulope, O., & Oyekale, A. F., (2017). Nutrient composition, functional, and pasting properties of unripe cooking banana, pigeon pea, and sweet potato flour blends. *Food Sci. Nutr.*, *5*, 750–762.
Elisabeth, D. A. A, (2015). Added value improvement of taro and sweet potato commodities by doing snack processing activity. *Procedia Food Science 3, the First International Symposium on Food and Agro-Biodiversity*, 262–273.
Fadeyibi, A., (2012). Storage methods and some uses of cassava in Nigeria. *Continental J. Agri. Sci.*, *5*, 12–18.
Falade, K. O., & Akingbala, J. O., (2010). Utilization of cassava for food. *Food Rev. Inter.*, *27*, 51–83.
FAO, (2011). *Statistical Database*. http://faostat.fao.org/site/567/DesktopDefault.aspx?PageID=567#ancor (Accessed on 8 June 2019).
Fidrianny, I., Windyaswari, A. S., & Wirasutisna, K. R., (2013). DPPH scavenging activity of various extracts of sweet potatoes leaves with varying tubers colors. *Inter J. Res. Pharm. Nano Sci.*, *3*(2), 133–145.
Gelencsér, T., Gal, V., Hodsayi, M., & Salgo, A., (2008). "Evaluation of quality and digestibility characteristics of resistant starch enriched pasta." *Food Bioprocess Technol., An International Journal*, *1*(2), 171–179. doi: 10.1007/s11947–007–0040-z.
Giri, N. A., Sheriff, J. T., Sajeev, M. S., & Pradeepika, C., (2016). Development and physico-nutritional evaluation of sweet potato flour based gluten-free cookies. *J. Root Crops*, *42*(1), 74–81.
Goni, I., & Valentin-Gamazo, C., (2003). "Chickpea flour ingredient slows glycemic response to pasta in healthy volunteers." *Food Chem.*, *81*(4), 511–515.
Grace, M. H., Yousef, G. G., Gustafson, S. J., Truong, V. D., Yencho, G. C., & Lila, M. A., (2014). Phytochemical changes in phenolics, anthocyanins, ascorbic acid, and carotenoids associated with sweet potato storage and impacts on bioactive properties. *Food Chem.*, *145*, 717–724.
Grant, V., (2003). *Select Markets for Taro, Sweet Potato and Yam*. A report for the Rural Industries Research and Development Corporation (RIRDC). Publication No 0 3 /052 RIRDC project No UCQ-13A. Online: http://www.rirdc.gov.au (Accessed on 8 June 2019).
Green, P. H. R., & Cellier, C., (2007). Coeliac disease. *New Engl. J. Med.*, *357*, 1731–1743.
Guerin-Deremaux, L., Li, S., Pochat, M., Wils, D., Mubasher, M., Reifer, C., et al., (2011). Effects of NUTRIOSE_ dietary fiber supplementation on body weight, body composition, energy intake and hunger in overweight men. *Inter. J. Food Sci. Nutr.*, *62*, 628–635.
Hagenimana, V., Carey, E., Gichuki, S. T., Oyunga, M. A., & Imungi, J. K., (1992). Changes in carotenoid content after drying and processing sweet potato products. *Eco. Food Nutr.*, *37*, 450–473.
Haggblade, S., Djurfeldt, A. A., Nyirenda, D. B., Lodin, J. B., Brimer, L., Chiona, M., Chitundu, M., Chiwona-Karltun, L., Cuambe, C., & Dolislager, M., (2012). Cassava commercialization in Southeastern Africa. *J. Agribusi. Develop Emerging Eco.*, *2*, 4–40.
Herawati, H., & Widowati, S., (2009). Characteristic of pearl-rice from sweet potato (*Ipomea batatas*) (in Bahasa Indonesia). *Post Harvest Tech. Bulletin*, *5*(1), 39–46. 2002.79.2.265.
Hill, I. D., Dirks, M. H., Liptak, G. S., Colletti, R. B., Fasano, A., & Guandalini, S., (2005). Guideline for the diagnosis and treatment of celiac disease in children: Recommendations of the North American Society for pediatric gastroenterology, hepatology and nutrition. *J. Pediatr. Gastroenterol. Nutr.*, *40*, 1–19.

Himeda, M., Nicolas, Y. N., Fombang, E., Facho, B., Kitissou, P., Mbofung, C. M. F., & Scher, J., (2014). Chemical composition, functional and sensory characteristics of wheat-taro composite flours and biscuits. *J. Food Sci. Tech.*, *51*(9), 1893–1901.

Hsi-Mei, L., Shih-Tong, J., & Cheng-Yi, L., (1998). ONMR and DSC for studying quality of taro paste as affected by processing and storage. *LWT-Food Sci. Tech.,*, *31*, 57–63.

Hue, S. M., Boyce, A. N., & Somasundram, C., (2012). Antioxidant activity, phenolic and flavonoid contents in the leaves of different varieties of sweet potato (*Ipomoea batatas*). *Afri. J. Crop Sci.*, *6*(3), 375–380.

Ikpeme-Emmanuel, C. A., Osuchukwu, N. C., & Oshiele, L., (2010). Functional and sensory properties of wheat (*Aestium triticium*) and taro flour (*Colocasia esculenta*) composite bread. *Afr. J. Food Sci.*, *4*(5), 248–253.

Ishiguro, K., & Yoshimoto, M., (2005). Content of eye-protective nutrient lutein in sweet potato leaves. In: *2nd International Symposium on Sweet Potato and Cassava* (pp. 213–214). Kuala Lumpur, Malaysia.

Iwe, M. O., & Onuh, J. O., (1992). Functional and sensory property of soybean and sweet potato flour mixture. *LWT-Food Sci. Tech.*, *25*, 569–573.

Jagatheesh, K., Arumugam, V., Elangovan, N., & Pavan, K. P., (2010). Evaluation of the anti-tumor and antioxidant activity of *Amorphophallus paeonifolius* on DMBA induced mammary carcinoma. *Inter. J. Chem. Pharma. Sci.*, *1*(2), 40–50.

Jekayinfa, S., & Olajide, J., (2007). Analysis of energy usage in the production of three selected cassava-based foods in Nigeria. *J. Food Eng.*, *82*, 217–226.

Jiao, Y., Jiang, Y., Zhai, W., & Yang, Z., (2012). Studies on antioxidant capacity of anthocyanin extract from purple sweet potato (*Ipomoea batatas* L.). *African J. Biotech.*, *11*(27), 7046–7054,

Jirarat, T., Sukruedee, A., & Persuade, P., (2006). Chemical and physical properties of flour extracted from taro (*Colocasia esculenta*) grown in different regions of Thailand. *Sci. Asia*, *32*, 279–284.

Jothy, A. N., Moorthy, S. N., & Vimala, B., (2003). Physico-chemical and functional properties of starch extracted from two species of Curcuma. *Inter J. Food Prop.*, *6*(1), 135–145.

Jyothi, G. K., Renjusha, M., Padmaja, G., Sajeev, M. S., & Moorthy, S. N., (2011). Nutritional and functional characteristics of protein fortified pasta from sweet potato. *Food Nutr. Sci.*, *2*, 944–955.

Jyothi, G. K., Renjusha, M., Padmaja, G., Sajeev, M. S., & Moorthy, S. N., (2012). Evaluation of nutritional and physicomechanical characteristics of dietary fiber enriched sweet potato pasta. *Euro Food Res. Tech.*, *234*, 467–476.

Katan, M. B., & De Roos, N. M., (2004). Promises and problems of functional foods. *Critical Rev. Food Sci. Nutr.*, *44*, 369–377.

Kaur, M., Kaushal, P., & Sandhu, K. S., (2011). Studies on physicochemical and pasting properties of taro (*Colocasia esculenta* L.) flour in comparison with a cereal, tuber and legume flour. *J. Food Sci. Technol.*, *48*, 1–7. doi: 10.1007/s13197-010-0227-6.

Krishnan, J. G., Menon, R., Padmaja, G., Sajeev, M. S., & Moorthy, S. N., (2011). Nutritional and functional characteristics of protein-fortified pasta from sweet potato. *Food Nutr. Sci.*, *2*, 944–955.

Kundu, N., Campbell, P., Hampton, B., et al. (2012) Antimetastatic activity isolated from *Colocasia esculenta* (taro). *Anticancer Drugs*, *23*(2), 200–211. doi:10.1097/CAD.0b013e32834b85e8.

Kusano, S., & Abe, H., (2000). "Anti-diabetic activity of white-skinned sweet potato (*Ipomoea batatas* L.) in obese Zucker fatty rats." *Biology & Pharmacology Bulletin*, *23*, 23–26. doi: 10.1248/bpb.23.23.

Kwame, O. D., Yu-Jiao, L., & Zhi-En, P., (2017). Evaluating the nutritional and sensory quality of bread, cookies and noodles made from wheat supplemented with root tuber flour. *Bri. Food J., 119*(4), 895–908.

Limroongreungrat, K., & Huang, Y. W., (2007). Pasta products made from sweet potato fortified with soy protein. *LWT – Food Sci. Tech., 40*, 200–206.

Lisona, E., Pradeep, H. A., & Prakash, L. H., (2011). *In vitro* bioactivity studies of wild *Curcuma angustifolia* rhizome extract against (hela) human cervical carcinoma cells. *J. Clin. Diag. Rese., 3*(2), 136–139.

Liu, Q., Donner, E., Yin, Y., Huang, R. L., & Fan, M. Z., (2006). "The physicochemical properties and *in vitro* digestibility of selected cereals, tubers, and legumes grown in China." *Food Chem., 99*(3), 470–477.

Lu, J., Wu, D. M., Zheng, Y. L., Cheng, B. H. W., & Zhang, Z. F., (2012). Purple sweet potato color attenuates domoic acid-induced cognitive deficits by promoting estrogen receptor-alpha-mediated mitochondrial biogenesis signaling in mice. *Free Radic. Biol. Med., 52*, 646–659.

Mareti, M. C., Grossmann, M. V. E., & Benassi, M. T., (2010). Physical and sensorial characteristics of cookies containing defatted soy flour and oat bran. *Food Sci. Technol., 4*, 878–883.

Mathew, J., (2007). Arrowroot. In: Peter, K. V., (ed.), *Underutilized and Underexploited Horticultural Crops* (p. 26). New Delhi, India: New India Publishing Agency.

Mbaeyi-Nwaoha, I. E., & Itoje, C. R., (2016). Quality evaluation of prawn crackers produced from blends of prawns and cassava (*Manihot esculenta*), pink and orange-fleshed sweet potato (*Ipomoea batatas* (L) Lam). *Starches, 7*(4), 051–059.

Menen, Z., Pragya, S., Alemzewed, C., & Yibeltal, J., (2015). Development of maize-based orange-fleshed sweet potato flatbread for lactating mothers at *Hawassa zuria woreda*, SNNPRS, Ethiopia. *Inter. J. Food Sci. Nutr. Eng., 5*(5), 183–190.

Mishra, A., Devi, M., & Jha, P., (2015). Development of gluten-free biscuits utilizing fruits and starchy vegetable powders. *J. Food Sci. Technol., 52*(7), 4423–4431.

Misra, R. S., & Sriram, S., (2001). Medicinal value and export potential of tropical tuber crops. In: *Recent Progress in Medicinal Plants, Crop Improvement and Commercialization* (Vol. 5, pp. 317–325).

Misra, R. S., Nedunchezhiyan, M., & Swamy, T. M. S., (2002). *Amorphophallus paeoniifolius*, a high-value cash crop for coastal areas. In: *National Conference on Coastal Agricultural Research* (pp. 6, 7). ICAR Research Complex, Goa.

Montagnac, J. A., Davis, C. R., Tanumihardjo, S. A., (2009). Nutritional value of cassava for use as a staple food and recent advances for improvement. Institute of food technologists. *Compre. Reviews Food Sci. Food Safety, 8*, 181–192.

Moorthy, R., (1994). Blanshard. *Influence of Gelatinization Characterization of Curcuma and Cassava Starch and Flour on Textural Properties of Some Food Products* (pp. 133–138), Columbia.

Nag, J. L., Shukla, N., Pararey, P. M., Soni, V. K., Netam, C. R., & Pandey, D. K., (2006). Effect of extraction methods on production of edible tikhur (Curcuma angustifolia Roxb.). *Abstracts Book, National Seminar on Medicinal, Aromatic & Spices Plants Perspective and Potential* (p. 185). IGKV, TCB, CARS, Bilaspur, Chhattisgarh.

Namsa, N. D., Mandal, M., Tangjang, S., & Mandal, S. C., (2011). Ethnobotany of the Monpa ethnic group at Arunachal Pradesh, India. *J. Ethnobiol. Ethnomed., 7*, 31–39.

NCDA, CS (North Carolina Department of Agriculture & Consumer Services), division of marketing, horticulture commodities, sweet potato, (2012). Available: http://www.ncagr.gov/markets/commodit/horticul/sweetpot (Accessed on 8 June 2019).

Nguimbou, R. M., Boudjeko, T., Njintang, N. Y., Himeda, M., Scher, J., & Mbofung, C. M. (2014). Mucilage chemical profile and antioxidant properties of giant swamp taro tubers. *J. Food Sci Tech., 51*(12), 3559–3567. doi:10.1007/s13197-012-0906-6.

Nip, W. K., Whitaker, C. S., & Vargo, D., (1994). Application of taro flour in cookie formulations. *Int. J. Food Sci. Tech., 29*(4), 463–468.

Nishaa, S., Vishnupriya, M., Sasikumar, J. M., Hephzibah, P., Christabel, G. V. K., et al., (2012). Antioxidant activity of ethanolic extract of *Maranta arundinacea* tuberous rhizomes. *Asian J. Pharma. Clini. Res., 5*(4), 86–88.

Njintang, N. Y., Mbofung, C. M. F., & Kesteloot, R., (2007). Multivariate analysis of the effect of drying method and particle size of flour on the instrumental texture characteristics of paste made from two varieties of taro (*Colocasia esculenta* L. Schott) flour. *J. Food Eng., 81*, 250–256.

Njintang, N. Y., Parker, M. L., Moates, G. K., Faulds, C. B., Smith, A. C., Waldron, K. W., Mbofung, C. M. F., & Scher, J., (2008). Microstructure and creep recovery characteristics of achu (a taro-based paste) made from freeze-dried taro chips as affected by moisture content and variety. *J. Food Eng., 87*, 172–180.

Njintang, Y. N., & Mbofung, C. M. F., (2006). Effect of precooking time and drying temperature on the physicochemical characteristics and in vitro carbohydrate digestibility of taro flour. *LWT-Food Sci. Tech., 3*(9), 684–691.

Njintang, Y. N., Boudjeko, T., Tatsadjieu, N. L., Nguema-One, E., Scher, J., & Mbofung, C. M. F., (2011). Compositional, spectroscopic and rheological analyses of mucilage isolated from taro (*Colocasia esculenta L. Schott*) corms. *J. Food Sci. Tech.* doi: 10.1007/s 13197-011-0580-0.

Oki, T., Masuda, M., Furuta, S., Nishiba, Y., Terahara, N., & Suda, I., (2002). Involvement of anthocyanins and other phenolic compounds in radical-scavenging activity of purple-fleshed sweet potato cultivars. *J. Food Sci., 67*, 1752–1756.

Onabanjo, O. O., & Ighere, D. A., (2014). Nutritional, functional and sensory properties of bread biscuit produced from wheat–sweet potato composite. *J. Food Technol. Res., 1*(3), 111–121.

Onwueme, I. C., (1994). *Tropical Root and Tuber Crops—Production, Perspectives and Future Prospects* (p. 228). FAO plant production & protection paper 126, FAO, Rome.

Oyewole, O., & Sanni, L., (1995). Constraints in traditional cassava processing-the case of fufu production. In: Agbor-Egbe, T., Brauman, A., Griffon, T., & Treche, S., (eds.), *Cassava Food Processing* (pp. 523–529). France, ORSTOM.

Paramapojn, S., & Gritsanapan, W., (2009). Free radical scavenging activity determination and quantitative analysis of curcuminoids in Curcuma zedaria rhizome extract by HPLC method. *Current Science, 97*(7), 1069–1073.

Pochapski, M. T., Fosquiera, E. C., Esmerino, L. A., Santos, E. B., Farago, P. V., Santos, F. A., et al., (2011). Phytochemical screening, antioxidant, and antimicrobial activities of the crude leaves' extract from *Ipomoea batatas* (L.) Lam. *Pharmacogn Mag., 7*, 165–170.

Preedy, V. R., Watsoa, R. R., & Patel, V. B., (2011). *Flour and Breads and Their Fortification in Health and Disease Prevention.* London: Academic Press.

Preetham, S., Lohith, K. D. H., & Chanda, D., (2007). Traditional knowledge on wild food plants. *Indian J. Traditi. Know., 3*, 233–234.

Raji, A., Ladeinde, O., & Dixon, A., (2008). Screening landraces for additional sources of field resistance to cassava mosaic disease and green mite for integration into the cassava improvement program. *J. Integr. Plant Bio., 50*, 311–318.

Rani, A., & Chawhaan, P. H., (2012). Extraction and scanning electron microscopic studies of Curcuma angustifolia Roxb. Starch. *Indian J. Nat. Prod. Resour.*, *3*, 407–410.

Rathod, R. P., & Annapure, U. S., (2016). Development of extruded fasting snacks by using vari rice, sweet potato and banana powder with applying response surface methodology. *Food Meas. Char.*, *10*, 715–725.

Ravindran, C. S., & George, J., (2008). Status of *Amorphophallus* in Kerala. In: *National Seminar on Amorphophallus: Innovative Technologies* (pp. 58–61). Rajendra Agricultural University, Patna, Bihar.

Reilly, N. R., & Green, P. H. R., (2012). Epidemiology and clinical presentations of celiac disease. *Semin. Immunopathol.*, 1–6.

Renjusha, Menon, G., Padmaja, & Sajeev, M. S. (2015). Ultrastructural and Starch Digestibility Characteristics of Sweet Potato Spaghetti: Effects of Edible Gums and Fibers, *Int. J. Food Prop.*, *18*(6), 1231–1247, DOI: 10.1080/10942912.2014.90326.

Rosarlo, M. D., Vinas, A., & Lorenz, K., (1999). Pasta products containing taro (*Colocasia esculenta*) and chaya (*Cnidoscolus chavamansa*). *J. Food Process Pres.*, *23*(1), 1–20.

Rubio-Tapia, A., & Murray, J. A., (2010). Coeliac disease. *Current Opinion in Gastroenterology*, *26*, 116–122.

Sarabhai, S., Indrani, D., Vijaykrishnaraj, M., Milind, A. V., & Prabhasankar, P., (2015). Effect of protein concentrates, emulsifiers on textural and sensory characteristics of gluten-free cookies and its immunochemical validation. *J. Food Sci. Technol.*, *52*(6), 3763–3772.

Sciarini, S. L., Ribotta, D. P., Leon, E. A., & Perez, T. G., (2008). Influence of gluten-free flours and their mixtures on batter properties and bread quality. *Food Bioprocess Technol.*, doi: 10.1007/s 11947-008-0098-2.

Shan, Q., Lu, J., Y., Zheng, J., Li, Z., Zhou, B., H., et al., (2009). Purple sweet potato color ameliorates cognition deficits and attenuates oxidative damage and inflammation in aging mouse brain induced by d-galactose. *J. Biomed. Biotechnol.*, 1–9.

Shittu, T., Dixon, A., Awonorin, S., Sanni, L., & Maziya-Dixon, B., (2008). Bread from composite cassava–wheat flour. In: *Effect of Cassava Genotype and Nitrogen Fertilizer on Bread Quality. Food Rese. Int.*, *41*, 569–578.

Shukla, D., Shankar, N. J. L., & Sahu, M. K., (2014). Study on preparation procedure and standardization of recipe for tikhur Barfi. *Inter. J. Proc. Post. Harv. Technol.*, *5*, 156–164.

Shukla, S. K., Susmita, S., Vijaya, K., & Mishra, S. K., (2007). "In vitro propagation of tikhur (Curcuma angustifolia Roxb.): A starch yielding plant." *Indian J. Biotech.*, *6*, 274–276.

Singh, A., & Wadhwa, N., (2012). Osmotic dehydration of *Amorphophallus paeoniifolius* slices and its phytochemical investigation. *Inter. J. Phar. Life Sci.*, *3*(7), 1797–1801.

Singh, S., Charanjit, S. R., Amrinder, S. B., & Saxena, D. C., (2004). Sweet potato-based pasta product: Optimization of ingredient levels using response surface methodology. *Int. J. Food Sci. Tech.*, *39*, 191–200.

Siyumbwa, S., Nomishan, N. B., & Okechukwu, P. N., (2014). Antioxidant and gastroprotective activity of ethanolic rind extract of *Manihot esculenta* Crantz. *Inter. J. Phar. Rese. Scho.*, *3*, 1–2.

Spennemann, D. H. R., (1992). Arrowroot production in the Marshall Islands—past, present and future. *Palawija News*, *9*(1), 1–2.

Srikaeo, K., Mingyai, S., & Sopade, P. A., (2011). Physicochemical properties, resistant starch content and enzymatic digestibility of unripe banana, edible canna, taro flours and their rice noodle products. *Int. J. Food Sci. Technol., 46*, 2111–2117. doi:10.1111/j.1365-2621.2011.02724.x.

Srinivas, T., (2007). Industrial demand for cassava starch in India. *Starch- Stärke, 59*, 477–481.

Sushama, S., Murthy, S. R., Sujan, G., Sivakamisundhari, P. S., & Preetham, J., (2011). Pharmacognostic and phytochemical evolution of Curcuma angustifolia. *Inter. J. Pharmacognosy Phytoche. Res., 7*(4), 820–824.

Taiwo, K. A., (2006). Utilization potentials of cassava in Nigeria: The domestic and industrial products. *Food Reviews Int., 22*, 29–42.

Teow, C., Truong, V., McFeeters, R., Thompson, R., Pecota, K., & Yencho, G., (2007). Antioxidant activities, phenolic and beta-carotene contents of sweet potato genotypes with varying flesh colors. *Food Chem., 103*, 829–838.

The Wealth of India: Raw Materials, (1962). New Delhi: Council of Scientific and Industrial Research (Vol. VI., pp. 303, 304).

Tilman, J. C., Colm, M. O. B., Denise, M. C., Anja, D., & Elke, K. A., (2003). Influence of gluten-free flour mixes and fat powder on the quality of gluten-free biscuits. *Eur. Food Res. Technol., 216*, 369–376.

Truong, V. D., & Avula, R., (2010). Sweet potato purees and powders for functional food ingredients. In: Ray, R. C., & Tomlins, K. I., (eds.), *Sweet Potato: Post Harvest Aspects in Food, Feed and Industry* (pp. 117–161). Nova Science Publishers Inc., New York.

Ubalua, A., (2007). Cassava wastes: Treatment options and value addition alternatives. *African J. Biotech., 6*, 2065–2073.

USDA (U. S. Department of Agriculture), Agricultural Research Service, (2009). USDA National Nutrient Database for standard reference, release 22. *Nutrient Data Laboratory Home Page*, http://www.ars.usda.gov/ba/bhnrc/ndl (Accessed on 8 June 2019).

Van Hal, M., (2000). Quality of sweet potato flour during processing and storage. *Food Reviews Inter., 16*(1), 1–37. doi: 10.1081/ Fri-100100280.

Van-Den, T., & Avula, R. Y., (2010). Sweet potato purees and dehydrated powders for functional food ingredients. In: Ray, R. C., & Tomlins, K. I., (eds.), *Sweet Potato: Post Harvest Aspects in Food.* ISBN 978-1-60876-343-6. © Nova Science Publishers, Inc.

Villarreal, M. E., Ribotta, P. D., & Iturriaga, L. B., (2013). "Comparing methods for extracting amaranthus starch and the properties of the isolated starches." *LWT - Food Sci. Tech., 51*(2), 441–447.

Wei, L. S., Wee, W., Siong, J. Y. F., & Syamsumir, D. F., (2011). *Antimicrobial, Antioxidant, Anticancerous Property and Chemical Composition of Different Parts (Corm, Stem, Leaves) of Colocasia Esculenta Extract* (Vol. XXIV, 3(1)). Annales-Medical University Lubin-Polonia.

Yi, B., Hu, L., Mei, W., Zhou, K., Wang, H., Luo, Y., Wei, X., Dai, H., et al., (2010). Antioxidant phenolic compounds of cassava (*Manihot esculenta*) from Hainan. *Molecules, 16*, 10157–10167.

Yoshimoto, M., Kurata, R., Okuno, S., Ishiguro, K., Yamakawa, O., Tsubata, M., & Mori, S., (2005). Takagaki. Nutritional value of and product development from sweet potato leaves. In: 2nd *International Symposium on Sweet Potato and Cassava* (pp. 183–184). Kuala Lumpur, Malaysia.

Zhang, Z. F., Lu, J., Zheng, Y. L., Wu, D. M., Hu, B., Shan, Q., et al., (2013). Purple sweet potato color attenuates hepatic insulin resistance via blocking oxidative stress and endoplasmic reticulum stress in high-fat- diet- treated mice. *J. Nutr. Biochem., 24*, 1008–1018.

CHAPTER 10

Bioethanol Fermentation: The Path Forward for Eco-Friendly and Sustainable Development

BINITA DEV,[1] ABDELRAHMAN SALEH ZAKY,[2] and R. JAYABALAN[1]

[1]*Food Microbiology and Bioprocess Laboratory, Department of Life Science, National Institute of Technology, Rourkela 769008, Odisha, E-mail: jayabalanr@nitrkl.ac.in*

[2]*Department of Microbiology, Faculty of Agriculture, Cairo University, Giza, 12613, Egypt*

ABSTRACT

Bioenergy reformation through fermentation has achieved considerable attention with the advent of global climatic change, population rise, the uncertainty of non-renewable sources, and high demand for alternative fuel. Bioethanol, the most potent fermented transportation fuel, due to its less carbon-dioxide emissions, renewable, and several environmental benefits can substitute fossil dependent fuels with the global rise in production from 120 billion liters (2016) to 137 billion liters (2026). The core of bioethanol production depends upon fermentation, in which simple sugars are fermented by a variety of microorganisms to bioethanol. Various renewable biomasses like agricultural wastes, industrial wastes, and algal-based raw materials are used for bioethanol production. *Saccharomyces cerevisiae* is the most availed yeast used for industrial fermentation. Up to 2–5% of the gasoline-ethanol blend can be used without engine modification, and this accounts for the necessity of production of huge quantities of bioethanol. This chapter focuses on the economic feasibility and environmental viability of bioethanol production.

10.1 INTRODUCTION

The world energy market is strongly influenced by fossil-derived fuels worth around 1.5 trillion dollars (Goldemberg, 2006) and can satisfy energy demands by 84% until 2030 (Shafieeand Topal, 2009). Unrestricted consumption of these fossil-derived fuels over the years has resulted in increased pollution and a drastic rise in atmospheric levels of Greenhouse Gases (GHG) (Ballesteros et al., 2006). The global population would be reaching around 9 billion or more in 2050 along with facing confrontation in a food shortage, biodiversity loss, freshwater scarcity, and political and economic imbalance and more significantly increased demand for energy consumption (Ingrao et al., 2016). As stated in fifth assessment report of United Nations Intergovernmental Panel on Climate change (IPCC, 2013), the main reason of drastic weather change in recent years is due to human activity which has resulted in increased global average temperature and likely to increase more by 5–6% at the end of the century. The panel emphasizes the reduction of GHG emissions by at least 50% until 2050 and limits the global average temperature not more than 2%. GHG emissions from industries, transportation encompass for more than 600 million transport units (cars, buses, and trucks) which are likely to reach 2.5 billion transport units by 2050 (Manzetti and Andersen, 2015). The global transportation sector solely consumes half of the fossil fuels (oil) with emissions of 19% carbon dioxide and more than 70% of carbon monoxide (Goldemberg, 2008). Global warming is a serious issue around the world since 1900 (Boluk and Mert, 2014). The combustion of fossil fuels emits about 90% of total global CO_2as per the recent report of the European Union Joint Research Center (Olivier et al., 2012) and is expected to rise approximately 57 Gt in 2050 from 31 Gt in 2011 (Dimitriou et al., 2015). This tragic scenario converged for the urgent establishment of green or low carbon sustainable technologies based on renewable biomass. Bioethanol, bio-diesel, and bio-hydrogen are some of the renewable biomass-based fuels that have the potential of mitigating CO_2 emissions. With the rapid growth of renewables as a world energy source, it is estimated to cover 14% of the total energy use by 2035 (Boluk and Mert, 2014).

Several state policies have been formulated by countries across the globe with the motive of efficient and economic utilization of biomass for satisfying CO_2 reduction targets as stated in Kyoto protocol and simultaneously for limiting dependence on fossil fuels (Sarkar et al., 2012). With carbon capture, storage, and utilization technologies, CO_2 emissions can be mitigated by turning into value-added products- chemicals and fuels (Cuellar-Franca et al., 2015). Bioethanol has been stated as a premier high-performance substitute of automotive fossil fuel with several promising environmental benefits and potential of tackling the energy crisis. Bioethanol market has entered in exponential phase globally with widespread importance at national and international level (Sarkar et al., 2012).

The biologically produced ethanol from renewable biomass is called fermentative ethanol or bioethanol. Bioethanol (C_2H_5OH) has been attributed as an upcoming valuable energy source over gasoline (C_7H_{17}) in the transport fuel market (Wyman, 1996) and has very less or no net CO_2 emissions when used as fuel. Around 73% to 83% of energy is afforded by 1 gallon of E85 and 96.7% from 1 gallon of E10 of the energy of one gallon of gasoline (US Department of energy, 2017). In comparison to gasoline, ethanol has higher octane no (106–110 vs. 91–96), high compression ratio, shorter burning time, greater evaporation enthalpy (1177 kJ/kg vs. 348 kJ/kg at 60°c) ensuring several benefits over gasoline and can potentially replace gasoline in transport fuel market (Sarkar et al., 2012; Manzetti and Andersen, 2015; Zabed et al., 2017b).

10.2 ECO-FRIENDLY

Ethanol has been portrayed as "the fuel of the future" by Henry Ford in 1925. He further stated that these would be derived from apples, weeds, sawdust – almost anything and today his prediction can be easily seen. Bioethanol with 35% oxygen has been earmarked as an environment-friendly fuel that possesses15% higher combustion efficiency than that of gasoline, which lacks oxygen. This results in reducing the NOx and particulate emissions (Karand, 2006). Sulfur dioxide (SO_2), a potent carcinogen is a precursor of atmospheric sulfuric acid (H_2SO_4) and sulfate aerosol. It is produced as a by-product during the burning of fossil fuels and is a major reason for acid rain. An ethanol and gasoline blend reduces the amount of sulfur in fuel as well as sulfur oxide emissions (Woods, 2008). Methyl tertiary butyl ether (MTBE), an oxygenated organic compound, is used as a gasoline additive in fuels for increasing the oxygen content of gasoline. Its use is controversial as it leads to groundwater contamination. Ethanol has been recommended as a safer substitute for MTBE and used first by California in 2003 (McCarthy and Tiemann, 2006). Bioethanol either in pure form or blend with gasoline (gasohol) can be used. The most popular blends are E15 (15% ethanol and 85% gasoline) and E85 (85% ethanol and 15% gasoline). In Brazil, bioethanol is used in its purest form or as gasohol comprising a blend of ethanol (24%) with gasoline (76%) (Dias De Oliveira et al., 2005) whereas, in the United States, gasohol or E10 consisting of 10% ethanol and 90% gasoline is used. Bioethanol –gasoline blending programs are applied as E5 in India and Sweden; E7 in Paraguay; E10 in Canada, Australia, Thailand, China, Columbia, and Peru; for Flexible-Fuel Vehicles (FFV) E85 in Canada and Sweden (Kadiman, 2005). Blending offers several advantages such as improved gasoline combustion, reduced exhaust emissions, enhanced engine torque, brake power, engine efficiency and thermal efficiency (Yuksel and Yuksel, 2004; Najafi et al., 2009; Bayraktar, 2005; Yoon and Lee, 2011). Studies revealed that the gasoline-ethanol blending reduces emissions of CO, NOx, and HC. There is a 20–30% reduction of CO emissions when both low and high-level blends are used. CO_2 emissions are markedly reduced when E85 blend is used. NOx emissions also reduce from 5–20% when E10 to E85 blends are used (Balat, 2009). Ethanol–diesel blend E15 also decreases NOx particulate matter emission by 41% (Subramanian et al., 2005; Chan et al., 2007). Aldehyde emissions are more with ethanol than gasoline, but it is less hazardous to the environment than by the polynuclear aromatic emissions from gasoline engines (Rice et al., 1991). Bioethanol reduces precursors of ozone by 20–30% and significantly reduces harmful gases worldwide. The complete combustion of different ethanol gasoline blend potentially reduces the CO and UHC (unburned hydrocarbon) emissions in engines (Bata and Roan, 1989). Blending also reduces the effect of anti-corrosive compounds, thereby enhancing engine running qualities (Manzetti and Andersen, 2015). Various combination of the ethanol-gasoline blend has been used in engines to which 10% ethanol reduce CO emission by 30% with a 5% increase in engine power output. The octane number also increases to 5% for each time, more 10% ethanol is added (Palmer, 1986). Similarly, when different blends of 10%, 20%, 30%, 40% ethanol were used in the variable compression ratio engine, the octane number rises with a decrease in heating value. Among all the combinations, 10% ethanol with 90% gasoline was obtained as the optimum one (Abdel et al., 1997).

India, like other nations, has a separate ministry for renewable energy development. A report issued by the Indian planning commission in April 2003 strengthened bioethanol program setting targets for blending, cost analysis, substrate availability, and expanding bioethanol blends from 5–10%. The bioethanol blending to 10% was made mandatory in the 11thfive-year plan (2007–2012) which is scaled to E20 by 2017 as imposed by National Policy on Biofuels (Government of India, 11thfive-year plan 2007–2012). An ethanol blending program (EBP) was implemented in January 2003 in nine Indian states and four Union territories (Bandyopadhyay, 2015). An innovative approach is taken by Biochemical Engineering Research Center;

IIT Delhi in 1980 for lignocellulose to ethanol demonstrated the production of 50L bioethanol per day. With the recent approach of Make in India and Swachh Bharat Abhiyan, the joint venture of DBT-ICT Center for energy bioscience at ICT Mumbai and Department of Biotechnology, Ministry of Science and Technology has successfully presented bioethanol production from lignocellulosic biomass for the first time. On 22nd April 2016, India's first 10 L/day capacity 2G or cellulosic ethanol plant was set up in Kashipur, Uttarakhand where any biomass such as rice straw, wheat straw, bamboo, bagasse, cotton stalk, etc. can be converted into ethanol in lesser than 24 h. Scaling up further can lead India to a global technology provider with an imminent reduction in carbon emissions (Jahnavi et al., 2017; Bandyopadhyay, 2015).

10.3 GLOBAL BIOETHANOL PRODUCTION

The United States ranks as the largest producer of Bioethanol with the production of approximately 15.8 billion gallons in 2017 in total and Brazil ranked second with the total production of more than 7 gallons (Statista, 2018). Together the United States and Brazil account for 89% of global bioethanol production (Renewable Fuel Association, 2018; Singh et al., 2016).

10.4 SUSTAINABLE DEVELOPMENT

Rapid industrialization, improved living standards, environmental pollution, global warming, energy security, diminished fossil fuels, and environmental safety have inspired the emergence of alternative, sustainable, eco-friendly, and clean green biofuels (Azhar et al., 2017; Shafiee and Topal, 2009). Bioethanol and biodiesel are the most potent alternative fuels that can substitute the use of petroleum-derived fuels in internal combustion engines (Thangavelu et al., 2016). Bioethanol, an eminent biofuel is preferred more than biodiesel due to the environmental sustainability of the feedstock, its cleaner emission due to oxygenation capacity, cost-saving production and promising evolution (Demirbas, 2007, 2009, 2011; Balat and Balat, 2009; Mussatto et al., 2010).

Ethanol is obtained from catalytic hydrolysis of ethylene, which is a non-renewable petrochemical component.

$$C_2H_4 + H_2O \rightarrow C_2H_5OH$$

Bioethanol is obtained from the fermentation of carbohydrate-based renewable and sustainable feedstock by certain fermenting yeast (e.g., *Saccharomyces cerevisiae*) and bacteria (e.g., *Zymomonasmobilis*) under anaerobic conditions.

$$C_6H_{12}O_6 \rightarrow 2C_2H_5OH + 2\ CO_2 + \text{heat}$$

The production expenses of bioethanol are more than petrochemically derived fossil fuels (Niven, 2005; Banerjee et al., 2010). There is always a search for exploring cheap and sustainable renewable carbohydrate sources worldwide for bioethanol production (Mohanty et al., 2009). These feedstocks are mostly sugar and starch-rich biomass, lignocellulosic biomass, and algae which are categorized under first, second, and third generation respectively. The conversion of this different biomass into bioethanol differs significantly by sugar obtainment.

10.5 FIRST GENERATION BIOETHANOL

First generation biofuel mostly comprises sugar-rich crops such as sugar cane, sugar beet, sweet sorghum, whey, fruits, and starch-rich crops such as corn, wheat, rice, potato, cassava, sweet potato, and barley. Sugarcane and corn are the most potent feedstocks used in Brazil and the United States, respectively, for bioethanol production (Lopes et al., 2016). Sugarcane and corn use C4 mode of photosynthesis that enables them to efficiently fix more atmospheric CO_2 and water into polysaccharides such as starch, cellulose, and hemicellulose (Zabed et al., 2017b; Lopes et al., 2016). Burning of sugarcane and corn-derived ethanol releases CO_2 which returns to the atmosphere and reabsorbed by the growing plants through photosynthesis in the following season, thereby completing the cycle. Therefore, sugarcane does not result in a significant release of CO_2 (Goldemberg et al., 2008) and reduces GHG emissions between 40 and 60% in comparison to gasoline (Lopes et al., 2016).

Sugar derived ethanol is generally called as Brazilian fuel as sugarcane is mostly used as a substrate for bioethanol in Brazil which accounts for 79% of the bioethanol production and this leads it to the second largest ethanol producer in the world from sugarcane followed by US corn ethanol production. The sugarcane juice contains organic nutrients and fermentable sugars that have rendered it as an appropriate feedstock for bioethanol production (Zabed et al., 2017b).

Sugar-rich crops such as sugarcane, sugar beet containing sucrose, glucose, and fructose are the major fermentable sugars which can be extracted by milling, grinding or crushing then followed by fermentation to yield ethanol (Bai et al., 2008). This contributes to the feasibility of the bioethanol production process from sugarcane in comparison to corn, which requires saccharification before fermentation.

Grains such as corn, wheat, and barley belong to starchy crops. Due to their worldwide availability, easy conversion, long-term storage capability, and high ethanol yield, starch-based feedstocks are extensively used for bioethanol production. Corn is the dominant starchy crop grown globally for bioethanol production. The United States is the leading producer of corn ethanol with the production of 14.3 billion gallons in 2014, and about 825 million gallons is exported to 51 countries worldwide (Global Going, 2015). This is further targeted to reach 36 billion gallons by 2022, using corn and corn Stover (Schnepf et al., 2010). Being the largest producer United States has successfully commercialized corn ethanol production (Dien et al., 2002). Bioethanol production from starch-based crop requires more energy requirement as starch cannot be directly fermented into ethanol, so it demands the use of hydrolytic enzymes for saccharification of complex sugars into simple readily fermentable sugars (Lin and Tanaka, 2006). In India, molasses are the main feedstock for bioethanol production with 330 distilleries that contribute over 4.0 billion litters annually (Raju et al., 2012).

To meet the current consumption demand of one trillion gallons of fossil fuel, corn, and sugarcane-derived ethanol is not sufficient. Also for their edible value, these are not preferred for sustainable bioethanol production as it raises debate over its food security, land issues (Bell and Attfield, 2009; Singh et al., 2016) and water intake (Podkuiko et al., 2014). So there is a necessity to derive ethanol from inedible feedstocks such as agro-residue, municipal waste, lignocellulosic, and algal biomass which does not compete with human consumable food sources (Lim and Kee, 2012). Ethanol production depends on the composition of the substrate and involves three steps: (1) to obtain fermentable sugars, (2) to ferment the sugars into ethanol, and (3) purification of ethanol and generation of co-products for animal feed, fertilizer, and other by-products.

10.5.1 BIOETHANOL PRODUCTION FROM FIRST GENERATION FEEDSTOCK

Sugar and starch containing substrates are used for the production of first-generation bioethanol. Sugar based feedstock mostly comprises of traditional crops such as sugarcane, sugar beet, sweet sorghum; sugar refinery wastes such as canes and molasses; and fruits such as watermelon and dates. Starch-based feedstock includes cereal crops such as corn, barley, and wheat; tuber crops such as potato, cassava, and Jerusalem artichoke (Nwufo, 2016).

Generally the process of bioethanol production from sugar crops (Figure 10.1) starts with the extraction of juice by crushing or milling the stalks or tuber in a specialized roller followed by addition of a source of lime, preferably lime milk (calcium hydroxide $Ca(OH)_2$) or calcium saccharate for fiber and sludge precipitation. This evades the requirement of hydrolysis. Sedimentation removes the precipitate and then filtration is done to remove the filter cake from the juice. The sugary filtrate is then evaporated and concentrated to sugar syrup termed as blackstrap molasses (BSM), which is further supplemented with nitrogen sources such as ammonium sulfate. This was also subjected to sterilization followed by the pH optimization and adjustment of sugar levels to a tolerable range (14–18%) for the subsequent addition of fermenting yeast, *Saccharomyces cerevisiae*, at an ambient growth temperature (33–35°C) and desired cell density (8–17% (v/v)). The leftover solid part such as bagasse, filter cake, and liquid part - vinasse serves several functions. The bagasse can be used for lignocellulosic bioethanol production and also for the generation of electricity. The vinasse and filter cake can be used as fertilizer in the agricultural field (Zabed et al., 2017b; Vohra et al., 2014).

The process of bioethanol production from starch-based feedstock (Figure 10.1) demands a saccharification step using amylolytic enzymes for converting large starch molecules into simpler monomers usually called as glucose syrup that is fermented, distilled, and dehydrated into anhydrous ethanol. This makes sugar ethanol cheaper than starch ethanol. However, the contribution of sugar-rich feedstock such as sugarcane and sugar beet in bioethanol production is only 40%, whereas that of starch-based ethanol is 60% (Mussatto et al., 2010). Two steps are involved in the refinement of starch into sugars-wet milling and dry grinding (Bothast and Schlicher, 2005).

In the dry grinding process, the ground grains and tubers are mixed with water to form a slurry then subsequently cooked and liquefied with thermostable α-amylase for starch conversion into dextrin at 110–115°C for an hour. The cooled liquefied slurry is then mixed with glucoamylase enzymes, which leads to the breakdown of dextrin into glucose and subsequently fermented with yeast into ethanol and carbon dioxide. The fermentation liquid or beer formed is fed to the distillation column, and purified ethanol is separated from the whole stillage. The stillage with remaining protein, oil, and fiber are processed into high protein-rich feed for animals called distillers dried grain with solubles (DDGS) after drying it to 27% (Zabed et al., 2017a; Vohra et al., 2014; Taherzadeh, 2013).

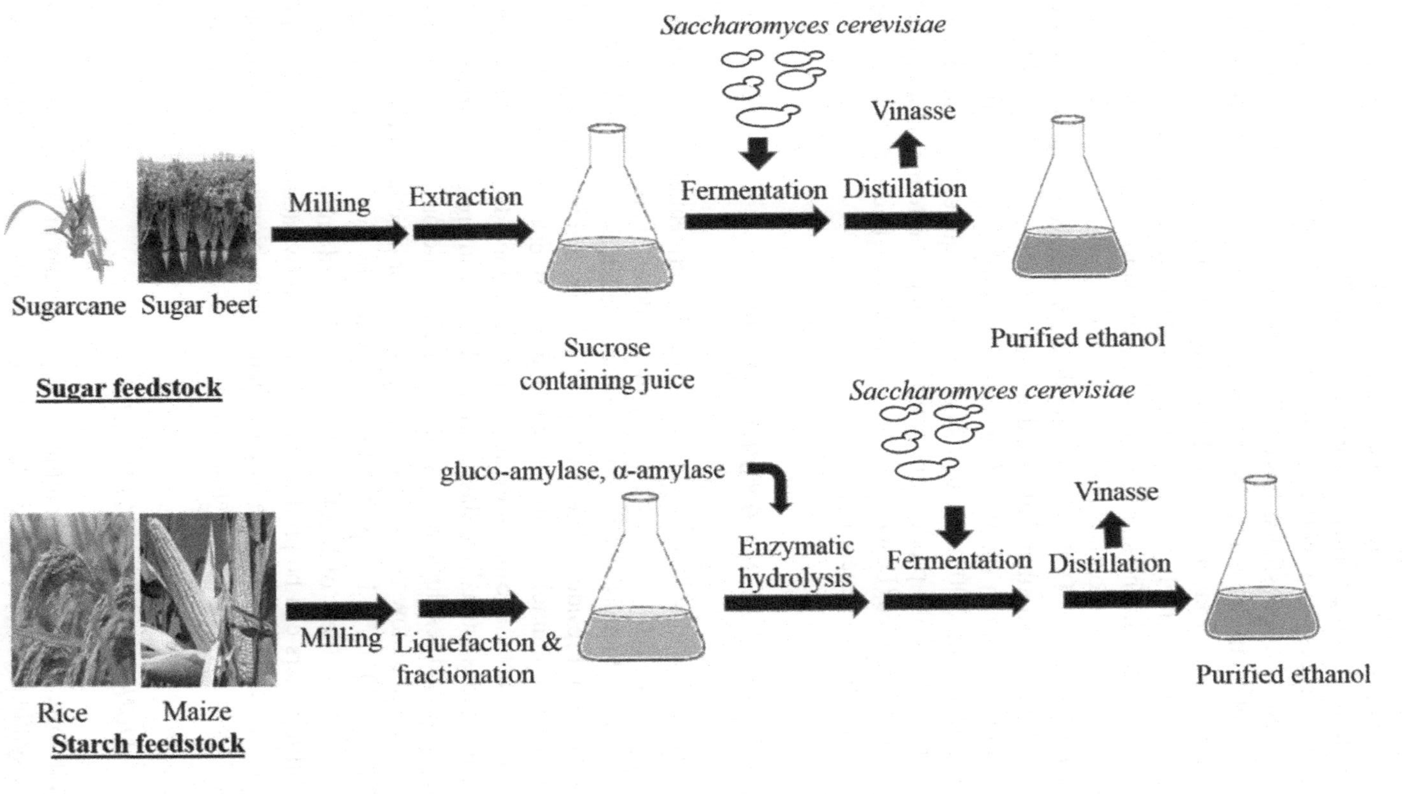

FIGURE 10.1 (See color insert.) Bioethanol production from first generation feedstock. (*Source*: Zabed et al., 2017b.)

Wet milling requires huge capital investment but produces high valued co-products such as high-fructose corn syrup (HFCS), dextrose, and glucose syrup. In the process, the shelled grains are first cleaned from an unwanted material such as cobs, sticks, husks, and stones. The cleaned grains are soaked in dilute sulfuric acid for 24–48 hours in steep tanks at 52°C. Steeping separates the protein layer around the starch by softening the kernel. The grain is then ground finely that break the kernel and removes the germ from the gluten and starch. The starch is subsequently subjected to liquefaction, enzymatic saccharification for breaking it into fermentable sugars which are fermented into ethanol. Both milling processes produce ethanol, but the dry grinding method is commercially used in ethanol industries as it produces more ethanol from corn (0.395 L/Kg) than wet milling process (0.372 L/kg) (Shapouri et al., 2002). Further dry grinding requires less equipment and investment with the production of another major protein product called dry grain with solubles (DDGS), making it more attractive and widely used method of bioethanol generation.

10.5.2 SECOND GENERATION BIOETHANOL

Second generation bioethanol is derived from the fermentation of the most abundant carbon form on earth, the lignocellulosic biomass. It includes several groups of biomass such as forest woody feedstocks – softwood, hardwood; agricultural residues – cereal straws, stovers, and bagasse and forest residues; organic municipal solid wastes and energy crops- poplar, switchgrass, and miscanthus. Since they are available in the adequate amount of 10–50 billion tons annually (Classen et al., 1999) and are mostly left agricultural waste and residues, they do not trigger debate on food/land issues and are sustainable, low-cost renewable feedstock for bioethanol production (Agbor et al., 2011). It is reported to be available in so huge amount that about 442 billion liters of bioethanol can be derived annually and a total of 491 billion liters can be produced annually from wasted crops and crop residue which counts to 16 times higher than current bioethanol production (Kim and Dale, 2004; Gamage et al., 2010).

10.5.2.1 LIGNOCELLULOSE STRUCTURE

Lignocellulose, also called photomass, constitute about half of the plant biomass synthesized by photosynthesis. It is the most available feedstock for biofuel. It is a complex composite material consisting of polysaccharides–cellulose and hemicellulose which are covalently cross-linked to the aromatic polymer-lignin in the plant cell wall forming a lignocellulosic matrix (Perez et al., 2002; Sanchez, 2009) This makes lignocellulose highly lignified, recalcitrant crystalline material that needs intensive energy and cost input during processing (Himmel et al., 2007).

Cellulose, hemicellulose, and lignin are available in varying proportion of 40–60%, 20–40%, and 10–25% (Mckendry, 2002; Kang et al., 2014) respectively, in the primary cell wall of plants. However, the composition varies depending on the substrate and harvesting season (Tutt et al., 2013). Cellulose and hemicellulose constituted the main substrate for bioethanol production and based on its hydrolysis into soluble sugars, the capacity of different lignocellulosic biomass for bioethanol production is determined.

Cellulose and hemicellulose are macromolecular polysaccharides synthesized from different sugar molecules and represent two-thirds of the biomass dry weight (Gírio et al., 2010) whereas lignin is rich in aromatic subunits synthesized from phenylalanine via phenylpropanoid pathway. Cellulose is an organic homopolymer with the structural formula ($C_6H_{10}O_{5n}$ and an average molecular weight of 100,000 Da (Saxena et al., 2009). It comprises linear strands of hundreds to thousands of D-glucopyranose units joined via β-1,4– linkage forming microfibrils that are linked by inter and intramolecular van der Waal forces and hydrogen bonds. The cross-linking among cellulose molecules renders the cellulose crystalline, compact, rigid, and resistant to biological attack and depolymerization (Rocha–Meneses et al., 2017; Mielenz, 2001). The cellulose content in cotton fiber can reach 95–97% and are the purest sources of cellulose (Chen, 2014) whereas less than 50% of the cellulose is present in softwood and hardwood (Kang et al., 2014).

Cellulose is usually wrapped with a low molecular weight 30,000 Da heteropolymer, called hemicellulose $(C_5H_8O_4)_n$ that remains connected with it forming a cellulose-hemicellulose complex. This covering around cellulose hinders the accessibility of enzymes, the digestion of cellulose, the liberation of fermentable sugars and the rate of hydrolysis (Harmsen et al., 2010; Zabed et al., 2016). It has a heterogeneous structure and is made up of different hexoses such as β–D-glucose, α–D-galactose and β–D-mannose; pentoses such as (β–D-xylose and α– L-arabinose) and also acetylated sugars such as α–D–glucuronic, α–D–galacturonic andα–D–4–O–methylgalacturonic acid (Agbor et al., 2011; Limayem and Ricke, 2012; Yu et al., 2017). The backbone frame of hemicellulose is principally composed of xylan including L- arabinose (10%) and D-xylose (90%) which are linked together through β–1,4 linkages. In comparison to recalcitrant cellulose, hemicellulose is less resistant and can be easily hydrolyzed with acid or base pretreatment (Gamage et al., 2010). Xylan and glucomannan are the two most commonly known hemicelluloses. Hardwood, forest waste, agricultural residues, municipal, and industrial debris are rich in xylan while softwoods mostly consist of glucomannan.

Lignin is the third most abundant polymer available. It is a rigid biopolymer with a molecular weight of 10000 Da attached covalently to the xylan component of the hemicellulose and confers rigidity and compactness to the plant cell wall thereby protecting against microbial attack and oxidative stress (Hendriks and Zeeman, 2009). It functions as a cementing material that holds the cellulose and hemicellulose together. It confers to the recalcitrance of lignocellulosic biomass and is a major deterrent to enzymatic saccharification and biomass degradation into fermentable sugars (Gamage et al., 2010; Agbor et al., 2011). This rigid phenylpropanoid derivative consists of three phenolic monomers- sinapyl, coumaryl, and coniferyl alcohol. Woods contain a greater amount of lignin with major quantity available in softwoods (30–60%) and hardwoods (30–55%) whereas the low amount of lignin is present in grasses (10–30%) and agricultural residues (3–15%) (Zabed et al., 2016; Limayem and Ricke, 2012).

10.5.2.2 BIOETHANOL PRODUCTION FROM SECOND-GENERATION FEEDSTOCK

Second generation feedstock consists of the lignocellulosic biomass of which lignin is tightly bound with hemicellulose and cellulose to form a recalcitrant lignocellulosic matrix and as such lignin resists the saccharification and fermentation of cellulose and hemicellulose. This leads to more difficulty in the conversion of lignocellulosic material into bioethanol than that of sugar or starch-based feedstock (Ruane et al., 2010). Generation of lignocellulosic biomass into bioethanol requires delignification and conversion of lignocellulose into fractions of lignin, cellulose, and hemicellulose by suitable pretreatment method. Once pretreatment is done, the biomass containing liberated cellulose and hemicellulose are saccharified enzymatically or chemically into simple monomers which are finally microbially fermented into ethanol (Figure 10.2). The ethanol obtained is purified by distillation to yield fuel grade ethanol. Therefore, the basic biochemical platform involves pretreatment, saccharification, fermentation, and product recovery (Zabed et al., 2016).

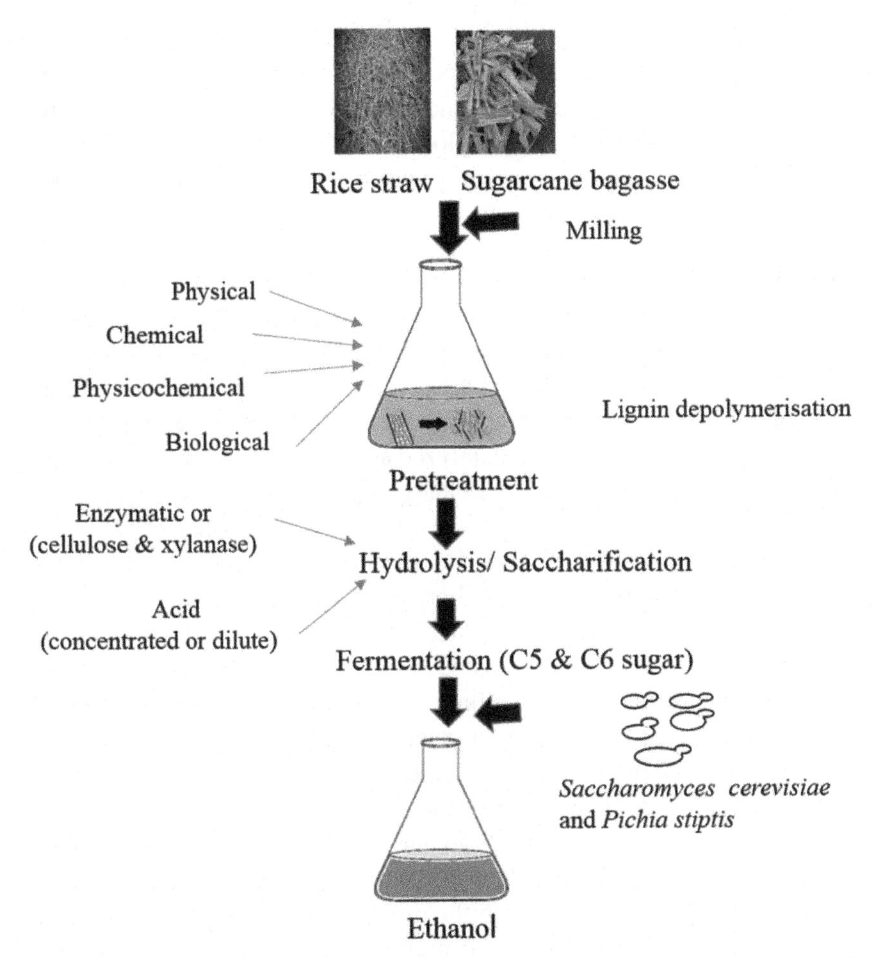

FIGURE 10.2 (See color insert.) Bioethanol production from the second-generation feedstock. (*Source*: Zabed et al., 2017b.)

10.5.2.2.1 Biomass Pretreatment Process

The pretreatment is the first step of lignocellulosic bioethanol production, which basically involves cellulose unwinding from hemicellulose and lignin. It causes changes in the macroscopic, microscopic, and submicroscopic structures of lignocellulosic biomass. It involves delignification, depolymerization of hemicellulose, exposure of the cellulose microfibrils for enzymatic or acid saccharification, a decrease of cellulose crystallinity, increase in surface area and porosity of the material (Wyman et al., 2005). As a result, the cellulose becomes amenable to cellulolytic enzymes and hydrolyzing agents, thus making the process cost-effective (Jacobsen and Wyman, 2000). The process of lignocellulosic biomass pretreatment may release some amount of cellulose and hydrolyzes the hemicellulose into monomeric fractions such as D-xylose and D–arabinose (Bhatia et al., 2012). Pretreatment can be done in a variety of ways-chemical, physical, biological, or combination of them.

1. **Physical Pretreatment:** It includes eco-friendly pretreatment such as grinding, milling that do not involve any harmful chemical residues and the generation of inhibitors. Several types of mechanical grinding such as ball milling, wet disk milling, colloid milling, and hammer milling are used for the pretreatment. Apart from grinding, thermolysis, uncatalyzed steam explosion, liquid hot water pretreatment, different high-energy irradiations such as gamma rays, electron beam, and microwave are also used (Zheng et al., 2009). Physical pretreatment aims to diminish the particle size, reduce cellulose polymerization and crystallinity, lignin depolymerization, and to increase the surface area. But it has certain drawbacks as it requires energy input, possess safety concerns, releases inhibitory compounds in some fermentation experiments, expensive, and is not feasible for full-scale commercial production (Huang et al., 2013).
2. **Chemical Pretreatment:** Lignocellulosic biomass can be dignified and hydrolyzed using chemicals as a catalyst such as acid and alkali. Some organic acids (such as fumaric and maleic acids) and inorganic acids (such as H_2SO_4, HCl, H_3PO_4, and HNO_3) are used as a catalyst. Such acids are used in the diluted or concentrated form. In the process of concentrated acid pretreatment, the concentration range of between 70–77% is used at 40–100°C for yielding monomeric sugars (Zabed et al., 2017b; Aditya et al., 2016) with small amount of inhibitors produced such as furfural from xylose and hydroxylmethyl furfural (HMF) from glucose and acetic acid that inhibits the growth of microorganisms and are to be removed from the residues before further biochemical reactions (Kurian et al., 2013). Concentrated acid causes corrosion of the equipment and must be recovered and hence are not preferred (Refaat, 2012; Laureano-Perez et al., 2005). Dilute acid pretreatment is considered to be most cost-effective for solubilizing hemicellulosic part of the biomass through its degradation and sufficiently increasing the enzyme accessibility to cellulose (Sims et al., 2010). It results in the same outcome as concentrated acid pretreatment without any requirement

of acid recovery. A concentration of below 4% is used in dilute acid pretreatment, and the biomass can be pretreated at high temperature (180°C) for short duration or at low temperature (120°C) for a long duration (30–90 min).

While acid pretreatment degrades hemicellulose fraction of the lignocellulosic biomass, the alkaline pretreatment causes structural lignin alteration by breaking ester and glycosidic linkages. It also enhances the internal surface area by swelling of cellulose fibers and partially decrystallizing cellulose increases cellulose digestion and porosity by saponification of the ester bonds of crosslinked xylan hemicelluloses (Cheng et al., 2010; Ibrahim et al., 2011; McIntosh and Vancov, 2010). NaOH, $Ca(OH)_2$ and NH_4OH are the most commonly used alkaline solutions for lignocellulosic biomass pretreatment. The less severe alkali pretreatment is generally conducted at room temperature with reaction duration from seconds to days.

Ionic liquids (ILs) have also been used for pretreatment of lignocellulosic biomass. These are large organic cations and small inorganic anionic salts that tend to remain in a liquid state at room temperature and over a wide temperature range of less than 100°C. They have the extensive potential of breaking intermolecular hydrogen bonds in polysaccharides and dissolving cellulose. ILs are called as green solvents as they don't form any toxic or explosive gases. 1-allyl-3-methylimidazolium chloride and 1-butyl3-methyl-imidazolium chloride are the most commonly used IL solvents (Vitz et al., 2009). Organosolv is another popular pretreatment derived from a paper industry where lignin and hemicelluloses are solubilized and extracted using a mixture of aqueous organic solvents such as methanol, ethanol, acetone, and ethylene glycol with an inorganic acid such as (Zabed et al., 2017b).

3. **Physicochemical Pretreatment:** As the name suggests employing a combination of both physical and chemical pretreatment. It includes different approaches such as a steam explosion, ammonia fiber explosion (AFEX), thermo hydrolysis also known as liquid hot water, ammonia recycling percolation (ARP), soaking aqueous ammonia (SAA), wet oxidation, CO_2 explosion, microwave pretreatment, ultrasound pretreatment and so on. This different physicochemical pretreatment exerts their effect on biomass by enhancing the surface area, removing hemicellulose and lignin and decreasing cellulose crystallinity (Zabed et al., 2017b).

4 **Biological Pretreatment:** Biological methods employ microbes such as white rot, brown rot, soft rot fungi and their enzymes for degradation of lignin and hemicellulose fraction of lignocellulosic biomass with low energy input, no waste generation, less inhibitors production and in an eco-friendly manner (Keller et al., 2003; Kumar et al., 2009; Shi et al., 2009). Brown rot fungi act on and degrade cellulose, whereas white and soft rot fungi act on and degrade both cellulose and lignin (Prasad et al., 2007). Among these, white rot fungi such as *Phanerochaetechrysosporium, Ceriporiopsissubvermispora, Phlebiasubseries,* and *Pleurotusstreams* have been most efficiently used for the delignification of lignocellulosic biomass (Keller et al., 2003). White rot fungus has lipolytic enzymes such as lignin peroxidases, polyphenoloxidases, manganese-dependent peroxidases, and laccases that cause oxidation of the lignin and disrupts it (Perie and Gold, 1991). Apart from these certain bacteria such as *Bacillus macerans, Cellulomonascartae, Cellulomonasuda, and Zymomonasmobilis* are also reported to have the delignification property of up to 50% (Singh et al., 2008). The major drawbacks associated with biological pretreatment are long incubation duration as it may require several weeks to months for high lignin degradation, sterile growth conditions, and slow rate of pretreatment.

10.5.2.2.2 Hydrolysis

Post pretreatment the holocellulose part of the pretreated biomass is to be converted into simple monomeric hexoses and pentoses for fermentation into bioethanol through hydrolysis. Hydrolysis is needed because the fermenting organisms can only digest simple sugars derived from complex carbohydrates. There are two methods commonly used- acid hydrolysis and enzymatic hydrolysis. Acid hydrolysis is the most utilized methods for conversion of pretreated biomass into fermentable sugars and can be accomplished in two approaches -dilute acid hydrolysis and concentrated acid hydrolysis.

Hydrolysis of the pretreated biomass with dilute acid is done in two steps so that both cellulose and hemicellulose are hydrolyzed at their optimal conditions as pentose sugars degrade rapidly than hexose sugars. A glucose yield of 50% was achieved with two-stage dilute acid hydrolysis (Hamelinck et al., 2005). The first step, similar to dilute acid pretreatment, is performed under mild conditions at low temperature to digest and depolymerize the hemicellulose as well as to recover pentose sugars. Whereas, the second stage is conducted to recover hexose sugars at a higher temperature and harsher conditions that result in the conversion of cellulose into glucose (Demirbas, 2008). According to NREL the first step of dilute acid hydrolysis is done at 190°C for 3 min and 0.7% sulfuric acid and the second step is conducted at 215°C for 3 minutes and 0.4% sulfuric acid (Lynd et al., 2002).

Concentrated acid hydrolysis can rapidly hydrolyze both hemicellulose and cellulose to 5- carbon sugars and glucose, respectively. The hydrolysis is done at low temperature (40°C) with the acid used in the range of between 41% to 100% and results in 90%yield of glucose (Balat, 2007; Zhang et al., 2007; Hamelinck et al., 2005). Since the acid concentrations are higher, it leads to the production of a few inhibitory compounds and corrosion of the equipment.

Enzymes can convert both hemicellulose and cellulose into fermentable sugars using substrate specific hemicellulase (xylanase) and cellulase under mild environmental conditions with a temperature between 40 to 50°Cand pH4–5 (Das Neves et al., 2007). These enzymes convert cellulose into glucose and hemicellulose into C5 sugars (xylose and arabinose) and C6 sugars (glucose, galactose, and mannose). The advantages of using enzymes over acids are they do not impart a corrosive effect to the reactor (Duffand Murray, 1996), there are less sugar degradation and fewer inhibitors production along with less maintenance cost and higher yields (Bhatia et al., 2012).

Cellulose is naturally degraded by cellulase possessing microorganisms-eubacteria and fungi under aerobic and anaerobic conditions (in the animal rumen). Cellulase producing bacteria belong to the genera *of Clostridium, Cellulomonas, Bacillus, Thermomonospora, Ruminococcus, Bacteroides, Erwinia, Acetovibrio, Microbispora,* and *Streptomyces* (Bisaria, 1998). *Clostridium phytofermentans, Clostridium thermocellum, Clostridium hungatei, and Clostridium papyrosolvens* are anaerobic bacteria producing cellulase with high specific activity (Duffand Murray, 1996; Bisaria, 1998). Fungal species known to produce cellulase are *Sclerotium rolfsii, Phanerochaete chrysosporium* and different species of *Trichoderma, Aspergillus, Schizophyllum* and *Penicillium* (Duffand Murray, 1996) of which cellulases from *Trichoderma ressei* and β-D-glucosidase from *Aspergillus niger* (Kaur et al., 2007) are produced for the commercial purpose.

A cellulase is a group of three major enzyme components which acts synergistically in the complete hydrolysis of cellulose. Typically it is a multienzyme complex consisting of endoglucanase that creates and expose free reducing and nonreducing ends by cleaving β–1, four glycosidic linkages in D-glucan chain, exoglucanase degrades the cellulose structure to liberate cellobiose units from these reducing and nonreducing ends and beta-glucosidases that breaks cellobiose into glucose (Lee, 1997).

Due to the heterogeneous chemical structure of hemicellulose, its degradation into monomers, dimers or oligomers requires multiple enzyme systems called xylanase or hemicellulase system. It includes endo–xylanase, exo–xylanase, ß–xylosidase, α–arabinofuranosidase, α–glucuronidase, acetyl xylan esterase, and feruloyl esterase (Saha, 2004). Xylan is present in an abundant amount and contributes to 70% of the total composition of hemicellulose. Endo and exoxylanase are the most important enzymes that hydrolyze the xylans into smaller chains. While ß–xylosidase cleaves from the nonreducing end of short xylooligosaccharides and xylobiose to yield xylose monomers, other debranching enzymes such as α-arabinofuranosidase and α-glucuronidase attack the xylan backbone to release arabinose and 4–o–methyl glucuronic acid respectively (Saha, 2000). Acetyl xylan esterase hydrolyzes xylan by breaking the ester bond between acetyl and xylose residues in xylans. Feruloyl esterase breaks the ester linkage between ferulic acid and arabinose (Saha, 2003; Howard et al., 2003).

10.5.2.2.3 Fermentation

The hexoses such as glucose, mannose, galactose, and pentoses such as xylose and arabinose obtained from hydrolysis are converted into ethanol by the metabolic activity of the fermenting microorganisms (yeast or bacteria) in a process called fermentation (Katahira et al., 2006). The overall reaction occurs to convert sugar into ethanol and carbon dioxide.

$$C_6H_{12}O_6 \rightarrow 2\ C_2H_5OH + 2\ CO_2$$

The theoretical ethanol yield is 0.51 Kg of bioethanol from1 Kg of glucose. However, practically some part of the glucose is used by the micro-organism for its development, and so the actual yield is always lesser than 100% (Demirbas, 2005). About 95% of sugars get converted into ethanol and carbon dioxide with 1% utilized for cellular growth and rest 4% into soluble products such as glycerol. An ideal fermenting organism should have high ethanol yield capacity, wide substrate utilization range, resistant to inhibitors liberated during hydrolysis and fermentation, tolerable to high sugar and ethanol concentration, low pH and high temperature (Banerjee et al., 2010). But practically all these features are seldom found together in one specific organism, and so there is a need to develop an organism with all these features for the industrial purpose (Sanchez and Cardona, 2008).

Saccharomyces cerevisiae and *Zymomonas mobilis* are the most utilized organism for industrial bioethanol production that can easily ferment hexose sugars (glucose, galactose & mannose) into ethanol but these are unable to ferment pentose sugars (xylose and arabinose) (Boulton et al., 1999). *Pichia stiptis*, *Candida sheathe,* and *Pachysolen tannophilus* are fermenting yeast strains capable of fermenting pentose sugars, but they are less efficient as they are sensitive to inhibitors, intolerable to high ethanol and low pH (Hahn-Hägerdal et al., 2007). There are different methods of conversion of lignocellulosic hydrolysates into bioethanol.

1. **Simultaneous Saccharification and Fermentation (SSF):** It carries out the combined enzymatic saccharification of the complex polysaccharides into free monomers and immediately their fermentation into ethanol by yeast in the same reaction reactor (Ikwebe and Harvey, 2011). The monosaccharides formed are fermented into ethanol without any separation process in between which keeps the glucose concentration low and avoids the excessive use of a β-glucosidase enzyme, thereby making the process cost saving. With SSF there is a diminished accumulation of inhibitors, increased hydrolysis rates, a higher yield of sugars and the ethanol in the broth dissuades the growth of undesirable microorganisms which makes the process much preferable approach than SHF. But the different pH and temperature requirements of enzymes used for hydrolysis and fermenting microbes possess difficulties in process optimization limitations (Gupta and, 2015; Dien et al., 2003; Chan et al., 2007; Aditya et al., 2016).
2. **Separate Hydrolysis and Fermentation (SHF):** In SHF, the hydrolysis and fermentation are carried out differentially with each process properly optimized. In the first reactor, the hydrolysate is fed for glucose fermentation. The ethanol formed is distilled later with the remaining hydrolysate fed in the second reactor for fermentation of the xylose. However, after hydrolysis, it leads to the generation of enzyme-inhibiting end products such as cellobiose and glucose which requires the addition of expensive β-glucosidase that overall slows down the hydrolysis rate and therefore lower ethanol yield (Ezhumalai and Thangavelu, 2010; Bhatia et al., 2012; Aditya et al., 2016). In addition to the above two common methods, some more integrated techniques are used.
3. **Simultaneous Saccharification and Co-Fermentation (SSCF):** It involves complete assimilation of whole sugars released after

lignocellulose pretreatment and hydrolysis by mixed cultures of different hexose and pentose fermenting microbes. As such, the microorganisms begin the fermentation process without any requirement of sugar separation and sterilization. Studies of bioethanol production using SSCF with corn stover, municipal solid waste, and sugarcane bagasse has proven to be efficient (Kleerebezem and van Loosdrecht, 2007; Thanakoses et al., 2003; Chan and Holtzapple, 2003; Fu and Holtzapple, 2010; Lynd et al., 2005; Vohra et al., 2014).

4. **Consolidated Bioprocessing (CBP):** It involves all the bioconversion steps from cellulase production to cellulose/hemicellulose saccharification to fermentation of hexose and pentose sugar into a single step in a single bioreactor with one or two microorganisms or one microbial community and hence also called as Direct Microbial Conversion (DMC). It is a cost saving technique as no capital investment is required for enzyme production, less energy input, fewer steps involved and better efficiency (Carere et al., 2008; Aditya et al., 2016). Several fungi such as *Trichoderma reesei*, *Fusarium oxysporium*, *Neurospora crassa*, and *Paecilomyces*s and bacteria such as *Clostridium thermocellum* are compatible for consolidated bioprocessing (CBP) (Sarkar et al., 2012).

10.5.3 THIRD GENERATION BIOFUEL

Algal feedstock microalgae and macroalgae are gaining global attention as a substitute for fossil fuel derived fuels and are categorized under third generation biofuel. First generation feedstock, when used for bioethanol production, raises questions of food, land, and water. These drawbacks are to some extent resolved by using waste and non-edible second-generation feedstock. However, the sophisticated and expensive technologies used for processing, pretreatment, and bioethanol production from second-generation bioethanol makes it challenging and not feasible for commercial production (Nigam and Singh, 2011; Alam et al., 2014; John et al., 2011; Daroch et al., 2013).

Algae serve as promising and best feedstock for bioethanol production because of higher yields and confer several advantages as a feedstock than other bioethanol feedstocks. They have a high photosynthetic efficiency, which enables them to tolerate high CO_2 and therefore fixes 5–7 times more CO_2 than terrestrial plants (Kim et al., 2015). Due to high CO_2 fixation capacity, they produce 5–10 times more biomass than woody terrestrial plants (Subhadra and Edwards, 2010; Packer, 2009). They grow ubiquitously in nature even in environmental conditions such as saline, industrial, and municipal wastewater that are unfavorable for growth of other crops (Ahmad et al., 2011; Surendhiran and Vijay, 2012; Xue et al., 2010; Georgianna and Mayfield, 2012) and hence reduces fresh water demand (Dutta et al., 2014). They do not alter the human food chain and are buoyant, which avoids the necessity of biopolymer supporting matrix like lignin and hemicellulose as in land plants. This results in enhanced saccharification efficiency and fermentation yield, making the process of bioethanol production simpler and easier by evading the requirement of enzymatic hydrolysis and pretreatment (John et al., 2011). This overall reduce the cost of bioethanol production (Sun and Cheng, 2002).

10.5.3.1 BIOETHANOL PRODUCTION FROM ALGAE

The bioethanol production from algae depends on the fermentation of different types of algal polysaccharides such as starch, cellulose, and sugars. The cell-bound carbohydrates are hydrolyzed into free sugars under optimized conditions of pretreatment and hydrolysis, which on fermentation are converted to ethanol. Microalgae such as *Chlorella, Dunaliella, Chlamydomonas, Scenedesmus, Spirulina* accumulate more than 50% of the dry weight of polysaccharides such as starch and glycogen (Chen et al., 2010). Certain species of microalgae can increase their carbohydrate content to 70% under specific conditions (Branyikova et al., 2011) as reserved food. Therefore microalgae can be ideally used for bioethanol production which can estimate to be around 5000–15,000 gal of ethanol/acre/year. The yield is much higher than the yield from different feedstocks (Nguyen, 2012).

The ethanol production from microalgal starch (Figure 10.3) is similar to the process of ethanol production from starch-based feedstock mostly consisting of two processes-saccharification and fermentation (Rubin, 2008; Matsumoto et al., 2003). Saccharification using acid or enzymes (α-amylase or glucoamylase) hydrolyzes starch into simple sugars which are fermented into ethanol using suitable yeast strain (John et al., 2011). Besides starch, several microalgae contain cellulose in their inner cell wall, which can be saccharified using cellulase to release simple sugars and fermented to yield bioethanol.

Similarly, carbohydrate content in macroalgae ranges from 25–50%, 30–60% and 30–50% in green algae, red algae, and brown algae respectively. Structurally macroalgae consist of cellulose, hemicellulose, and other polymers such as alginate, mannitol, ulvan, carrageenan, glucan, and laminarin.

Conversion of alginate and mannitol into free sugars requires pretreatment and saccharification. Laminarin can be hydrolyzed into glucose using β-1,3-glucanases orlaminarinases (Ozçimen and Inan, 2015; Behera et al., 2015).

Pretreated and saccharified macroalgal biomass (Figure 10.3) can be converted into ethanol with a different group of microorganisms such as yeast, fungi, and bacteria. *Saccharomyces cerevisiae* and *Zymomonas mobilis* are the most potent bioethanol fermenting microorganisms, but due to narrow substrate range, these organisms cannot ferment mannitol. *Zymobacter palmae* can yield ethanol from mannitol containing seaweeds under aerobic conditions. The yeast *Pichia angophorae* can ferment both substrates mannitol and laminarin for ethanol production (Horn et al., 2000; Nguyen, 2012).

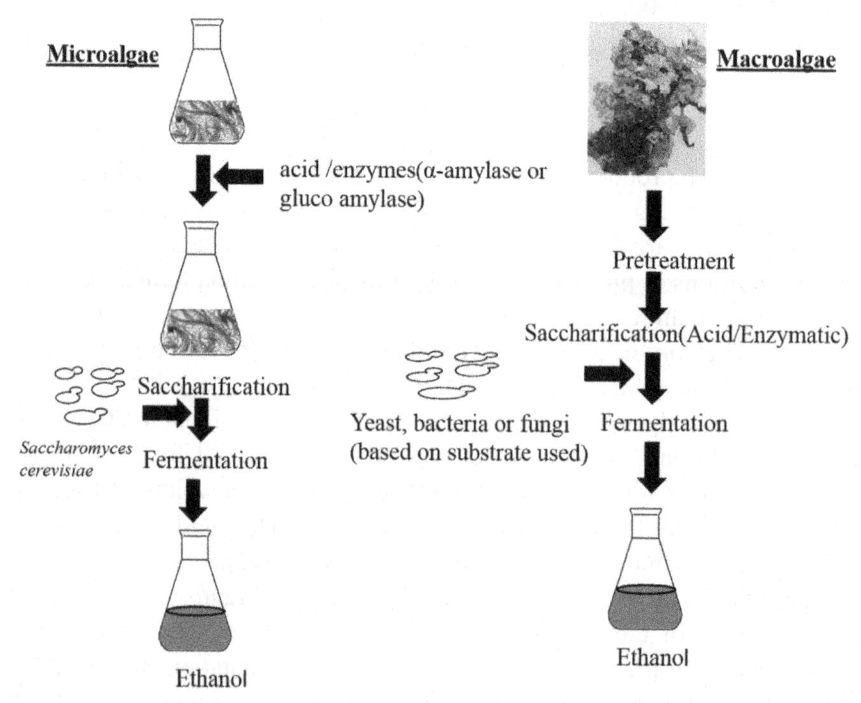

FIGURE 10.3 (See color insert.) Bioethanol production from third generation feedstock. (*Source*: Liu et al., 2014)

10.6 GROWING RISK OF WATER AND BIOFUEL- WATER SECURITY AND ENERGY SECURITY

Water plays an imperative role in our day-to-day life in agriculture, industrial sectors, and domestic purposes. Depletion of fresh water at an alarming rate due to overpopulation and climate change has sensitized the global researchers and environmentalists for its conservation and efficient use. With increasing demographics, growing energy requirements, industrialization, and urbanization, the world is on the edge of an unprecedented water shortage that could trigger the next great global crisis leading to third world war.

According to the United Nations World Water Development Report (2016), of the total fresh water available worldwide, the global industries including energy-based consume about 19% to 23% of the fresh water which is quite huge and impart strains on freshwater availability. In Indian scenario according to Central Pollution Control Board (CPCB) about 6% of the total available fresh water is used for industrial purpose which is equivalent to consumption of about 500 billion cubic meters fresh water annually and this is going to increase drastically in 2020 and 2050 by 8.5 and 10.1 percent of the total world water available from the current level of 6%.

The National Water Policy, 2012 (NWP, 2012) strongly urged for a freshwater reduction in industries by implementing new economical, sustainable, eco-friendly, and socially sustainable technologies that could result in reduced usage of fresh water. Biomass processing plants particularly lignocelluloses based biorefinery consume massive quantities of fresh water around 1.9–5.9 m^3 water per m^3 (Fang et al., 2015) of biofuel for large-scale operations but its unsustainable in the long run due to fresh water shortage. Keeping given the ecological footprint, a nonpotable water source appears to be promising reaction media. Over the past few years, seawater has received growing attention due to its abundance, around 96.5% on earth and has emerged as an alternative medium for enzymatic and fermentation purpose in biorefineries. Seawater with 97% similarity to that of fresh water and rest 3% as minerals have a salinity around 25–35 practical salinity units (PSU) depending on geographical areas with pH range from 7.6 to 8.5. Several works in recent years have been reported using seawater as the reaction medium, and its application has shown promising outcome (Table 10.1), and therefore this can be further implemented for producing bioethanol from all generation feedstocks (Figure 10.4).

Seawater, as reaction media, has been used in the chemoenzymatic production of HMF from glucose (Grande et al., 2012). Several of the fermentative approaches were conducted using seawater for obtaining enzymes, acids such as L-glutaminase production with *Beauveria* species (Sabu et al., 2000) and *Vibrio costicola* (Prabhu and Chandrasekaran, 1997); chitinase production from prawn waste using marine fungus *Beauveria bassiana* (Suresh and Chandrasekaran, 1998); succinic acid production using *Actinobacillus succinogenes* from wheat hydrolysates (Lin et al., 2011) and itaconic acid production using *Ustilago maydis* (Klement et al., 2012). Chemocatalytic

TABLE 10.1 Utilization of Seawater as a Reaction Medium for Biochemical Conversion

Substrate used	Medium used	Microorganism/ enzyme used	Method	Outcome	References
Mixed substrates consisting of mostly cellulose, starch, lipids, and proteins	Artificial seawater	*Recombinant and wild-type Halomonas campaniensis* strain LS21	Fermentation	PHB production 70% for recombinant and 30% for wild-type	Yue et al., 2014
Amorphous celluloses (Sigmacell-20 and Sigmacell-101) and microcrystalline cellulose (Avicel 150)	Different concentrations of seawater (1X, 2X, 4X)	Commercially available Accellerase-1500 a "cocktail	Enzymatic hydrolysis	Slightly diminished hydrolytic rates (90%) than those observed for control reactions using citrate buffer	Grande and De Maria, 2012
Pretreated cellulose and hydrolyzed hemicellulose part of pretreated beech wood	Seawater	*Ustilago maydis*	Fermentation	Itaconic acid production	Klement et al., 2012
Lignocellulosic biomass-rice straw, sugar cane bagasse, eucalyptus, pine, wheat straw, Corncob	Seawater	Commercially available cellulase	Ionic liquid (IL) pretreatment and enzymatic hydrolysis	Lack of any adverse effects on the efficiency of IL pretreatment and increased activity of commercial enzymes at the seawater pH	Ren et al., 2016
Avicennia marina	Seawater	Sludge	Fermentation	Biomethane production	Almardeai et al., 2016
Date palm Residues	Artificial seawater	NA	Hydrothermal Pretreatment	The lower crystallinity of cellulose in pretreated date palm	Fang et al., 2015
Wheat	Seawater	*Actinobacillus succinogenes*	Fermentation	Succinic acid production	Lin et al., 2011

TABLE 10.1 *(Continued)*

Substrate used	Medium used	Microorganism/ enzyme used	Method	Outcome	References
Corn cob	Seawater	NA	Acetic acid steam and $FeCl_3$ mediated hydrolysis	Furfural production and high delignification efficiency	Mao et al., 2013
Glucose	Seawater	NA	Chemo-enzymatic conversion	HMF production	Grande et al., 2012
Polystyrene beads	Seawater	*Vibrio costicola*	Chemo-enzymatic conversion	L-glutaminase production	Prabhu and Chandrasekaran, 1997
Polystyrene beads	Seawater	*Beauveria* spp.	Fermentation	L-glutaminase production	Sabu et al., 2000
Prawn waste	Seawater	*Beauveria bassiana* BTMF S10	Fermentation	Chitinase production	Suresh and Chandrasekaran, 1998

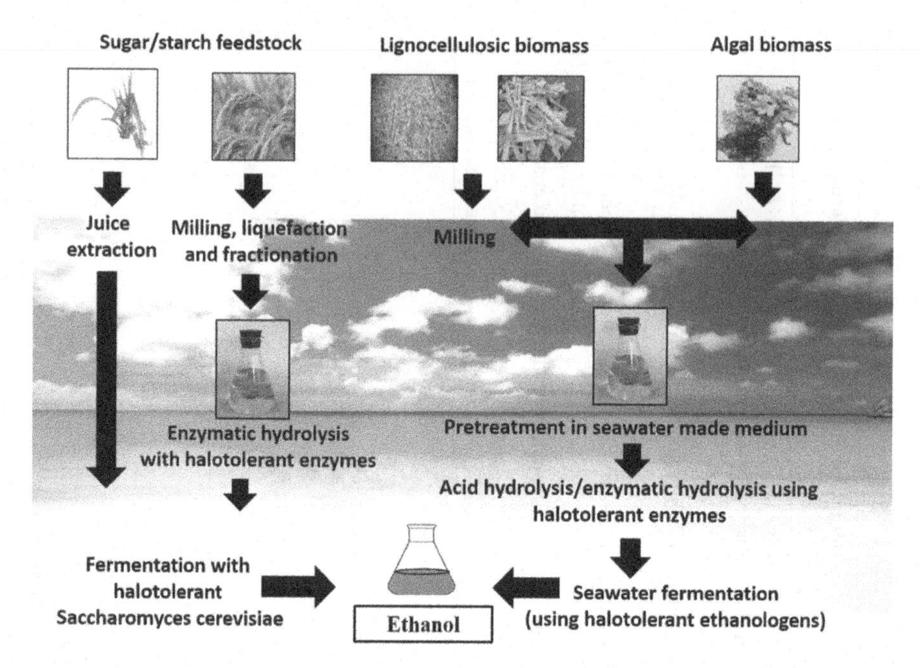

FIGURE 10.4 (See color insert.) Bioethanol production from 1st generation, 2nd generation, and 3rd generation feedstock using seawater as the reaction medium. (*Source*: Ren et al., 2016)

operations were also performed in concentrated seawater using malic acid for the depolymerization of cellulose (vom Stein et al., 2010) and furfural formation from corncob residues in seawater using $FeCl_3$ as catalyst (Mao et al., 2013). Some of the reports on seawater as a reaction medium for ligno-cellulose pretreatment are also available where seawater based hydrothermal pretreatment of the date palm leaflets was performed which resulted in higher total glucan production and significant ethanol yield having a negligible difference in ethanol yield when treated with fresh water. Also, no inhibition of *Saccharomyces cerevisiae* was noticed that might be because of increased tolerance and capable of metabolizing salts (Fang et al., 2015). Salt tolerant marine yeasts can be used to ferment first generation and second-generation substrates in seawater media. Seawater consists of various minerals that support the growth of fermentative ethanologens (Lin et al., 2011). Recently one of the isolated marine strain S. cerevisiae AZ65produced 52.2 g/L bioethanol from sugarcane molasses after 48 hours in 15 L batch fermentations with seawater based medium (Greetham et al., 2018). The enzymatic depolymerization of microcrystalline cellulose in concentrated seawater with commercially available cocktail Accellerase-1500 was reported for the first time with slightly lower hydrolytic rates (90%) than in pure citrate buffer

(Grande and De Maria, 2012). The study by Ren et al., (2016) with seawater as a reaction medium for IL pretreatment of lignocellulosic biomass showed no negative effect on pretreatment efficiency. Additionally, they conducted enzymatic hydrolysis of the IL pretreated biomass, where the commercial cellulolytic enzymes were functionally active in seawater pH (Ren et al., 2016). These results depict the promising aspects of using seawater as a reaction medium that can be feasible for use in biorefineries.

10.7 CONCLUSION

With several environmental and economic benefits, the demands for bioethanol, CO_2 neutral transportation fuel is growing worldwide. The fermentative production of bioethanol counteracts by reducing the dependence on crude oil imports, diminishing greenhouse gas emissions, contributing to the national economy and energy security. Also, the huge consumption of fresh water in bioethanol production puts additional pressure on limited freshwater sources. The use of seawater can curb the exploitation of freshwater in biorefineries.

10.8 ACKNOWLEDGMENT

Binita Dev acknowledges CSIR UGC for the Junior Research Fellowship.

KEYWORDS

- **ammonia fiber explosion**
- **ammonia recycling percolation**
- **blackstrap molasses**
- **distillers dried grain with solubles**
- **dry grain with soluble**
- **ethanol blending program**
- **flexible-fuel vehicles**
- **greenhouse gases**
- **high-fructose corn syrup**
- **hydroxyl-methyl furfural**

REFERENCES

Abdel-Rahman, A. A., & Osman, M. M., (1997). Experimental investigation on varying the compression ratio of SI engine working under different ethanol-gasoline fuel blends. *International Journal of Energy Research, 21*(1), 31–40.

Aditiya, H. B., Mahlia, T. M. I., Chong, W. T., Nur, H., & Sebayang, A. H., (2016). Second generation bioethanol production: A critical review. *Renewable and Sustainable Energy Reviews, 66*, 631–653.

Agbor, V. B., Cicek, N., Sparling, R., Berlin, A., & Levin, D. B., (2011). Biomass pretreatment: Fundamentals toward application. *Biotechnology Advances, 29*(6), 675–685.

Ahmad, A. L., Yasin, N. M., Derek, C. J. C., & Lim, J. K., (2011). Microalgae as a sustainable energy source for biodiesel production: A review. *Renewable and Sustainable Energy Reviews, 15*(1), 584–593.

Alam, F., Mobin, S., & Chowdhury, H., (2015). Third generation biofuel from algae. *Procedia Engineering, 105*, 763–768.

Almardeai, S., Javid, U., BBastidas-Oyanedel, J. R., & Schmidt, J. E., (2016). Seawater biorefinery of Avicennia marina–biomethane potential at seawater conditions. *Int. J. of Sustainable Water & Environmental Systems, 8*(1), 00–00.

Azhar, S. H. M., Abdulla, R., Jambo, S. A., Marbawi, H., Gansau, J. A., Faik, A. A. M., & Rodrigues, K. F., (2017). Yeasts in sustainable bioethanol production: A review. *Biochemistry and Biophysics Reports, 10*, 52–61.

Bai, F. W., Anderson, W. A., & Moo-Young, M., (2008). Ethanol fermentation technologies from sugar and starch feedstocks. *Biotechnology Advances, 26*(1), 89–105.

Balat, M., (2007). An overview of biofuels and policies in the European Union. *Energy Sources, Part B, 2*(2), 167–181.

Balat, M., (2009). Bioethanol as a vehicular fuel: A critical review. *Energy Sources, Part A, 31*(14), 1242–1255.

Balat, M., & Balat, H., (2009). Recent trends in global production and utilization of bio-ethanol fuel. *Applied Energy, 86*(11), 2273–2282.

Ballesteros, I., Negro, M. J., Oliva, J. M., Cabañas, A., Manzanares, P., & Ballesteros, M., (2006). Ethanol production from steam-explosion pretreated wheat straw. In: *Twenty-Seventh Symposium on Biotechnology for Fuels and Chemicals* (pp. 496–508). Humana Press.

Bandyopadhyay, K. R., (2015). *Biofuel Promotion in India for Transport: Exploring the Grey Areas* (p. 12). The Energy and Resources Institute (TERI): New Delhi, India.

Banerjee, S., Mudliar, S., Sen, R., Giri, B., Satpute, D., Chakrabarti, T., & Pandey, R. A., (2010). Commercializing lignocellulosic bioethanol: Technology bottlenecks and possible remedies. *Biofuels, Bioproducts and Biorefining: Innovation for a Sustainable Economy, 4*(1), 77–93.

Bata, R. M., & Roan, V. P., (1989). Effects of ethanol and/or methanol in alcohol-gasoline blends on exhaust emissions. *Journal of Engineering for Gas Turbines and Power, 111*(3), 432–438.

Bayraktar, H., (2005). Experimental and theoretical investigation of using gasoline-ethanol blends in spark-ignition engines. *Renewable Energy, 30*(11), 1733–1747.

Behera, S., Singh, R., Arora, R., Sharma, N. K., Shukla, M., & Kumar, S., (2015). Scope of algae as third generation biofuels. *Frontiers in Bioengineering and Biotechnology, 2*, 90.

Bell, P. J., & Attfield, P. V., (2009). *Breakthrough in Yeast for Making Bio-Ethanol from Lignocellulosics.*

Bhatia, L., Johri, S., & Ahmad, R., (2012). An economic and ecological perspective of ethanol production from renewable agro waste: A review. *Amb. Express*, *2*(1), 65.

Bisaria, V. S., (1998). Bioprocessing of agro-residues to value-added products. In: *Bioconversion of Waste Materials to Industrial Products* (pp. 197–246). Springer, Boston, MA.

Bölük, G., & Mert, M., (2014). Fossil & renewable energy consumption, GHGs (greenhouse gases) and economic growth: Evidence from a panel of EU (European Union) countries. *Energy*, *74*, 439–446.

Bothast, R. J., & Schlicher, M. A., (2005). Biotechnological processes for conversion of corn into ethanol. *Applied Microbiology and Biotechnology*, *67*(1), 19–25.

Boulton, R. B., Singleton, V. L., Bisson, L. F., & Kunkee, R. E., (1999). Yeast and biochemistry of ethanol fermentation. In: *Principles and Practices of Winemaking* (pp. 102–192). Springer, Boston, MA.

Brányiková, I., Maršálková, B., Doucha, J., Brányik, T., Bišová, K., Zachleder, V., & Vítová, M., (2011). Microalgae—novel highly efficient starch producers. *Biotechnology and Bioengineering*, *108*(4), 766–776.

Carere, C. R., Sparling, R., Cicek, N., & Levin, D. B., (2008). Third generation biofuels via direct cellulose fermentation. *International Journal of Molecular Sciences*, *9*(7), 1342–1360.

Chan, E. S., Rudravaram, R., Narasu, M. L., Rao, L. V., & Ravindra, P., (2007). Economics and environmental impact of bioethanol production technologies: An appraisal. *Biotechnology and Molecular Biology Reviews*, *2*(1), 14–32.

Chan, W. N., & Holtzapple, M. T., (2003). Conversion of municipal solid wastes to carboxylic acids by thermophilic fermentation. *Applied Biochemistry and Biotechnology*, *111*(2), 93–112.

Chen, C. Y., Zhao, X. Q., Yen, H. W., Ho, S. H., Cheng, C. L., Lee, D. J., & Chang, J. S., (2013). Microalgae-based carbohydrates for biofuel production. *Biochemical Engineering Journal*, *78*, 1–10.

Chen, H., (2014). Chemical composition and structure of natural lignocellulose. In: *Biotechnology of Lignocellulose* (pp. 25–71). Springer, Dordrecht.

Chen, P., Min, M., Chen, Y., Wang, L., Li, Y., Chen, Q., & Deng, S., (2010). Review of biological and engineering aspects of algae to fuels approach. *International Journal of Agricultural and Biological Engineering*, *2*(4), 1–30.

Cheng, Y. S., Zheng, Y., Yu, C. W., Dooley, T. M., Jenkins, B. M., & VanderGheynst, J. S., (2010). Evaluation of high solids alkaline pretreatment of rice straw. *Applied Biochemistry and Biotechnology*, *162*(6), 1768–1784.

Claassen, P. A. M., Van Lier, J. B., Contreras, A. L., Van Niel, E. W. J., Sijtsma, L., Stams, A. J. M., & Weusthuis, R. A., (1999). Utilization of biomass for the supply of energy carriers. *Applied Microbiology and Biotechnology*, *52*(6), 741–755.

Cuéllar-Franca, R. M., & Azapagic, A., (2015). Carbon capture, storage and utilization technologies: A critical analysis and comparison of their life cycle environmental impacts. *Journal of CO2 Utilization*, *9*, 82–102.

Daroch, M., Geng, S., & Wang, G., (2013). Recent advances in liquid biofuel production from algal feedstocks. *Applied Energy*, *102*, 1371–1381.

Das Neves, M. A., Kimura, T., Shimizu, N., & Nakajima, M., (2007). State of the art and future trends of bioethanol production. *Dynamic Biochemistry, Process Biotechnology and Molecular Biology*, *1*(1), 1–14.

Demirbaş, A., (2005). Bioethanol from cellulosic materials: A renewable motor fuel from biomass. *Energy Sources*, *27*(4), 327–337.

Demirbas, A., (2007). Progress and recent trends in biofuels. *Progress in Energy and Combustion Science*, *33*(1), 1–18.

Demirbas, A., (2008). The importance of bioethanol and biodiesel from biomass. *Energy Sources, Part B*, *3*(2), 177–185.

Demirbas, A., (2009). Biofuels securing the planet's future energy needs. *Energy Conversion and Management*, *50*(9), 2239–2249.

Demirbas, A., (2011). Competitive liquid biofuels from biomass. *Applied Energy*, *88*(1), 17–28.

Dias De Oliveira, M. E., Vaughan, B. E., & Rykiel, E. J., (2005). Ethanol as fuel: Energy, carbon dioxide balances, and ecological footprint. *AIBS Bulletin*, *55*(7), 593–602.

Dien, B. S., Bothast, R. J., Nichols, N. N., & Cotta, M. A., (2002). The US corn ethanol industry: An overview of current technology and future prospects. *International Sugar Journal*, *103*(1241), 204–208.

Dien, B. S., Cotta, M. A., & Jeffries, T. W., (2003). Bacteria engineered for fuel ethanol production: Current status. *Applied Microbiology and Biotechnology*, *63*(3), 258–266.

Dimitriou, I., García-Gutiérrez, P., Elder, R. H., Cuéllar-Franca, R. M., Azapagic, A., & Allen, R. W., (2015). Carbon dioxide utilization for production of transport fuels: Process and economic analysis. *Energy & Environmental Science*, *8*(6), 1775–1789.

Duff, S. J., & Murray, W. D., (1996). Bioconversion of forest products industry waste cellulosic to fuel ethanol: A review. *Bioresource Technology*, *55*(1), 1–33.

Dutta, K., Daverey, A., & Lin, J. G., (2014). Evolution retrospective for alternative fuels: First to fourth generation. *Renewable Energy*, *69*, 114–122.

Ezhumalai, S., & Thangavelu, V., (2010). Kinetic and optimization studies on the bioconversion of lignocellulosic material into ethanol. *Bioresources*, *5*(3), 1879–1894.

Fang, C., Thomsen, M. H., Brudecki, G. P., Cybulska, I., Frankær, C. G., Bastidas-Oyanedel, J. R., & Schmidt, J. E., (2015). Seawater as alternative to freshwater in pretreatment of date palm residues for bioethanol production in coastal and/or arid areas. *Chem. Sus. Chem.*, *8*(22), 3823–3831.

Fu, Z., & Holtzapple, M. T., (2010). Anaerobic mixed-culture fermentation of aqueous ammonia-treated sugarcane bagasse in consolidated bioprocessing. *Biotechnology and Bioengineering*, *106*(2), 216–227.

Gamage, J., Lam, H., & Zhang, Z., (2010). Bioethanol production from lignocellulosic biomass, a review. *Journal of Biobased Materials and Bioenergy*, *4*(1), 3–11.

Georgianna, D. R., & Mayfield, S. P., (2012). Exploiting diversity and synthetic biology for the production of algal biofuels. *Nature*, *488*(7411), 329.

Gírio, F. M., Fonseca, C., Carvalheiro, F., Duarte, L. C., Marques, S., & Bogel-Łukasik, R., (2010). Hemicelluloses for fuel ethanol: A review. *Bioresource Technology*, *101*(13), 4775–4800.

Global, G., (2015). Ethanol industry outlook Washington DC. *The Renewable Fuels Association USA (RFA)*.

Goldemberg, J., (2006). The promise of clean energy. *Energy Policy*, *34*(15), 2185–2190.

Goldemberg, J., (2008). Environmental and ecological dimensions of biofuels. In: *Proceedings of the Conference on the Ecological Dimensions of Biofuels* (Vol. 10). Washington, DC.

Goldemberg, J., Coelho, S. T., & Guardabassi, P., (2008). The sustainability of ethanol production from sugarcane. *Energy Policy*, *36*(6), 2086–2097.

Government of India, (2007). *11th Five-Year Plan: 2007–2012*. Planning Commission, New Delhi. http://planningcommission.nic.inplans/planrel/fiveyr/>11th/11_v2/11th_vol2.pdf (Accessed on 9 June 2019).

Grande, P. M., & De Maria, P. D., (2012). Enzymatic hydrolysis of microcrystalline cellulose in concentrated seawater. *Bioresource Technology*, *104*, 799–802.

Grande, P. M., Bergs, C., & Domínguez de María, P., (2012). Chemo-enzymatic conversion of glucose into 5-hydroxymethylfurfural in seawater. *Chem. Sus. Chem.*, *5*(7), 1203–1206.

Greetham, D., Zaky, A., Makanjuola, O., & Du, C. (2018). A brief review on bioethanol production using marine biomass, marine microorganism and seawater. *Current Opinion in Green and Sustainable Chemistry, 14,* 53–59.

Gupta, A., & Verma, J. P., (2015). Sustainable bio-ethanol production from agro-residues: A review. *Renewable and Sustainable Energy Reviews*, *41*, 550–567.

Hahn-Hägerdal, B., Karhumaa, K., Fonseca, C., Spencer-Martins, I., & Gorwa-Grauslund, M. F., (2007). Towards industrial pentose-fermenting yeast strains. *Applied Microbiology and Biotechnology*, *74*(5), 937–953.

Hamelinck, C. N., Van Hooijdonk, G., & Faaij, A. P., (2005). Ethanol from lignocellulosic biomass: Techno-economic performance in short-, middle-and long-term. *Biomass and Bioenergy*, *28*(4), 384–410.

Harmsen, P. F. H., Huijgen, W., Bermudez, L., & Bakker, R., (2010). *Literature Review of Physical and Chemical Pretreatment Processes for Lignocellulosic Biomass* (No. 1184). Wageningen UR-Food & Biobased Research.

Hendriks, A. T. W. M., & Zeeman, G., (2009). Pretreatments to enhance the digestibility of lignocellulosic biomass. *Bioresource Technology*, *100*(1), 10–18.

Himmel, M. E., Ding, S. Y., Johnson, D. K., Adney, W. S., Nimlos, M. R., Brady, J. W., & Foust, T. D., (2007). Biomass recalcitrance: engineering plants and enzymes for biofuels production. *Science*, *315*(5813), 804–807.

Horn, S. J., Aasen, I. M., & Østgaard, K., (2000). Production of ethanol from mannitol by Zymobacterpalmae. *Journal of Industrial Microbiology and Biotechnology*, *24*(1), 51–57.

Howard, R. L., Abotsi, E. L. J. R., Van Rensburg, E. J., & Howard, S., (2003). Lignocellulose biotechnology: Issues of bioconversion and enzyme production. *African Journal of Biotechnology*, *2*(12), 602–619.

Huang, X. F., Santhanam, N., Badri, D. V., Hunter, W. J., Manter, D. K., Decker, S. R., & Reardon, K. F., (2013). Isolation and characterization of lignin-degrading bacteria from rainforest soils. *Biotechnology and Bioengineering*, *110*(6), 1616–1626.

Ipcc, W. G. I., (2013). Contribution to the IPCC 5th Assessment Report. *Climate Change*, 36.

Ibrahim, M. M., El-Zawawy, W. K., Abdel-Fattah, Y. R., Soliman, N. A., & Agblevor, F. A., (2011). Comparison of alkaline pulping with steam explosion for glucose production from rice straw. *Carbohydrate Polymers*, *83*(2), 720–726.

Ikwebe, J., & Harvey, A. P., (2011). Intensification of bioethanol production by simultaneous saccharification and fermentation (SSF) in an oscillatory baffled reactor (OBR). In: *World Renewable Energy Congress-Sweden* (No. 057, pp. 381–388). Linköping, Sweden. Linköping University Electronic Press.

Ingrao, C., Bacenetti, J., Bezama, A., Blok, V., Geldermann, J., Goglio, P., & Zabaniotou, A., (2016). Agricultural and forest biomass for food, materials and energy: bio-economy as the cornerstone to cleaner production and more sustainable consumption patterns for accelerating the transition towards equitable, sustainable, post-fossil-carbon societies. *Journal of Cleaner Production*, *117*, 4–6.

Jacobsen, S. E., & Wyman, C. E., (2000). Cellulose and hemicellulose hydrolysis models for application to current and novel pretreatment processes. In: *Twenty-First Symposium on Biotechnology for Fuels and Chemicals* (pp. 81–96). Humana Press, Totowa, NJ.

Jahnavi, G., Prashanthi, G. S., Sravanthi, K., & Rao, L. V., (2017). Status of availability of lignocellulosic feedstocks in India: Biotechnological strategies involved in the production of bioethanol. *Renewable and Sustainable Energy Reviews, 73*, 798–820.

John, R. P., Anisha, G. S., Nampoothiri, K. M., & Pandey, A., (2011). Micro and macroalgal biomass: A renewable source for bioethanol. *Bioresource Technology, 102*(1), 186–193.

Kadiman, O. K., (2005). Crops: Beyond foods. In: *Proceedings First International Conference of Crop Security* (pp. 20–23). Malang, Indonesia.

Kang, Q., Appels, L., Tan, T., & Dewil, R., (2014). Bioethanol from lignocellulosic biomass: Current findings determine research priorities. *The Scientific World Journal.*

Kar, Y., & Deveci, H., (2006). Importance of P-series fuels for flexible-fuel vehicles (FFVs) and alternative fuels. *Energy Sources, Part A, 28*(10), 909–921.

Katahira, S., Mizuike, A., Fukuda, H., & Kondo, A., (2006). Ethanol fermentation from lignocellulosic hydrolysate by a recombinant xylose-and cello oligosaccharide-assimilating yeast strain. *Applied Microbiology and Biotechnology, 72*(6), 1136–1143.

Kaur, J., Chadha, B. S., Kumar, B. A., Kaur, G., & Saini, H. S., (2007). Purification and characterization of ß-glucosidase from Melanocarpus sp. MTCC 3922. *Electronic Journal of Biotechnology, 10*(2), 260–270.

Keller, F. A., Hamilton, J. E., & Nguyen, Q. A., (2003). Microbial pretreatment of biomass. In: *Biotechnology for Fuels and Chemicals* (pp. 27–41). Humana Press, Totowa, NJ.

Kim, H. M., Wi, S. G., Jung, S., Song, Y., & Bae, H. J., (2015). Efficient approach for bioethanol production from red seaweed Gelidiumamansii. *Bioresource Technology, 175*, 128–134.

Kim, S., & Dale, B. E., (2004). Global potential bioethanol production from wasted crops and crop residues. *Biomass and Bioenergy, 26*(4), 361–375.

Kleerebezem, R., & Van Loosdrecht, M. C., (2007). Mixed culture biotechnology for bioenergy production. *Current Opinion in Biotechnology, 18*(3), 207–212.

Klement, T., Milker, S., Jäger, G., Grande, P. M., De María, P. D., & Büchs, J., (2012). Biomass pretreatment affects Ustilago maydis in producing itaconic acid. *Microbial Cell Factories, 11*(1), 43.

Kumar, P., Barrett, D. M., Delwiche, M. J., & Stroeve, P., (2009). Methods for pretreatment of lignocellulosic biomass for efficient hydrolysis and biofuel production. *Industrial & Engineering Chemistry Research, 48*(8), 3713–3729.

Kurian, J. K., Nair, G. R., Hussain, A., & Raghavan, G. V., (2013). Feedstocks, logistics and pre-treatment processes for sustainable lignocellulosic biorefineries: A comprehensive review. *Renewable and Sustainable Energy Reviews, 25*, 205–219.

Laureano-Perez, L., Teymouri, F., Alizadeh, H., & Dale, B. E., (2005). Understanding factors that limit enzymatic hydrolysis of biomass. *Applied Biochemistry and Biotechnology, 124*(1–3), 1081–1099.

Lee, J., (1997). Biological conversion of lignocellulosic biomass to ethanol. *Journal of Biotechnology, 56*(1), 1–24.

Lim, S., & Lee, K. T., (2012). Implementation of biofuels in Malaysian transportation sector towards sustainable development: A case study of international cooperation between Malaysia and Japan. *Renewable and Sustainable Energy Reviews, 16*(4), 1790–1800.

Limayem, A., & Ricke, S. C., (2012). Lignocellulosic biomass for bioethanol production: Current perspectives, potential issues and future prospects. *Progress in Energy and Combustion Science, 38*(4), 449–467.

Lin, C. S., Luque, R., Clark, J. H., Webb, C., & Du, C., (2011). A seawater-based biorefining strategy for fermentative production and chemical transformations of succinic acid. *Energy & Environmental Science, 4*(4), 1471–1479.

Lin, Y., & Tanaka, S., (2006). Ethanol fermentation from biomass resources: Current state and prospects. *Applied Microbiology and Biotechnology, 69*(6), 627–642.

Lopes, M. L., De Lima Paulillo, S. C., Godoy, A., Cherubin, R. A., Lorenzi, M. S., Giometti, F. H. C., & De Amorim, H. V., (2016). Ethanol production in Brazil: A bridge between science and industry. *Brazilian Journal of Microbiology, 47*, 64–76.

Lynd, L. R., Van Zyl, W. H., McBride, J. E., & Laser, M., (2005). Consolidated bioprocessing of cellulosic biomass: an update. *Current Opinion in Biotechnology, 16*(5), 577–583.

Lynd, L. R., Weimer, P. J., Van Zyl, W. H., & Pretorius, I. S., (2002). Microbial cellulose utilization: Fundamentals and biotechnology. *Microbiology and Molecular Biology Reviews, 66*(3), 506–577.

Manzetti, S., & Andersen, O., (2015). A review of emission products from bioethanol and its blends with gasoline. Background for new guidelines for emission control. *Fuel, 140*, 293–301.

Mao, L., Zhang, L., Gao, N., & Li, A., (2013). Seawater-based furfural production via corncob hydrolysis catalyzed by FeCl 3 in acetic acid steam. *Green Chemistry, 15*(3), 727–737.

Matsumoto, M., Yokouchi, H., Suzuki, N., Ohata, H., & Matsunaga, T., (2003). Saccharification of marine microalgae using marine bacteria for ethanol production. *Applied Biochemistry and Biotechnology, 105*(1–3), 247–254.

McCarthy, J. E., & Tiemann, M., (2006). MTBE in Gasoline: Clean Air and Drinking Water Issues. *Congressional Research Service Reports.* 26.

McIntosh, S., & Vancov, T., (2010). Enhanced enzyme saccharification of Sorghum bicolor straw using dilute alkali pretreatment. *Bioresource Technology, 101*(17), 6718–6727.

McKendry, P., (2002). Energy production from biomass (part 1): Overview of biomass. *Bioresource Technology, 83*(1), 37–46.

Mielenz, J. R., (2001). Ethanol production from biomass: Technology and commercialization status. *Current Opinion in Microbiology, 4*(3), 324–329.

Mohanty, S. K., Behera, S., Swain, M. R., & Ray, R. C., (2009). Bioethanol production from mahula (Madhucalatifolia L.) flowers by solid-state fermentation. *Applied Energy, 86*(5), 640–644.

Mussatto, S. I., Dragone, G., Guimarães, P. M., Silva, J. P. A., Carneiro, L. M., Roberto, I. C., & Teixeira, J. A., (2010). Technological trends, global market, and challenges of bio-ethanol production. *Biotechnology Advances, 28*(6), 817–830.

National Water Policy, (2012). *Ministry of Water Resources*. Government of India. http://mowr.gov.in/sites/default/files/NWP2012Eng6495132651_1.pdf (Accessed on 8 June 2019).

Najafi, G., Ghobadian, B., Tavakoli, T., Buttsworth, D. R., Yusaf, T. F., & Faizollahnejad, M., (2009). Performance and exhaust emissions of a gasoline engine with ethanol-blended gasoline fuels using artificial neural network. *Applied Energy, 86*(5), 630–639.

Nguyen, T. H. M., (2012). Bioethanol production from marine algae biomass: Prospect and troubles. *Journal of Vietnamese Environment, 3*(1), 25–29.

Nigam, P. S., & Singh, A., (2011). Production of liquid biofuels from renewable resources. *Progress in Energy and Combustion Science, 37*(1), 52–68.

Niven, R. K., (2005). Ethanol in gasoline: Environmental impacts and sustainability review article. *Renewable and Sustainable Energy Reviews, 9*(6), 535–555.

Nwufo, O. C., Nwafor, O. M. I., & Igbokwe, J. O., (2016). Effects of blends on the physical properties of bioethanol produced from selected Nigerian crops. *International Journal of Ambient Energy, 37*(1), 10–15.

Olivier, J. G., Janssens-Maenhout, G., Muntean, M., & Peters, J. A. H. W., (2012). *Trends in Global CO2 Emissions, 2012 Report*. PBL Netherlands Environmental Assessment Agency, Institute for Environment and Sustainability of the European Commission's Joint Research Centre.

Özçimen, D., & İnan, B. (2015). An overview of bioethanol production from algae. *Biofuels-Status and Perspective,* 141–162.

Packer, M., (2009). Algal capture of carbon dioxide, biomass generation as a tool for greenhouse gas mitigation with reference to New Zealand energy strategy and policy. *Energy Policy*, *37*(9), 3428–3437.

Palmer, F. H., (1986). Vehicle performance of gasoline containing oxygenates. In: *International Conference on Petroleum-Based Fuels and Automotive Applications. IMECHE Conference Publications 1986–11. Paper No C319/86.*

Pérez, J., Munoz-Dorado, J., De la Rubia, T. D. L. R., & Martinez, J., (2002). Biodegradation and biological treatments of cellulose, hemicellulose and lignin: An overview. *International Microbiology*, *5*(2), 53–63.

Périé, F. H., & Gold, M. H., (1991). Manganese regulation of manganese peroxidase expression and lignin degradation by the white rot fungus dichomitussqualens. *Applied and Environmental Microbiology*, *57*(8), 2240–2245.

Podkuiko, L., Ritslaid, K., Olt, J., & Kikas, T., (2014). Review of promising strategies for zero-waste production of the third generation biofuels. *Agronomy Research*, *12*(2), 373–390.

Prabhu, G. N., & Chandrasekaran, M., (1997). Impact of process parameters on L-glutaminase production by marine Vibrio costicola in solid state fermentation using polystyrene as an inert support. *Process Biochemistry*, *32*(4), 285–289.

Prasad, S., Singh, A., & Joshi, H. C., (2007). Ethanol as an alternative fuel from agricultural, industrial and urban residues. *Resources, Conservation and Recycling*, *50*(1), 1–39.

Raju, S. S., Parappurathu, S., Chand, R., Joshi, P. K., Kumar, P., & Msangi, S., (2012). *Biofuels in India: Potential, Policy and Emerging Paradigms*.

Refaat, A. A., (2012). *Biofuels from Waste Materials*.

Ren, H., Zong, M. H., Wu, H., & Li, N., (2016). Utilization of seawater for the biorefinery of lignocellulosic biomass: Ionic liquid pretreatment, enzymatic hydrolysis, and microbial lipid production. *ACS Sustainable Chemistry & Engineering*, *4*(10), 5659–5666.

Renewable Fuel Association, (2018). http://www.ethanolrfa.org/industry/locations/ (Accessed on 8 June 2019).

Rice, R. W., Sanyal, A. K., Elrod, A. C., & Bata, R. M., (1991). Exhaust gas emissions of butanol, ethanol, and methanol-gasoline blends. *Journal of Engineering for Gas Turbines and Power*, *113*(3), 377–381.

Rocha–Meneses, L., Raud, M., Orupõld, K., & Kikas, T., (2017). Second-generation bioethanol production: A review of strategies for waste valorization. *Agronomy Research*, *15*(3), 830–847.

Ruane, J., Sonnino, A., & Agostini, A., (2010). Bioenergy and the potential contribution of agricultural biotechnologies in developing countries. *Biomass and Bioenergy*, *34*(10), 1427–1439.

Rubin, E. M., (2008). Genomics of cellulosic biofuels. *Nature*, *454*(7206), 841.

Sabu, A., Keerthi, T. R., Kumar, S. R., & Chandrasekaran, M., (2000). L-Glutaminase production by marine Beauveria sp. under solid state fermentation. *Process Biochemistry*, *35*(7), 705–710.

Saha, B. C., (2000). α-L-Arabinofuranosidases: Biochemistry, molecular biology and application in biotechnology. *Biotechnology Advances*, *18*(5), 403–423.

Saha, B. C., (2003). Hemicellulose bioconversion. *Journal of Industrial Microbiology and Biotechnology, 30*(5), 279–291.

Saha, B. C., (2004). Lignocellulose biodegradation and applications in biotechnology. In: *ACS Symposium Series* (pp. 2–35). Washington, DC, American Chemical Society, 1999.

Sánchez, C., (2009). Lignocellulosic residues: Biodegradation and bioconversion by fungi. *Biotechnology Advances, 27*(2), 185–194.

Sanchez, O. J., & Cardona, C. A., (2008). Trends in biotechnological production of fuel ethanol from different feedstocks. *Bioresource Technology, 99*(13), 5270–5295.

Sarkar, N., Ghosh, S. K., Bannerjee, S., & Aikat, K., (2012). Bioethanol production from agricultural wastes: An overview. *Renewable Energy, 37*(1), 19–27.

Saxena, R. C., Adhikari, D. K., & Goyal, H. B., (2009). Biomass-based energy fuel through biochemical routes: A review. *Renewable and Sustainable Energy Reviews, 13*(1), 167–178.

Schnepf, R., & Yacobucci, B. D., (2010). Renewable fuel standard (RFS): Overview and issues. In: *CRS Report for Congress* (No. R40155).

Shafiee, S., & Topal, E., (2009). When will fossil fuel reserves be diminished? *Energy Policy, 37*(1), 181–189.

Shapouri, H., Duffield, J. A., & Wang, M. Q., (2002). *The Energy Balance of Corn Ethanol: An Update* (No. 34075). United States Department of Agriculture, Economic Research Service.

Shi, J., Sharma-Shivappa, R. R., Chinn, M., & Howell, N., (2009). Effect of microbial pretreatment on enzymatic hydrolysis and fermentation of cotton stalks for ethanol production. *Biomass and Bioenergy, 33*(1), 88–96.

Sims, R. E., Mabee, W., Saddler, J. N., & Taylor, M., (2010). An overview of second generation biofuel technologies. *Bioresource Technology, 101*(6), 1570–1580.

Singh, P., Suman, A., Tiwari, P., Arya, N., Gaur, A., & Shrivastava, A. K., (2008). Biological pretreatment of sugarcane trash for its conversion to fermentable sugars. *World Journal of Microbiology and Biotechnology, 24*(5), 667–673.

Singh, R., Srivastava, M., & Shukla, A., (2016). Environmental sustainability of bioethanol production from rice straw in India: A review. *Renewable and Sustainable Energy Reviews, 54*, 202–216.

Statista, (2018). *Fuel Ethanol Production Worldwide in 2017 by Country*. https://www.statista.com/statistics/281606/ethanol-production-in-selected-countries/ (Accessed on 8 June 2019).

Subhadra, B., & Edwards, M., (2010). An integrated renewable energy park approach for algal biofuel production in the United States. *Energy Policy, 38*(9), 4897–4902.

Subramanian, K. A., Singal, S. K., Saxena, M., & Singhal, S., (2005). Utilization of liquid biofuels in automotive diesel engines: An Indian perspective. *Biomass and Bioenergy, 29*(1), 65–72.

Sun, Y., & Cheng, J., (2002). Hydrolysis of lignocellulosic materials for ethanol production: A review. *Bioresource Technology, 83*(1), 1–11.

Surendhiran, D., & Vijay, M. (2012). Microalgal biodiesel-a comprehensive review on the potential and alternative biofuel. *Research Journal of Chemical Sciences ISSN*, 2231, 606X.

Suresh, P. V., & Chandrasekaran, M., (1998). Utilization of prawn waste for chitinase production by the marine fungus Beauveria bassiana by solid-state fermentation. *World Journal of Microbiology and Biotechnology, 14*(5), 655–660.

Taherzadeh, M. J., Lennartsson, P. R., Teichert, O., & Nordholm, H., (2013). Bioethanol production processes. *Biofuels Production*, 211–253.

Thanakoses, P., Black, A. S., & Holtzapple, M. T., (2003). Fermentation of corn stover to carboxylic acids. *Biotechnology and Bioengineering, 83*(2), 191–200.

Thangavelu, S. K., Ahmed, A. S., & Ani, F. N., (2016). Review on bioethanol as alternative fuel for spark ignition engines. *Renewable and Sustainable Energy Reviews, 56*, 820–835.

Tutt, M., Kikas, T., & Olt, J., (2013). Influence of harvesting time on biochemical composition and glucose yield from hemp. *Agronomy Research, 11*(1), 215–220.

U.S Department of Energy, (2017). *Alternatives Fuel Data Center: Fuel Properties Comparison*. https://www.afdc.energy.Gov/fuels/fuel_comparison_chart.pdf (Accessed on 8 June 2019).

Vitz, J., Erdmenger, T., Haensch, C., & Schubert, U. S., (2009). Extended dissolution studies of cellulose in imidazolium based ionic liquids. *Green Chemistry, 11*(3), 417–424.

Vohra, M., Manwar, J., Manmode, R., Padgilwar, S., & Patil, S., (2014). Bioethanol production: Feedstock and current technologies. *Journal of Environmental Chemical Engineering, 2*(1), 573–584.

Vom Stein, T., Grande, P., Sibilla, F., Commandeur, U., Fischer, R., Leitner, W., & De María, P. D., (2010). Salt-assisted organic-acid-catalyzed depolymerization of cellulose. *Green Chemistry, 12*(10), 1844–1849.

Woods, J., (2008). *Sustainable Biofuels: Prospects and Challenges*. The Royal Society, London Google Scholar.

WWAP (United Nations World Water Assessment Programme), (2016). The United Nations World Water Development Report, Water and Jobs. Paris, UNESCO.

Wyman, C., (1996). *Handbook on Bioethanol: Production and Utilization*. CRC Press.

Wyman, C. E., Dale, B. E., Elander, R. T., Holtzapple, M., Ladisch, M. R., & Lee, Y. Y., (2005). Coordinated development of leading biomass pretreatment technologies. *Bioresource Technology, 96*(18), 1959–1966.

Xue, F., Gao, B., Zhu, Y., Zhang, X., Feng, W., & Tan, T., (2010). Pilot-scale production of microbial lipid using starch wastewater as raw material. *Bioresource Technology, 101*(15), 6092–6095.

Yoon, S. H., & Lee, C. S., (2011). Lean combustion and emission characteristics of bioethanol and its blends in a spark ignition (SI) engine. *Energy & Fuels, 25*(8), 3484–3492.

Yu, J., Paterson, N., Blamey, J., & Millan, M., (2017). Cellulose, xylan and lignin interactions during pyrolysis of lignocellulosic biomass. *Fuel, 191*, 140–149.

Yue, H., Ling, C., Yang, T., Chen, X., Chen, Y., Deng, H., & Chen, G. Q., (2014). A seawater-based open and continuous process for polyhydroxyalkanoates production by recombinant Halomonas campaniensis LS21 grown in mixed substrates. *Biotechnology for Biofuels, 7*(1), 108.

Yüksel, F., & Yüksel, B., (2004). The use of ethanol-gasoline blend as a fuel in an SI engine. *Renewable Energy, 29*(7), 1181–1191.

Zabed, H., Sahu, J. N., Boyce, A. N., & Faruq, G., (2016). Fuel ethanol production from lignocellulosic biomass: An overview on feedstocks and technological approaches. *Renewable and Sustainable Energy Reviews, 66*, 751–774.

Zabed, H., Boyce, A. N., Sahu, J. N., & Faruq, G., (2017a). Evaluation of the quality of dried distiller's grains with solubles for normal and high sugary corn genotypes during dry–grind ethanol production. *Journal of Cleaner Production, 142*, 4282–4293.

Zabed, H., Sahu, J. N., Suely, A., Boyce, A. N., & Faruq, G., (2017b). Bioethanol production from renewable sources: Current perspectives and technological progress. *Renewable and Sustainable Energy Reviews, 71*, 475–501.

Zhang, Y. H. P., Ding, S. Y., Mielenz, J. R., Cui, J. B., Elander, R. T., Laser, M., & Lynd, L. R., (2007). Fractionating recalcitrant lignocellulose at modest reaction conditions. *Biotechnology and Bioengineering, 97*(2), 214–223.

Zheng, Y., Pan, Z., & Zhang, R., (2009). Overview of biomass pretreatment for cellulosic ethanol production. *International Journal of Agricultural and Biological Engineering, 2*(3), 51–68.

CHAPTER 11

Traditional Fermentation by the Rabha-Hasong, Mishing, and Karbi Communities of Assam and Prospects of Value Addition for Enhancement of Nutritional Qualities in Ethnic Foods

HRIDIP KUMAR SARMA[1] and DEEP PRAKASH PARASAR[1]

[1]*Microbial Communication and Fungal Biology Group, Department of Biotechnology, Gauhati University, Guwahati–781014, Assam, India*

ABSTRACT

Assam is a state in Northeast India, situated south of the eastern Himalayas along the Brahmaputra and Barak River valleys is home to many ethnic communities who have been residing for hundreds of years. They represent multi-ethnic, multi-linguistic and multi-religious societies unique in their own. The communities depend primarily on fermented food products that differ in terms of their names, ingredients used and methods of preparation. Each community follow their own traditional knowledge of food fermentation that has been passed on from generation to generation. This customary is associated with societal makeup. Three major ethnic communities of the region are the Rabha Hasongs, the Mishings and the Karbis. They inherit a rich tradition of indigenous fermentation for preparation of ethnic food. The starter cultures are prepared from rice or other cereals like millets, maize and lesser known ethnic fruits. The use and addition of plant extracts and condiments have been observed that are known to enhance the quality traits of the fermented products. Recent studies have confirmed the identity of the microbial consortia responsible for ethnic fermentation. These primarily include yeasts, lactic acid bacteria and some fungi. Both *Saccharomyces* and non-*Saccharomyces* yeasts like *Pichia anomala, Wickerhamomyces, Saccharomycopisis fibuligera, Rhodotorula,*

Candia tropicalis and *Candida glabrata* have also been observed. The effects of secondary metabolites produced during fermentation are not in the literature. The role of plant phytochemicals in microbial sustenance and fermentation efficiency has never been studied. It is envisaged that value addition of these indigenous food materials can provide industrially important products of commercial importance.

11.1 INTRODUCTION

The great Indian Himalayas extending from the Indus Trench of Nanga Parbat (8125 masl) in the west to the Yarlung Tsangpo-Brahmaputra gorge below Namche Barwa (7756 masl) in the east are a sacred place for millions of Buddhist and Hindu people and the center of a rich diversity of cultures and biological resources. The Himalayan arc extending between latitudes 26°20′ and 35°40′ north and longitudes 74°50′ and 95°40′ east (Ives, 2006) include hilly and mountainous terrains of northeast India where hundreds of indigenous ethnic communities evolved and flourished in the last 5000 years.

The northeastern region of India (NEI) comprises of eight sister states including Assam that fall within the Himalayan Biodiversity Hotspot and are rich in flora, fauna, and indigenous ethnic human communities. Northeast India constitutes about 8% of the total geographical area of India, having a population of approximately 40 million, which is about 3.1% of the total Indian population. Agriculture is the main occupation of the tribes, who practice 'Jhum or shifting' cultivation for their agricultural needs. The region is characterized by the presence of diverse physiographic hills, valleys, plains, and mountains. It has the richest reservoir of plant diversity in India and is one of the 'biodiversity hotspots' of the world, supporting about 50% of India's biodiversity (Mao and Hynniewta, 2000; Chatterjee et al., 2006). The region is also known as the 'Cradle' of flowering plants and is a habitat of many botanical curiosities and rarities that promise tremendous potential in ethnic medicine. Families of some of the plants such as the Nepenthaceae, Liliaceae, Clethraceae, Ruppiaceae, Siphonodontaceae, and Tetracentraceae are unique to the world, not found anywhere but only in this region. The northeastern region has drawn attention for its high biodiversity and is a priority for leading conservation agencies across the globe. This area is inhabited largely by tribal people, who make up 75% of the population of the region (Agrahar-Murungkar and Subbulakshmi, 2006). Approximately 225 out of the 450 tribes of India reside in this region, representing different ethnic groups with distinct cultural entities (Chatterjee et al., 2006; Mao et al., 2009). The people of this region have a very rich reserve of traditional knowledge owing to their livelihood in the tough hilly terrains and coarse tropical plains. The people possess great knowledge of the environment and depend on the forests, plants, and plant products for food and other purposes (Jaiswal, 2010). Food plays a very important role in defining the identity of the ethnic groups. It is noteworthy that in recent times there is a high demand for food products of the ethnic people among the urban-dwelling societies due to probiotic and nutritional contents inherent in traditional foods (Asati and Yadav, 2003; Medhi et al., 2013).

11.2 FERMENTED FOODS OF NORTHEAST INDIA

Northeast India is the center of a diverse food culture comprising fermented and non-fermented ethnic foods and alcoholic beverages (Sathe and Mandal, 2016). Most of the people of this region carry their own methods of fermenting food materials for the purpose of preservation and enhancement of desirable qualities and to achieve this, they have been improvising the traditional methods through time immemorial. Indigenous fermented foods are an intrinsic part of diet of the ethnic tribes in the Himalayan belt of India, testifying the oldest and most economical methods for development of a diversity of aromas, flavors, and textures, as well as food preservation measures through inculcation of biological enrichment of food products attained by the processes of microbial manipulation practiced through conventional means (Sekar and Mariappan, 2007). Ethnic fermented foods, alcoholic beverages, and drinks have been consumed by the ethnic people of northeast India for more than 2500 years. All the fermented foods are region-specific and have unique substrates and preparation methods. Locally available materials such as milk, vegetable, bamboo, soybean, meat, fish, and cereal are commonly fermented (Das and Deka, 2012; Tamang et al., 2012). More than 250 different types of familiar and less-familiar ethnic fermented foods and alcoholic beverages are prepared and consumed by the different ethnic people of northeast India. Ethnic fermented foods of northeast India include a variety of eatables like soybean and non-soybean legume foods, fermented vegetables (gundruk, sinki, and anishi) and bamboo shoots (soibum and mesu), fermented cereal and pulse foods (kinema, bhatootu, marchu, chilra, and tungrymbai), fermented, and smoked fish products (ngari and hentak), preserved meat products, milk beverages (kadi, churpa, and nudu), non-food mixed amylolytic starters and alcoholic beverages (ghanti, jann, and daru)

(Tamang et al., 2012). The food-processing procedure makes use of various technologies to convert perishable, relatively bulky, and typically inedible raw materials into more useful palatable foods or potable beverages with increased self-life. Processing of food contributes to the food security by minimizing the wastage and loss of various ethnic food by increasing their availability and market value (Rolle and Satin, 2002).

11.3 ASSAM AND THE HISTORY OF ITS ETHNIC PEOPLE

Assam, the pride of Northeast India, is situated south of the eastern Himalayas along the Brahmaputra and Barak river valleys covering an area of 78,438 km^2 (30,285 sq miles) and is home to more than 40 ethnic communities who have been residing for hundreds of years. They represent multi-ethnic, multi-linguistic, and multi-religious societies unique in their own. The ethnic communities depend primarily on fermented food products and alcoholic beverages that differ to quite an extent in terms of their names, ingredients used, and methods of preparation. Each community has its own traditional knowledge of preparing fermented food and alcoholic beverages that has been passed on from generation to generation. This customary has its roots in their socio-cultural life and is found to be associated with many occasions like merrymaking, ritual ceremonies, festivals, marriages, and funerals. These substances have been reported to have less significant health effects which may be accredited to the medicinal properties of the plant parts and herbs used in the preparation of starter cultures. Rice is the common substrate used for fermentation whilst other cereals like millets, Job's tears, maize, and lesser-known ethnic fruits are also used for fermentation.

The major tribal communities of the region include the Bodos, Rabhas, Karbis, Dimasas, Deories, Sonowal-Kacharis, Kacharis, Mishings, Ahoms, Morans, Motoks, etc. All these ethnic communities harbor their own traditional knowledge for food fermentation and preparation of their indigenous alcoholic beverages. Rice beer is an integral part of their livelihood and is called with different names such as *jumai* by the Bodos, *suze* by the Deoris, *morpo* by the Mikirs, *chu* by the Garos, *jou* by the Dimasas, *zu* by the Tiwas, *apong* by the Mishings, *laopani*, and *mod* by the Ahoms, Kacharis, and some tribal communities. Three of the major ethnic communities residing in the plains and foothills of the Himalayas within the state of Assam are the Rabha-Hasongs, the Mishings and the Karbis and the processes of fermented food preparation is exceptional in case of these communities which is elaborated in the succeeding pages. The three communities are regarded as descendants of the Tibeto-Burmese tribe.

11.4 ETHNIC FERMENTATION BY THE RABHA-HASONG COMMUNITY

The Rabhas are one of the largest aboriginal tribal communities of Assam and are descendants of the same clan that represent the other Kachari communities like Bodo-Kachari, Mishing-Kachari, Karbi-Kachari, and Sonowal-Kacharis. Rabhas are also known as Rabha-Hasongs and are a widely scattered community with maximum densities in the undivided Goalpara, Kamrup, and Darrang districts of Assam. Rice is the staple food of the tribe, and a diversity of ethnic rice varieties are cultivated by the community which are widely used for the preparation of rice beer and other fermented products. Rabha-Hasongs are generally regarded as inhabitants of the plains of the southern bank of the Brahmaputra River in lower Assam and bear resemblances with the Garo and Jaintia communities of the hilly terrain in Meghalaya. They have a rich tradition of ethnic fermentation and follow their own practices of rice beer preparation called, i.e., *chokot*, a traditional alcoholic preparation. Preparation is through old age process during which parts of plants like *Plumbagozeylenica* L. are used as a key ingredient for starter preparation. The Rabhas add concoctions of plant herbs during the preparation of rice-beer which are believed to have medicinal properties.

Rice-beer cake is popularly known as *bakhor*, *surachi* or *phap* among the Rabhas while rice-beer is known as *choko* or *jonga-mod*. A considerable amount of rice-beer cake is mixed along with plants materials for the preparation of fresh rice-beer cakes. Some round flats globules (each around 50 gm) are prepared from the grounded mixture. Generally, different parts of 10 specific plants species viz., *Ananascomosus (L.)*, *Artocarpusheterophyllus*, *Calotropis gigantean*, *Capsicum frutescens*, *Clerodendrumviscosum Vent.*, *Dennstaeditascabra* (Wall.) *T. Moore*, *Ochthochloacoracana Edgrew*, *Plumbagoindica*, *Saccharumofficinarum*, and *Scopariadulcis L* are added to *phaps* for preparing the rice-beer cake. To prepare rice-beer *choko*, tightly cooked fresh rice (using less amount of water) is used. After cooking, rice is scattered on a broad mat made of bamboo and cooled. Then a particular amount of rice-beer cake (generally two pieces for boiled rice prepared from 2 kg of fresh rice) is powdered and mixed with the cooked rice. A special type of cylinder made of bamboo net, known among Rabhas as *janthi*, is placed inside an earthen pitcher, known as *jonga*. The mixture of rice and cake are kept inside *jonga* and outside *janthi* with an open mouth tightly sealed with banana leaves that are pre-warmed on before use. Fermentation of *choko* or rice beer takes 4–5 days during summer and 7–8 days during winter. The produce is collected and stored in a dried shell of the fruit

Lagenariasicerariastandl by pouring the *choko* from the *jonga* by making a hole in the dried shell of the fruit. The stored produce is further fermented by adding water and rice-beer cake. The final produce is collected after 3–4 days of fermentation and distilled through an indigenous process using earthen or metallic pitcher-like pots/handi called *luduki*, whereby three pots are placed over one another with the uppermost pots having holes at the bottom while the collecting pot is sealed. This entire filtering and distilling apparatus is made airtight with jute *(Cochoruscapsularis L.)* fiber and mud at the junctions. The resulting drink is strong liquor, known as *fotika*, and the Rabhas believe that it has a curative effect on psychiatric patients.

The traditional drink '*choko*' is often prepared from an indigenous local variety of Bodorice, which can be preserved for 6 to 12 months. This fermented beverage is strongly alcoholic, appears reddish brown in color and is believed to be sweet in taste better known as '*sachotinamod*' by the Rabhas. It is also observed that the Rabhas often consume endemic glutinous rice with rice-beer as a delicacy, popularly known as '*juguli*.' Traditional know-how of the Rabhas denotes that '*juguli*' works as a tonic, diminishes sleeping disorders and has a laxative effect. Traditional belief also states that these alcoholic beverages possess high medicinal significance and do relieve migraine headache, body ache, and inflammation of body parts, cure diarrhea and urinary problems, including tract infections. Rice-beer is believed to have antihelminthic properties too (Figure 11.1).

11.5 ETHNIC FERMENTATION BY THE MISHING COMMUNITY

The *Mishing* community of Assam primarily is the descendants of the Mongoloid race and is closely related to the Austro-Asiatic races of India (Singh et al., 1996). They are one of the tribal communities of Assam residing in the plains and live in far-flung areas of the Brahmaputra valley adjacent to the hilly terrains of Arunachal Pradesh and Nagaland, bounded by the districts of Dhemaji, Lakhimpur, Sonitpur, and Sibsagar. Mishings have distinct entities from the rest of the tribes of Assam with their special unique culture and tradition (Baruah and Kalita, 2007). The community has been residing for hundreds of years in the region practicing their traditional way of livelihood options. Rice beer prepared by the *Mishings* is known as *Apong*, and the starter cake is called as *aopopitha*. *Apong* forms an integral part of the culture, traditions, and rituals of this tribe. Two forms of *Apong* are prepared for consumption and cultural use by fermenting rice with *E'pob* (starter cakes). These are *Nogin Apong* and *Po:ro Apong (saimod*) *(saimod*) that are produced either by fermenting rice or by fermenting mixture of rice (usually glutinous varieties) with the ash of paddy husk and straw. The community also prepares a lot of fermented food and pickles from indigenous sources. The starter material for fermentation is a homely prepared yeast culture (*e'pob*) maintained in a starchy (rice) medium mixed with a concoction of powdered plant/ plant parts that are thought to contribute to the medicinal properties and aroma to the finished product. *Po:ro Apong* is used in various ailments and believed to cure dysentery, body ache, and stomach disorders. Such fermented food produce are also recorded to increase the perspiration rate, thereby expelling the extra heat from the body, and hence this drink is used preferably during hot summer humid conditions.

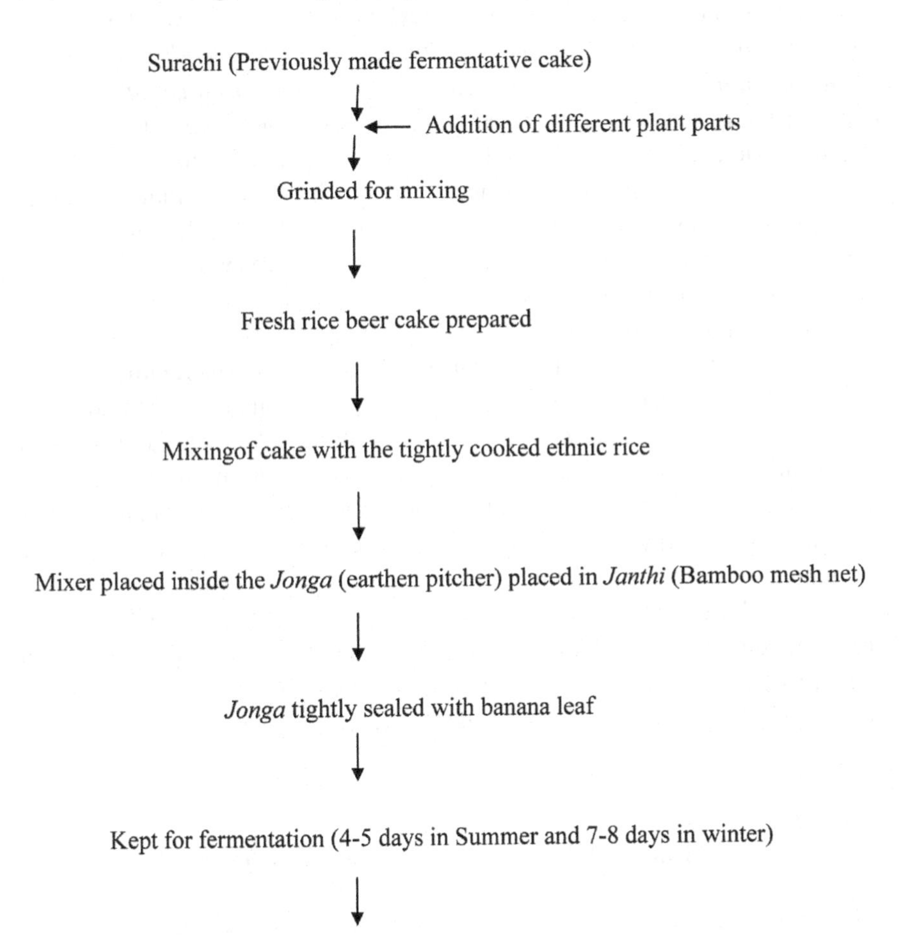

FIGURE 11.1 Indigenous method of Choko preparation by the Rabha-hasong ethnic community of Assam.

The Mishings observe a festival called "*Ali-Aye Ligang*" where the tradition of offering fermented beverages to the spirits of their forefathers has been in place for hundreds of years. The recipes of fermented brews are rolled down through generations with or without modification and as and when required. As a whole, the steps for processing of brews are complicated and require a lot of exercises.

The *Mishings* add several natural ingredients from nature to facilitate the process of spontaneous fermentation. The process of cake preparation is vital to ensure good quality fermentation. Leaves of several plants that harbor medicinal properties are added as concoctions during the preparation of *apo-pitha*. These include leaves of *Centellaasiatica, Hydrocotylesibthorpioides Oldenlandiacorymbosa, Saccharumofficinarum, Clerodendrumviscosum, CyclosoruslensaIpomea*sp, *Scopariadulcis, Drymeriacordata, Capsicumannuum, Ananascomosus, and Lygodiumflexuosum*. The collected leaves are cleaned, dried in the sun on bamboo mats called '*opoh*' and crushed to a powder before use. They are also used fresh at times. The primary substrate used for fermentation is soaked rice that are cultivated locally and kept for the entire year. Dough of the boiled or soaked rice is prepared as oval-shaped balls of approximately 6cm × 3cm and are dried in the sun. Fermentation is carried out in an earthen pot called '*killing*' which is fumigated by placing it on a '*torap*' (a bamboo frame constructed over the fireplace) until the pot turns blackish. Boiled or soaked rice is spread on banana leaves and allowed to cool. To this, powdered *apo-pitha (*starter culture cake materials) is added in the ratio of 1:1 kg of *apo-pitha*: rice and the whole mixture is kept for fermentation inside the '*killing*.' The pots are covered with banana leaves and allowed to ferment for five days. A little water is often added to the fermented product which is then filtered to retrieve the *apong*.

The Mishings also prepare a traditional alcoholic drink called '*saimod*' from half-burned hay and husk. The ash is mixed with an equal amount with boiled rice and *apo-pitha*. The mixture is compactly packed in *killing* and fermented for about 15 days and finally filtered to get the finished product (Figure 11.2).

11.6 ETHNIC FERMENTATION BY THE KARBI COMMUNITY

The Karbis are one of the major tribes of the state of Assam and are settled mostly in the districts of KarbiAnglong and North Cachar Hills. They inherit a rich tradition of indigenous fermentation for preparation of ethnic food and a traditional alcoholic beverage called *hor-alank*. This beverage is used as a refreshing drink and also bears significance in many social ceremonies and events. *Hor-alank* is prepared from yeast starter cultures called *thap* perpetuated in rice starter materials. The preparation is often associated with the addition of condiments and plant extracts like leaves of *Croton joufra, Artocarpusheterophyllus, Phlogocanthusthysiflorus, Solanumviarum,* and barks of *Acacia pennata*. Condiments are prepared as pastes that are flattened into small disc-shaped cakes of about 6 cm in diameter and 0.5 cm in thickness. These are overlaid with powder of previous years "*thaps*," kept in a bamboo sieve called "*ingkrung*" and dried for about three days under the sun or above a fireplace. Cakes prepared thus are stored for about a year for further use. Alcoholic rice beer is prepared by boiling newly harvested rice that are allowed to cool. It is followed by the addition of powdered *thaps* in a ratio of 5 Kg rice: 7 *thaps* or cakes. The whole mixture is allowed to ferment through traditional means for a period of 2 days at room temperature (Figures 11.3 and 11.4).

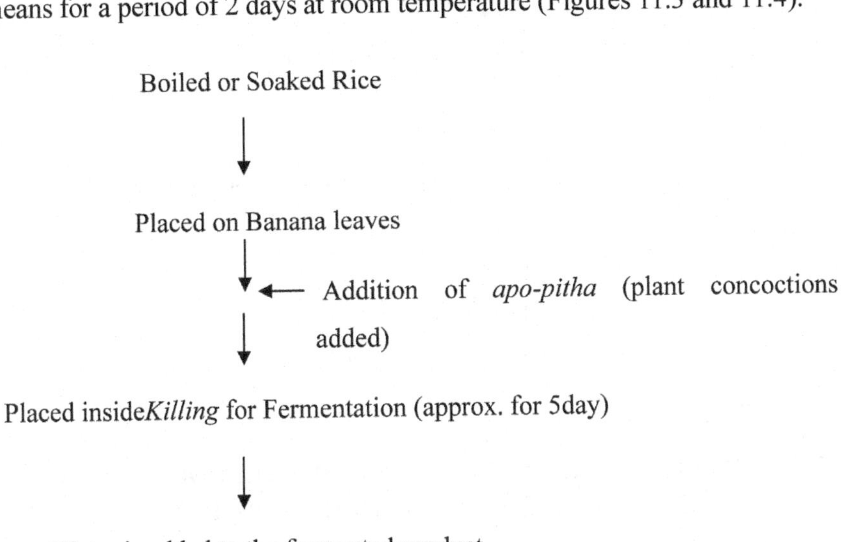

FIGURE 11.2 Indigenous method of Apong preparation by the Mishing ethnic community of Assam.

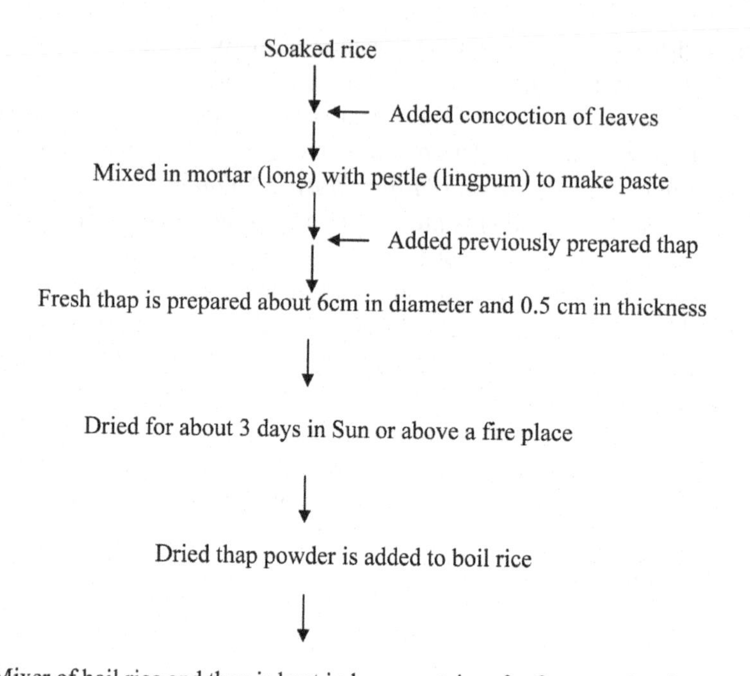

FIGURE 11.3 Indigenous method of Hor-alank preparation by the Karbi ethnic community of Assam.

11.7 PROBIOTIC VALUE AND THE NUTRITIONAL ANALYSIS OF THE RICE BEER PREPARED BY THESE COMMUNITIES

Recently the identity of microbial consortia responsible for ethnic fermentation practiced by the tribes and communities of northeast India has been deciphered. These primarily include yeasts and other fermenting microorganisms like lactic acid bacteria (LAB) and certain other fungi. Some studies have identified the presence of non-*Saccharomyces* yeasts like *Pichiaanomala, Wickerhamomyces, Saccharomyco pisisfibuligera, Rhodotorula, Candia tropicalis*, and *Candida glabrata* in the perpetuated fermentative cakes. It is apparent that the effects of secondary metabolites produced during fermentation have not been thoroughly appreciated. Additionally, the role of plant phytochemicals like phytohormones, terpenoids, carotenoids, phytosterols, glucosinolates, polyphenols, flavonoids, isoflavonoids, and anthocyanin's in the sustenance of microbes and process of fermentation has not been discussed. The possibility of value addition of these indigenous food materials holds good scope for commercialization provided they are studied in details and their deleterious traits are removed. It is now understood that the old age practice of fermentation could provide enormous scope for identifying products and molecules that are of pharmacological, medicinal, nutritional, and commercial significance.

FIGURE 11.4 (See color insert.) Fermentative cake and the plant parts added as ingredients by the Rabha-Hasong community during the preparation of alcoholic starter cake.
(A) Fermentative cake
(B) *Saccharum officinarum*
(C) *Ananascomosus (L.) merr* of Rabha Community
(D) *Capsicum frutescens*
(E) *Artocarpus heterophyllus*
(F) *Cynodondactylon*

It is evident from published research that significant quantities of several B vitamins are available in fermented beverages, including commercial beer.

Mayer et al., (2001) had demonstrated the worth of beer as a source of folic acid, leading to a decreased homocysteine content in the blood thereby decreasing a significant risk factor for vascular diseases. While, chronic alcoholism leads to the obverse effect, traditional beers significantly lower serum concentrations of homocysteine than did those consuming wine or spirits (Cravo et al., 1996). In present days, it is also understood that the microbes present in the fermentative rice beers produced worldwide act as a good probiotic drink with tremendous health benefits. Unfortunately, only a few significant research outcomes explain the variety of proposed beneficial health effects of probiotics in traditional drinks, a loophole to back up the claims. Clinical symptoms that have been reportedly treated or have the potential to be treated with probiotics include diarrhea, gastroenteritis, irritable bowel syndrome, and inflammatory bowel disease (IBD; Crohn's disease and ulcerative colitis), cancer, depressed immune function, inadequate lactase digestion, infant allergies, failure-to-thrive, hyperlipidaemia, hepatic diseases, *Helicobacter pylori* infections, and others (Bengmark, 2000; Benchimol and Mack, 2004; Brown and Valiere, 2004). The possible application or use of traditional fermentative beverages from Assam promises great potential as probiotic agents with a very strong traditional know-how inherent and propagated through ages. Microorganisms including yeasts that are domesticated through traditional fermentation are widely used in the preparation of ethnic foods that inhabit digestive tract of humans that have been shown to improve the quantity, availability, and digestibility of some dietary nutrients. Traditional fermented food rich in LAB increases folic acid in yogurt, bifidus milk and kefir (Rajalakshmi and Vanaja, 1967; Shahani and Chandan, 1979; Deeth and Tamime, 1981; Alm, 1982). Similarly, niacin and riboflavin levels in yogurt are increased with fermentation (Deeth and Tamime, 1981; Alm, 1982). These LABs have been found to be associated with yeasts during traditional fermentation. It is needless to say that a lot of desirable characteristics are inherent in traditional drinks that need to be revisited with a holistic endeavor. Removal of undesirable characteristics and incorporation of desirable ones like flavors, aroma, texture, and taste in traditional drinks will improve the value addition and commercial value of these ethnic fermented products.

11.8 CONCLUSION

From the investigations, it is evident that the three ethnic communities of Assam inherit vast sources of traditional knowledge, which is associated with the use of different medicinal plant and microbial diversity for the preparation of alcoholic fermentative cakes. These communities follow similar methods of rice beer preparation, which must have undergone horizontal divergence during the process of perpetuation contributed by geographic separation, unique climate, and the availability of the resources. The plant materials and the diversity of microflora promise great potential for value-added products. It is apparent that the effects of secondary metabolites produced during fermentation have not been thoroughly appreciated from the region. Additionally, the role of plant phytochemicals like phytohormones, terpernoids, carotenoids, phytosterols, glucosinolates, polyphenols, flavonoids, isoflavonoids, and anthocyanin's expressed by the microbes during fermentation has not been deciphered. The possibility of value addition of these indigenous food materials holds good scope for commercialization provided they are studied in details and their deleterious traits are removed. It is now understood that the old age practice of fermentation could provide enormous scope for identifying products and molecules that are of pharmacological, medicinal, nutritional, and commercial significance.

KEYWORDS

- **inflammatory bowel disease**
- **lactic acid bacteria**
- **Northeastern Region of India**

REFERENCES

Agrahar, M. D., & Subbulakshmi, G., (2006). Preparation techniques and nutritive value of fermented foods from the Khasi tribes of Meghalaya. *Ecol. Food Nutr., 45*, 27–38.

Agranoff, B. W., & William, H., (2000). Unwitting neurochemist? *Neurochem. Res., 25*, 1431–1434.

Alm, L., (1982) Effect of fermentation on lactose, glucose, and galactose content in milk and suitability of fermented milk products for lactose-deficient individuals. *J. Dairy. Sci., 65*, 346–352.

Asati, B. S., & Yadav, D. S., (2003). Diversity of horticultural crops in northeastern region. *ENVIS Bulletin: Himalayan Ecol., 12*, 1–10.

Baruah, M., & Kalita, D., (2007). Ethnomedicine used by *Mishings* tribes of Dibrugarh district, Assam. *Indian Journal of Traditional Knowledge, 6*(4), 595–598.

Benchimol, E. I., & Mack, D. R., (2004). Probiotics in relapsing and chronic diarrhea. *J. Pediatr. Hematol. Oncol., 26*, 515–517.

Bengmark, S., (2000). Colonic food: Pre- and probiotics. *Am. J. Gastroenterol., 95*, S5–S7.
Brown, A. C., & Valiere, A., (2004). Probiotics and medical nutrition therapy. *Nutr. Clin. Care, 7*, 56–68.
Chatterjee, S., Saikia, A., Dutta, P., Ghosh, D., Pangging, G., & Goswami, A. K., (2006). *Background Paper on Biodiversity Significance of North East India for the Study on Natural Resources, Water and Environment Nexus for Development and Growth in North Eastern India*. New Delhi: WWF-India.
Cravo, M. L., Gloria, L. M., Selhub, J., Nadeau, M. R., Camilo, E., Rosendi, M. P., et al., (1996). Hyperhomocysteinemia in chronic alcoholism: Correlation with folate, vitamin B12 and vitamin B6. *Am. J. Clin. Nut., 63*, 220–224.
Das, A. J., & Deka, C., (2012). Fermented foods and beverages of Northeast India. *Int. Food Res. J., 19*, 377–392.
Deeth, H. C., & Tamime, A. Y., (1981). Yogurt, nutritive and therapeutic aspects. *J. Food Prot., 44*, 78–86.
Fernandes, C. F., Shahani, K. M., & Amer, M. A., (1987). Therapeutic role of dietary lactobacilli and lactobacillic fermented dairy products. *FEMS Microbiol. Rev., 46*, 343–356.
Ives, J. D. (2004). Himalayan perceptions: Environmental change and the well-being of mountain peoples. *Himalayan Journal of Sciences, 2*(3): 17–19.
Jaiswal, V., (2010). Culture and ethnobotany of Jaintia tribal community of Meghalaya, Northeast India – a mini review. *Indian J. Trad. Knowl., 9*, 38–44.
Kilara, A., & Shahani, K. M., (1975). Lactase activity of cultured and acidified dairy products. *J. Dairy Sci., 59*, 2031–2035.
Leavitt, J., Barrett, J. C., Crawford, B. D., & Tso, P. O. P., (1978). Butyric acid suppression on the in vitro neoplastic state of Syrian hamster cells. *Nature, 271*, 262.
Leopold, C. S., & Eileler, D., (2000). Basic coating polymer for the colon-specific drug delivery in inflammatory bowel disease. *Drug Dev. Ind. Pharm., 26*, 1239–1246.
Mack, D. R., Michail, S., & Wet, S., (1999). Probiotics inhibit enteropathogenic *E. coli* adherence *in vitro* by inducing intestinal mucin gene expression. *Am. J. Physiol., 276*, 941–950.
Mallett, A. K., Bearne, C. A., & Rowland, I. R., (1989). The influence of incubation pH on the activity of rat and human gut flora enzymes. *J. Appl. Bacteriol., 66*, 433–437.
Mao, A. A., Hynniewta, T. M., & Sanjappa, M., (2009). Plant wealth of Northeast India with reference to ethnobotany. *Indian J. Trad. Know., 8*, 96–103.
Mao, A. A., & Hynniewta, T. M., (2000). Floristic diversity of North East India. *J. Assam Sci. Soc., 41*, 255–266.
Martini, M. C., Kukielka, D., & Savalano, D. A., (1991). Lactose digestion from yogurt: Influence of a meal and additional lactose. *Am. J. Clin. Nutr., 53*, 1253–1258.
Mayer, O., Simon, J., & Roslova, H., (2001). A population study of beer consumption on folate, and homocysteine concentrations. *Eur. J. Clin. Nut., 55*, 605–609.
Medhi, P., Kar, A., & Borthakur, S. K., (2013). Medicinal uses of wild edible plants among the *Ao Nagas* of Mokokchung and its vicinity of Nagaland, India. *Asian Reson., 2*, 64–67.
Rajalakshmi, R., & Vanaja, K., (1967). Chemical and biological evaluation of the effects of fermentation on the nutritive value of foods prepared from rice and grams. *Br. J. Nutr., 21*, 467–473.
Rolfe, R. D., (2000). The role of probiotic cultures in the control of gastrointestinal health. *J. Nutr., 130*, 396S–402S.
Rolle, R., & Satin, M., (2002). Basic requirements for the transfer of fermentation technologies to developing countries. *Int. J. Food Microbiol., 75*, 181–187.

Rombeau, J. L., Kripke, S. A., & Settle, R. G., (1990) Short-chain fatty acids. Production, absorption, metabolism and intestinal affects. In: Kritchesvsky, D., (ed.), *Dietary Fiber: Chemistry, Physiology, and Health Effects* (pp. 317–337). New York and London: Plenum Press.

Sathe, G. B., & Mandal, S., (2016). Fermented products of India and its implication: A review. *Asian J. Dairy Food Res., 35*, 1–9.

Sekar, S., & Mariappan, S., (2007). Usage of traditional fermented products by Indian rural folks and IPR. *Indian J. Trad. Know., 6*, 111–120.

Shahani, K. M., & Chandan, R. C., (1979). Nutritional and healthful aspects of cultured and culture-containing dairy foods. *J. Dairy Sci., 62*, 1685–1694.

Singh, J., Bhuyan, T. C., & Ahmed, A., (1996). Ethnobotanical studies on the Mishing tribes of Assam with special reference to food and medicinal plans-I. *Journal Economic Taxonomy Botany, 12*, 350–356.

Tamang, J. P., Tamang, N., Thapa, S., Dewan, S., Tamang, B., Yonzan, H., Rai, A. P., Chettri, R., Chakrabarty, J., & Kharel, N., (2012). Microorganism and nutritive value of ethnic fermented foods and alcoholic beverages of North-East India. *Indian J. Trad. Know., 11*, 7–25.

CHAPTER 12

Rice Protein: Properties, Extraction, and Applications in Food Formulation

NEERAJ GHANGHAS,[1] M. T. MUKILAN,[1] PRAMOD K. PRABHAKAR,[1] and NITIN KUMAR[2]

[1]*Food Science and Technology Lab, Department of Food Science and Technology, National Institute of Food Technology Entrepreneurship and Management, Sonepat, HR, India, E-mail: pkprabhakariitkgp@gmail.com, pramodkp@niftem.ac.in*

[2]*Department of Food Engineering, National Institute of Food Technology Entrepreneurship and Management, Kundli, Sonepat, HR, India*

ABSTRACT

Rice (*Oryza sativa*) belongs to the family of grasses, and it is a major industrial cereal crop. Rice feeds more than half of the total world population and is cultivated in more than 100 countries across the World. India and China alone contribute 50% (approx.) of the total global trade (USDA, 2018). Brown Rice is obtained after de-husking paddy and is followed by the removal of bran to obtain white rice. Bran is abundant in oil, protein, and minerals, while the endosperm contains a high amount of starch. Parboiling of rice improves the nutrient contents inside the rice kernel. Starch and protein are the two main components of rice and rice protein unique hypoallergenic. Rice protein is also rich in lysine relative to other cereals. Rice proteins are found in two types of protein bodies (PB-1 and PB-2). These PB provide required nitrogen at the time of germination. Proteins can be extracted from rice by exploiting the physiochemical properties of the different protein present in rice. Rice consists of four different types of protein, and their solubility vary in different solvents; albumin (water-soluble), globulin (salt soluble), glutelin (alkali-soluble), and prolamin (alcohol-soluble). By dissolving them in an appropriate solvent followed by mechanically separating the supernatant and then adjusting the pH to its isoelectric point, which results in precipitation of protein. The rice protein has a potential application in food products such as bread, biscuits, edible films, etc.

12.1 INTRODUCTION

Rice (*Oryza sativa*), an important crop belongs to the kingdom Plantae, Division Angiosperms, Class Monocots, Order Poales, Family Poaceae genus Oryza, and Species *O. sativa* (Sharif et al., 2014; Verma et al., 2015), is considered to be the staple food for millions of people around the globe and is cultivated in more than 100 countries across the World. India and China alone contribute 50% (approx) of the total global trade (USDA, 2018). Rice is mainly consumed in cooked/boiled form. The composition, granular structure and unique functional properties of rice such as flavor carrying capacity, water and oil absorption, digestibility, solubility, and hypoallergenic etc. make the grain useful in various value-added products such as gluten-free bread, beverages, meat products, low fat sauces, puddings, and salad dressings (Bhattacharyya et al., 2004).

12.2 COMPOSITION AND STRUCTURE OF RICE

The grain's chemical nature and composition vary depending on cultivar variety, soil, and the environment it is grown in. The mature rice grains are harvested as rough rice commonly known as paddy, it comprises a starchy white kernel covered by a bran layer coating, and caryopsis (brown rice) enclosed in a tough siliceous hull as shown in Figure 12.1. The rice grain comprises the hull (16–28% dry mass basis) and the caryopsis (Zhou et al., 2002). The mass distribution of the rice caryopsis (Hinton and Shaw, 1954) is pericarp, 1–2%; aleurone plus seed coat and nucleus, 4–6%; embryo, 2–3%; and starchy endosperm, 89–94%. The percentage of protein (Nx6.25), fat, available carbohydrate, fiber, and ash in brown rice are 7.3%, 2.2%, 64.3%, 0.8%, and 1.4%, respectively (Zhou et al., 2002).

Brown rice is obtained after the removal of the husk or hull by the process known as dehusking or dehulling. Dehusking is the first step in the milling operation of rice, when dehusking is done the bran layer is exposed to air, and due to the activity of endogenous lipase enzyme, brown rice develops off-flavor. Moreover, the brown rice is not appealing to the consumers due to its color, but its caryopsis is rich in nutrients (Sharif et al., 2014).

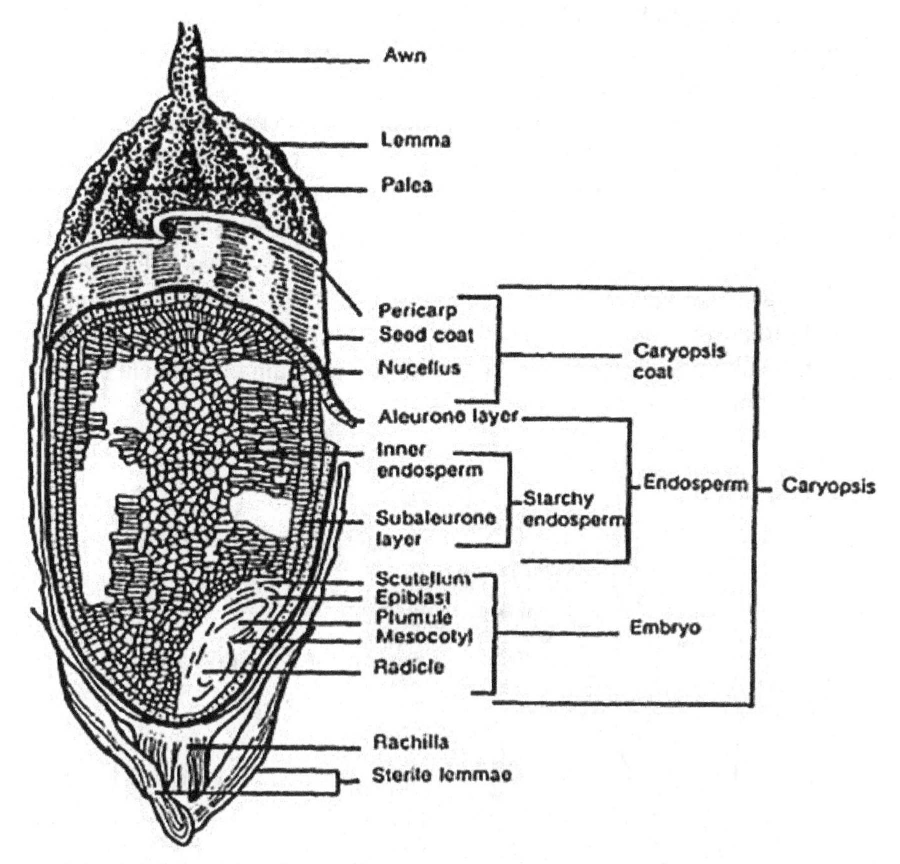

FIGURE 12.1 Structure of rice grains.
(*Source*: Reprinted with permission from Amagliani et al., 2017b. © 2017 Elsevier.)

Parboiling is a unit operation performed before dehulling to increase the nutrient concentration in the white starchy kernel. In parboiling, the paddy is exposed to steam and water, which transfers the water-soluble nutrients from the bran to the center of the rice. The caryopsis is subjected to abrasive and friction force by abrasion and friction type rollers to produce milled or white rice; the abrasive and friction force removes the bran that comprises the pericarp, seed coat, nucellus, aleurone, pulverized embryo, some starchy endosperm, and husk fragments thus making rice more appealing.

12.3 RICE PROTEIN SYSTEM

The second major component in rice is protein, next to starch and contributes approximately 8% and 80%, respectively. The nutritive quality of rice protein ranks high among the cereal proteins but considerably lower than animal and legume sources of protein (Amagliani et al., 2017; Ju et al., 2001). The rice protein is present as protein bodies (PB). The mature grain endosperm contains two protein body types PB-I & PB-II, and these bodies are present in higher quantity in the sub-aleurone layer as compared to the central part, while none of the PB were seen in the aleurone cells. PB-I and PB-II has different structure and composition, as shown in Figure 12.2. PB-I is spherical with a lamellar structure and is the site for the accumulation of prolamin while, PB-II has no lamellar structure and does not contain prolamin subunits. Glutelin and Globulin are the main component of PB-II (Tanaka et al., 1980).

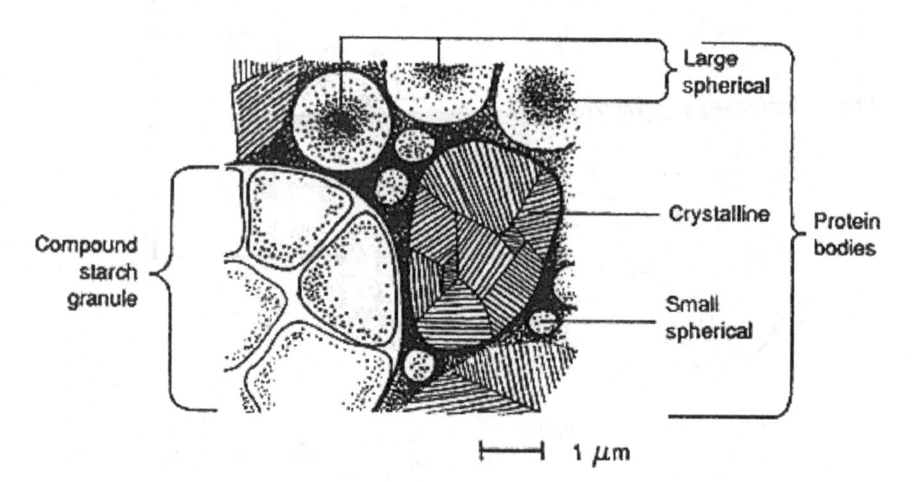

FIGURE 12.2 Schematic structure of rice protein bodies and compound starch granule in the endosperm subaleurone layer.
(*Source*: Reprinted with permission from Amagliani et al., 2017b. © 2017 Elsevier.).

Rice proteins are categorized into four fractions, and these classifications are based on the solubility of the different types of protein (Osborne, 1924). Further, Amagliani et al., (2017) reported that rice protein fractions are categorized as water-soluble albumin, salt soluble globulin, alkali & acid soluble glutelin, and alcohol soluble prolamin. Glutelin is the major protein fraction of the rice protein, while prolamin is present the minor fractions. According to Juliano (1985), the percentages of different protein fraction present in rice are albumin (5–17%), globulin (4–12%), glutelin (73–86%) and prolamin (2–5%). The total protein content and the different protein fraction depends upon the cultivar of the rice as depicted in Table 12.1, and the composition of amino acids in different protein fractions is shown in Table 12.2.

TABLE 12.1 Total Protein Content and Protein Fraction in Different Cultivar of Rice

Cultivar	Protein Content (Nx5.95)	Protein Fractions (%)			
		Albumin	Globulin	Prolamine	Glutelin
Tam	10.5	7.0	10.9	8.6	73.5
Du	9.2	8.8	9.4	7.8	74.0
Gie	8.8	10.5	8.3	7.6	73.6
Sai Duong	10.6	4.0	12.5	6.4	77.1
Nep Mua	10.8	8.2	10.5	6.0	75.3
Bau Chau Qui	9.7	7.1	7.6	6.5	78.8
Cuom Chau Qui	8.7	6.2	10.2	7.9	75.7
Tran Chau Lun	9.6	5.9	8.3	8.1	77.7
NN5	7.8	5.4	9.0	7.0	78.6
NN8	8.9	5.8	6.0	8.2	80.0
NN756	7.0	7.4	7.1	5.5	80.0
NR2151	10.6	8.9	11.2	9.5	70.4

Source: Adapted from Khoi et al., (1987).

TABLE 12.2 Amino Acid Composition of Basmati Dehraduni Rice Protein and Individual Fractions

Amino Acid	Rice Protein fractions				
	Albumin	Globulin	Prolamine	Glutelin	Total Protein
Aspartic Acid	9.13±0.22	6.91±0.09	7.34±0.05	9.14±0.42	7.30±1.40
Glutamic Acid	12.22±0.03	15.84±0.44	21.84±0.37	17.40±1.66	16.31±1.70
Serine	4.15±0.06	4.86±0.03	4.76±0.30	4.50±0.04	4.99±0.59
Glycine	5.96±0.21	6.23±0.08	2.96±0.18	3.86±0.36	4.42±0.76
Histidine*b	1.90±0.20	1.64±0.08	1.17±0.03	1.90±0.27	2.11±0.05
Arginine	7.75±0.30	9.87±0.09	6.14±0.07	9.27±1.08	8.68±0.41
Threonine*	3.12±0.30	2.24±0.18	2.05±0.19	2.69±0.25	3.16±0.45
Alanine	7.61±0.01	6.41±0.11	6.37±0.07	4.99±0.23	6.42±0.26
Proline	5.64±0.28	6.06±0.16	4.44±0.08	4.45±0.14	6.04±0.76
Tyrosine*	4.26±0.10	6.55±0.02	9.60±0.55	5.94±0.33	3.81±0.49
Valine	6.77±0.11	5.93±0.07	6.24±0.07	6.22±0.08	7.03±0.51
Isoleucine*	3.79±0.02	2.85±0.04	4.90±0.22	4.63±0.24	4.78±0.12
Leucine*	8.07±0.08	7.96±0.17	13.41±0.81	8.93±1.37	9.03±0.71

TABLE 12.2 *(Continued)*

Amino Acid	Rice Protein fractions				
	Albumin	Globulin	Prolamine	Glutelin	Total Protein
Phenylalanine*	3.77±0.08	3.82±0.03	6.35±0.36	6.39±0.01	5.64±0.32
Lysine*	10.01±0.61	4.33±0.11	0.50±0.17	6.37±0.17	5.32±0.50
Tryptophan*	2.68±0.06	2.30±0.00	0.76±0.06	1.55±0.28	1.63±0.01
		Distribution of Amino Acids			
Hydrophobic	47.46	47.76	46.61	42.79	48.32
Hydrophilic	11.53	13.67	16.41	13.13	11.96
Basic	19.66	15.84	7.81	17.54	16.11
Acidic	21.35	22.75	29.18	26.54	23.61
Limiting Essential Amino Acid	Threonine	Threonine	Threonine	Threonine	Threonine
E/T%[d]	44.37	37.64	44.98	44.62	42.51

[a] Measured as g/16g N+- standard error of the mean (SEM); [b,*] essential amino acids; [c] Based on the WHO/FAO (1985) Amino Acid requirement; [d] Proportion of total Essential Amino Acids and Total Amino Acid as calculated by WHO/FAO (1985).

Source: Adapted from Steenson et al., (1995).

1. **Glutelin:** It is defined as the protein fraction of rice protein which is soluble in dilute alkaline/acid solution and forms the main component of the protein body (PB-II) (Sawai et al., 1968). Glutelin contributes to large portion of the endosperm protein in rice and wheat. In rice, 86% of the endosperm protein is glutelin, and this may lead us to consider that it is synthesized as a reserve protein (Sawai et al., 1968). Glutelin has the second highest lysine content present in rice after albumin. In the total rice glutelin fraction, the most abundant amino acids are glutamine/glutamic acid, asparagine/aspartic acid, alanine, arginine, and glycine while methionine, tryptophan, and histidine are the least abundant amino acids in the total rice glutelin (Wen et al., 1985). The molecular mass peak was observed at a range of 11–52 kDa when analyzed using MALDI-TOF mass spectrometry (Adebiyi et al., 2009).
2. **Globulin:** Globulin is defined as the protein fraction of rice protein which is soluble in an aqueous solution of salt and is the second most abundant Osborne solubility group in rice (Amagliani et al., 2017). The globulin fraction is divided into α-globulin and polypeptides that have lower molecular weight than the α-globulin (Pan and Reeck, 1988). When analyzed using MALDI-TOF mass spectrometry. The molecular mass peaks shown by globulin ranges from 13 to 127 kDa

(Adebiyi et al., 2009). Among all the four types of proteins found in rice, Globulin is richest in sulfur amino acids, i.e., methionine, and cysteine, while it has a low level of lysine content as compared to albumin and globulin.

3. **Albumin:** It is defined as the protein fraction, which is soluble in water. The extraction of pure albumin fraction from rice is difficult as minerals are present in the rice bran that dissolve the globulin into the solvent (Amagliani et al., 2017). Albumin contains the highest lysine content among all other protein fractions of the rice protein. Albumin fraction gave many molecular mass peaks ranging from 11 to 76 kDa when analyzed through MALDI-TOF mass spectrometry (Adebiyi et al., 2009).
4. **Prolamin**: Rice prolamin is defined as the protein fraction, which is soluble in alcohol (Amagliani et al., 2017). Rice Prolamin consists of heterogeneous polypeptides with molecular masses of 12–17 kDa, and is similar to other cereal prolamins, also having high contents of glutamine and low level of lysine, histidine, cysteine, and methionine (Shyur et al., 1994). The rice has the highest leucine, phenylalanine, and tyrosine compared to other protein fractions. These fractions accumulate within PB, which are formed by direct dilation of the endoplasmic reticulum membrane (Shyur et al., 1994). Prolamine showed a high-intensity major peak at 19 kDa when analyzed through MALDI-TOF mass spectrometry (Adebiyi et al., 2009).

12.4 EXTRACTION OF RICE PROTEIN ISOLATE

1. **Alkali Extraction:** Alkali based extraction is the most common method used for economical extraction of protein from cereals, especially rice at the laboratory scale due to its convenience, ease, simplicity, and cost-effectiveness. The disadvantages of this method consist of large buffer use, time consumed, lower yield purity, and chances for exposure of the material to severe alkaline conditions that may deteriorate protein quality. De Souza et al., (2016) evaluated rice protein obtained by a fast alkaline extraction (ALK) method derived from broken rice. Broken rice has low economic value; therefore, it is advantageous to use broken rice obtained from rice milling to produce rice flour (RF) for protein extraction. The RF is used for protein extraction with the objective of reducing contact time required with the solvent, the traditional method involved soaking of

rice for a 24 hour period in the solvent, which was cumbersome and posed difficulty during extraction. The use of RF reduces the contact time required with solvent to 30 minutes, and this reduction in contact time is advantageous because lower contact time reduces the quality deterioration chances. The protein-rich and starch-rich fractions separate out by centrifugation. Protein concentrate with 79% protein is obtained from the protein-rich fraction by isoelectric precipitation, and also the protein profile of RF and isoelectric precipitation protein concentrate is found to be similar. In this method, 100 g of RF is added to 1500 mL of 0.18% NaOH and incubated at 30°C in a thermal bath which is coupled to a magnetic stirrer table. Now to separate the protein extract, the suspension is centrifuged at 3380 xg for 5 min. and the protein extract for precipitation of proteins is separated. The starchy residue in protein extract is washed with 250 mL of 0.18% NaOH and is again centrifuged at 3380 xg for 5 min, and this extract is added to the protein extract previously obtained. The protein is precipitated from protein extract by adjusting the protein extract pH to 4.8, i.e., isoelectric point by using 48 mL of 0.1 M HCl. The isoelectric precipitated protein concentrated (IPPC) is neutralized and is dried at 28 ± 2°C.

2. **Microwave-Assisted Extraction**: Microwave treatment is a well known, powerful, and convenient method. Microwave treatment has the extraction capabilities of solid-liquid interfaces; it can enhance the extraction from biological matter because of its ability to break down hydrogen bonds and also induces dipole rotation. Therefore it can be used to enhance the extraction methods to enable greater extraction yield along with the better efficiency and good quality. Phongthai et al., (2016) optimized conditions required to extract protein from rice bran. In MAE method 0.89g of defatted rice bran is mixed with the 10 ml distilled water. After adjusting the pH of the solution to 10 by using 3 M sodium carbonate, it is kept on the rotating plate of the microwave machine. In the microwave, the solution is exposed to 1000 W of microwave power for 90 seconds. Afterward, the slurry is cooled down, and for 10 minutes it is centrifuged at 10,000 g, and the temperature is set at 4°C. After centrifugation, the supernatant is collected and filtered followed by the isoelectric precipitation of the protein. Protein precipitation is performed by pH adjustment of the supernatant to about 4.5 to obtain IPPC. The derived IPPC is neutralized by pH adjustment to 7.0, followed by the freeze-drying process. The protein yield of MAE is found to be 1.54 fold greater

than that of the protein yield of the ALK method ($P<0.05$), while the protein extracts from both the methods have shown the similar protein digestibility. Also, MAE yields higher protein compared to the ALK method. Protein extracted by Microwave-Assisted Extraction is found to be superior than that of the soy protein concentrates.

3. **Enzymatic Extraction**: The enzymatic extraction method is commonly used to effect protein extraction and isolation, particularly in rice proteins it has historically lead to yields with 40% purity at most. Hence, a mass reproducible production method that could yield highly purified Rice protein Isolate was discussed by Morita and Kiriyama (1993), and promising results were obtained by using thermally stable a-amylase (Termamyl 120L). Thick slurry of RF is prepared by mixing 5 kg of flour in 10 L of distilled water containing 0.6% Termamyl the mixture is then indirectly heated in a water bath maintained at 97°C for 2 hr. This leads to the digestion and gelatinization of starch, and now the mixture is vacuum filtered through cheesecloth. The precipitate is washed three times using boiling water to remove sugar syrup and remaining enzyme followed by drying at room temperature. The preparation is Rice protein isolate. The protein recovery is also reported to be high (85.2–92.3%), which is about 10 times higher than that of original RF. Morita and Kiriyama (1993) reported that the rice protein isolate prepared by a heat stable α-amylase (Termamyl 120L) contains 82.1% protein, 0.6% lipids, 1.1% carbohydrate, 1.3% ash, 6.4% dietary fiber and 8.5% moisture. The dietary fiber amount was determined by the method referred in Prosky et al., (1988).
4. **Extraction of Rice Protein by Fractions:** Ju et al., (2001) described the process of extraction of the proteins from RF by using a combination of solvents. In this method, RF (100g) is firstly defatted using hexane (400mL) and then dried at 20°C for 24 hours. For protein extraction, RF is mixed with 400 mL of deionized water at 20°C for 24 hours using stirrer (300 rpm) to extract albumin. After the extraction of the albumin, the globulin is extracted using 5% NaCl followed by extraction of glutelin using 0.02% NaOH. The NaOH is used to adjust the pH to 11, and in both these steps, 20°C is maintained, and mixing is done for 30 min. The precipitation of the protein fractions (albumin, globulin, and glutelin) is done by pH adjustment of their respective supernatants to their isoelectric points. The isoelectric point of albumin, globulin, and glutelin is 4.1, 4.3, and 4.8, respectively. Prolamin extraction is a complex process due

to extensive cross-contamination with other rice protein fractions; therefore, needs to be extracted separately. In order to prevent cross-linking and to extract the prolamin fraction, the extraction process for albumin, globulin, and glutelin is to be repeated twice. The precipitation of the prolamin is done by adding a threefold volume of acetone to the supernatant. The precipitated protein fractions are neutralized by washing with deionized water.

12.5 RICE PROTEIN QUALITY

Protein ranks second in abundancy among the constituents of milled rice, followed by starch and may vary substantially depending on rice variety and process of milling. Rice protein content on an average can comprise nearly 8% of the grain, a relatively low amount but possessing high nutritional value. It is considered superior due to its unique amino acid composition and the balanced proportions of 8 essential amino acids (Verma and Srivastava, 2017). Its amino acid profile comprises of good amounts of both essential and nonessential amino acids, with the latter being present in larger quantities. It comprises of 9 amino acids like lysine, methionine, and threonine that are essential for humans. Lysine is also the limiting amino acid when it comes to rice. Rice protein is well known as a hypoallergenic that enables its use as an alternative nutrition choice in medical cases. Its nutritional quality is considered equal or greater than other cereals but relatively low in comparison with proteins derived from animal sources, legumes, and oilseed crops.

Rice albumin has also been noted to have a suppressive effect on the increase in blood glucose levels post oral consumption of glucose and starch. Wheat albumin is also known to have the same function, but only for starch loading, rice albumin is believed to work similar to dietary fiber at the small intestine by suppressing absorption, making it useful in diabetes treatment (Ina et al., 2016). The protein efficiency ratios (PER) of the rice fractions, namely rice bran, polish, germ, and bran sub-fractions enriched with protein from ALK, are very high for proteins derived from a cereal source, similar to ones seen on an amino acid chemical score. The biological availability of rice bran protein is, however, lower when compared to animal protein sources but shows high protein superiority in comparison to other vegetable protein sources.

Sanders (1985) discussed the PER values of protein from different milling byproducts of rice obtained through alkali extraction of rice bran. Protein enriched bran, and its sub-fractions are found to have the highest PER values among the byproducts, with values ranging from 2.31–2.49 and 2.24–2.38, respectively. Bran post ALK is seen to have values ranging 1.99–2.19, with normal and defatted bran having the lowest values of 1.61 and 1.9, respectively. Germ fractions are found to have values ranging around 1.74–1.9 depending on variety and slight changes within the method of extraction. Rice bran protein showed values of 72.26, 7.07, 3.77 and 2.29, respectively for Biological Value (BV), net protein utilization (NPU), net protein ratio (NPR), and PER during testing on animal subjects; these values are similar to ones found from proteins of animal sources (Table 12.3). The true digestibility (TD) value of rice bran protein is similar to that of Casein and is around 94.8% and is much higher compared to rice endosperm protein, Isolated Soy Protein and whey protein isolate whose values are 90.8%, 91.7%, and 92.8%, respectively. Han et al., (2015) discussed and compared various protein quality parameters such as BV, protein digestibility-corrected amino acid score (PDCAAS), TD, nitrogen intake and fecal nitrogen amounts obtained from different protein sources (casein, isolated soy protein, rice bran protein, rice endosperm protein and whey protein isolate). Nitrogen intake values ranged from 1.12 to 1.30 for all the sources, with casein and WPI representing the lowest and highest values, respectively. Fecal nitrogen values were found to be higher in diets consisting of REP, ISP, and WPI in decreasing order. The BV of RBP (72.6) is superseded by only WPI (78.8), Casein has the lowest value of 59.7 amongst the sources, further proving that rice protein is of very high quality (Table 12.3).

TABLE 12.3 Functional Properties, Total Lysine, Tryptophan, and Methionine Content of Different Rice Varieties

Varieties	BV[a]	TD[b]	NPU	UN	Crude Protein	Lysine	Tryptophan	Methionine
NFSM*	89	90	80	4.84	-	-	-	-
IS**	75	95	69	1.12	10.18	3.91	1.07	2.85
Pusa 2–21	75	92	68	1.20	10.99	3.83	1.26	2.30
Padma	79	94	74	1.06	9.21	4.27	1.08	2.21
Basmati- 370	69	95	65	0.96	9.06	3.72	1.01	1.87
Jaya	-	-	-	-	8.95	4.19	1.06	2.48

[a]BV, corrected for metabolic and endogenous losses of nitrogen; [b]TD, corrected for metabolic losses of nitrogen; NFSM* for Nonfat skimmed milk; IS** for Improved Sabarmati.

Source: Adapted from Chavan et al., (1978).

12.6 FUNCTIONAL PROPERTIES

Rice protein is lacking in certain functional properties such as its poor water solubility which leads to its underutilization in comparison to other plant protein products (Shih, 1996). Solubility is a critical parameter that directly influences its applicability as a functional ingredient in food systems. Glutelin is the major protein fraction of the rice protein and contributes highly towards the overall protein functionality of rice proteins. It is extracted from rice by using alkali or acid because it is soluble in both alkaline and acidic conditions.

1. **Emulsification:** Emulsifying properties are affected by the rate of protein adsorption at the Oil-Water interface, the extent of adsorption and reduction in interfacial tension achieved. Emulsifying properties of proteins are evaluated by measuring the distribution of droplet size post homogenization and storage. Rice proteins (mainly glutelins) are not great emulsifiers primarily because of their high molecular weight and poor solubility. In general, proteins with poor solubility in aqueous systems and high disulfide bonding make poor emulsifiers (Figure 12.3). Studies show that emulsions formed from glutelin fractions had an average particle size that was uncharacteristically higher compared to emulsions from whey proteins. Poor emulsifying capacity of Rice bran protein isolate is also attributed to the weaker protein and oil interactions conferred by the low surface hydrophobicity of rice bran (Wang et al., 1999).
2. **Foaming Characteristics and Solubility:** It was found to increase slowly between pH 4 to 7 and a rapid increase up to pH 9. Enzymatic pre-processing of glutelin fractions was shown to increase its solubility at higher pH values, the overall characteristics are in agreement with glutelins being soluble in dilute alkali (Agboola et al., 2005). Extremely acidic or alkali conditions can also improve solubility by virtue of dissociation of glutelin aggregates. Solubility of white rice was found to be lower than that of rice bran proteins in the pH range of 4–7; however, it was the inverse in pH ranges below and above (Figure 12.3).
3. **Foaming Capacity**: It describes the amount of interfacial area that proteins are able to stabilize per unit weight or concentration. Foam stability refers to the ability of proteins to stabilize the foam against

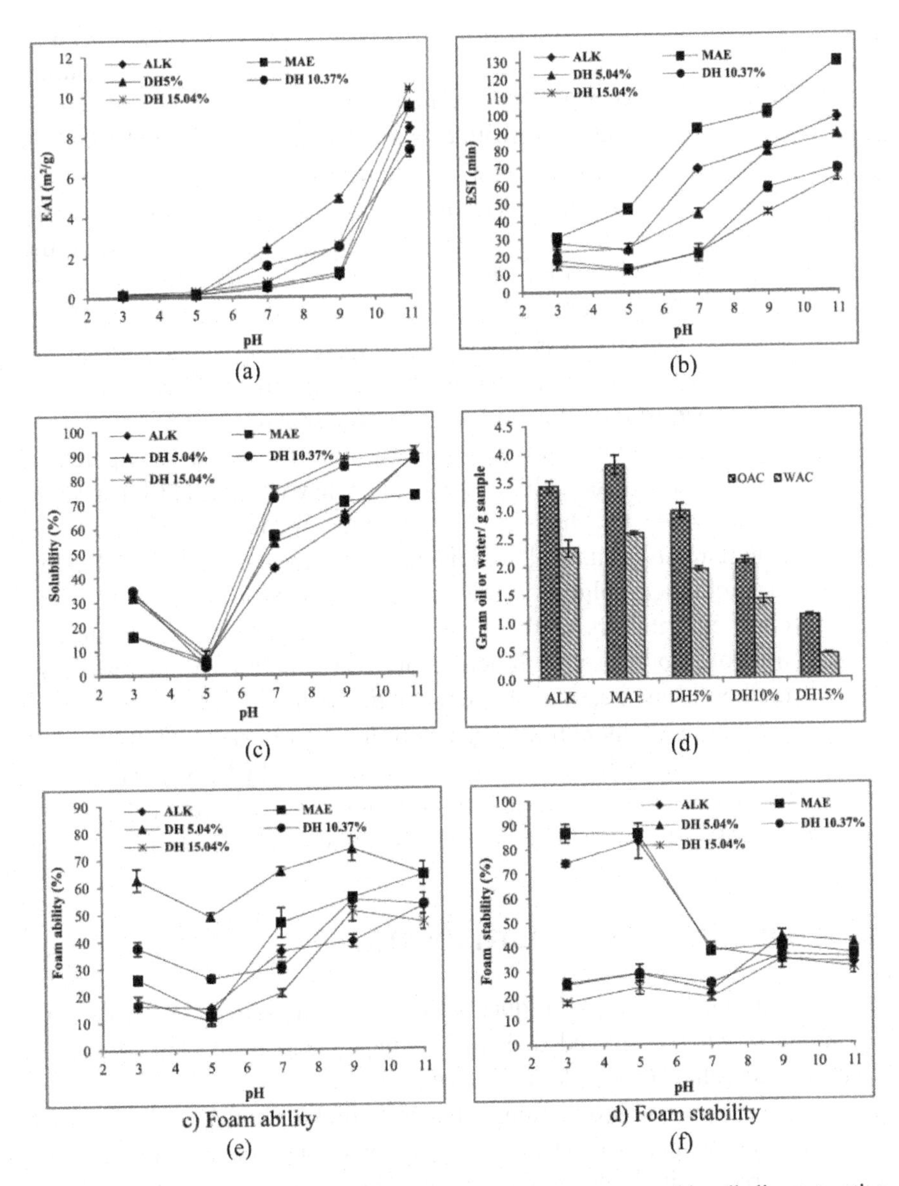

FIGURE 12.3 Functional properties of protein concentrate prepared by alkaline extraction (ALK), microwave-assisted extraction (MAE) (a) Emulsion activity index (EAI); (b) Emulsion stability index; (c) Protein solubility; (d) Oil and Water absorption capacity (OAC and WAC); (e) Foam ability; and (f) Foam stability.

(*Source*: Reprinted with permission from Phongthai et al., 2016. © 2016 Elsevier.)

gravitational and mechanical stress. These values are also found to be low, poor solubility and high disulfide linkages again contributing towards it. This can be overcome by modification of the proteins by enzymatic or chemical treatments to make it more favorable towards its usage as surface-active components. According to Amagliani et al., (2017), foaming capacity of rice bran increases greatly with an increase in the concentration of NaCl from 0.4%–2% while the foaming capacity of brown and white rice increases at NaCl concentrations of 0.4%–0.8%, greater concentrations do not have any effects. Foaming capacity was also found to increase significantly from pH 5–11 in white rice, rice bran, and brown rice proteins (Figure 12.3e).

4. **Gel Formation:** Glutelins are however shown to form stable gels at lower minimum gelation concentration (MGC) as compared to commercially used gelling agents like whey protein that form stable gels at 9%–10%, thus providing promise for its extension towards commercial use as well. Ironically, the high disulfide cross-linking that inhibited solubility and foaming characteristics is one of the chief contributors to the superior gelling property of the glutelin fraction. Food protein functionality can generally be increased by methods like heating and enzymatic hydrolysis, the degree and conditions of its application determining the extent of increase in functionality achieved. Limited or controlled hydrolysis is advised to preserve the overall quality of protein; extensive hydrolysis may improve solubility but may bring down emulsification and foaming.

12.7 APPLICATIONS IN FOOD INDUSTRY

The rice milling industries produce a huge amount of byproducts that are discarded or poorly utilized due to their physicochemical characteristics. Byproducts such as broken kernels, rice husk, and bran are used as feed for livestock, rice bran oil production and even as fuel for boilers. The potential for extracting greater utility from rice bran mostly remains untapped, especially in terms of its high protein quality. The degree of utilization of rice bran protein and rice bran itself is limited mainly due to the fiber content of the bran layer and the poor stability, which comprises a variety of hydrolyzing enzymes like lipases and oxidases that can develop deterioration through rancidity due to the breakdown of bran oil into glycerol and FFA. This deterioration is even high after milling due to the abrasion of the grain surface during the process. A large number of uses for rice and its protein fractions have been established through research but have not percolated into commercial practice; however, its usage in the preparation of specialty breads and other baked items is very well recognized.

1. **Food Formulation Potential:** Rice is commonly regarded as a food product that acts as a natural hypo allergen, it is among the first solid foods served to infants during their post-weaning transition and also used extensively in elimination diets for food allergy diagnostics in children and adults (Amagliani et al., 2017). Moreover, rice polish finds use in baby foods and infant formulations. Rice protein, with its inherent amino acid ratio, is a highly nutritious source of consumable protein, this combined with its hypoallergenicity and other qualities make it a highly valued ingredient for food formulations. Commercial allergen friendly protein food powders have also been introduced in the market as critical protein sources that contain all the essential amino acids that act as protein supplements and as an alternative protein ingredient in food formulations. There has been an increase in the utilization of rice protein ingredients as supplements by various athletes as an alternative to soy and whey proteins. Certain studies have also stated that rice protein can be employed as value-added ingredients in producing baked products such as bread and biscuits (Amagliani et al., 2017). Extensively hydrolyzed rice protein formulas (eRHFs) have been shown to be acceptable alternatives for infants and children diagnosed with cow's protein milk allergy (CMPA), this has been established through clinical trials of a test population with over 90% of trial subjects tolerating the rice protein alternative over a month-long period of study (Vandenplas et al., 2014). RF continues to be the most widely incorporated ingredient in products like bread and pasta that are gluten-free, as it adds towards overall quality and functionality.
2. **Edible Films Derived from Rice Protein:** Edible films from concentrates have also been proposed and developed in various research studies. Edible rice protein films prepared by combining rice protein with polysaccharides like pullulan and alginate to give rise to food grade films having appreciable tensile strength (TS), film strength and water vapor permeability (WVP) (Shih, 1996). Pure pullulan films or ones consisting low rice protein have been shown to have high relative TS of up to 30mPa but fragile under normal conditions; increased rice protein addition can lead to reduced brittleness and film strength. However, for films with greater than 50% protein

concentrate, water permeability increases. Films prepared from 53% rice protein concentrate in 1:1 ratio with pullulan is shown to provide optimal characteristics. Addition of small amount 1.5% (w/w) of propylene glycol alginate (PGA) and oil improved the film's cross-linking and overall barrier properties, which shall be used in the food packaging industry as edible food wrapping (Shih, 1996). Numerous other research conducted on the usage of rice protein isolate and concentrate to form edible biofilms for their applicability in edible packaging. Protein films formed from rice bran protein concentrate along with phenolic extracts obtained through solid state fermentation of rice bran with other additives also produced satisfactory results. The film may find their application as edible coatings on produce such as raw and minimally processed (MP) fruits and vegetables, as the coating can form a semi-permeable barrier impervious to gases and water vapor, maintaining quality of the produce and can also act as potential carriers for additives to improve product quality and preserve it. Schmidt et al., (2015), Gnanasambandam et al., (1997) and Shin et al., (2011) also elaborate the viability of producing edible food-grade films and packaging through rice bran protein and other additives like plasticizers, and their various mechanical, textural, and barrier properties have also been studied.

KEYWORDS

- **edible film**
- **functional properties**
- **hypoallergenic properties**
- **prolamin**
- **protein bodies**
- **rice protein**

REFERENCES

Adebiyi, A. P., Adebiyi, A. O., Hasegawa, Y., Ogawa, T., & Muramoto, K., (2009). Isolation and characterization of protein fractions from deoiled rice bran. *European Food Research and Technology*, *228*(3), 391–401.

Agboola, S., Ng, D., & Mills, D., (2005). Characterization and functional properties of Australian rice protein isolates. *Journal of Cereal Science*, *41*(3), 283–290.

Amagliani, L., O'Regan, J., Kelly, A. L., & O'Mahony, J. A., (2017a). Composition and protein profile analysis of rice protein ingredients. *Journal of Food Composition and Analysis*, *59*, 18–26.

Amagliani, L., O'Regan, J., Kelly, A. L., & O'Mahony, J. A., (2017b). The composition, extraction, functionality and applications of rice proteins: A review. *Trends in Food Science & Technology*, *64*, 1–12.

Bhattacharyya, P., Ghosh, U., Chowdhuri, U. R., Chattopadhyay, P., & Gangopadhyay, H. (2004). Effects of different treatments on physico-chemical properties of rice starch. *Journal of Scientific & Industrial Research*, *63*, 826–829.

Chavan, J. K., & Duggal, S. K., (1978). Studies on the essential amino acid composition, protein fractions and biological value (BV) of some new varieties of rice. *Journal of the Science of Food and Agriculture*, *29*(3), 225–229.

Childs, N., (2018). *Rice Outlook*. USDA-ERS, Washington DC.

De Souza, D., Sbardelotto, A. F., Ziegler, D. R., Marczak, L. D. F., & Tessaro, I. C., (2016). Characterization of rice starch and protein obtained by a fast alkaline extraction method. *Food Chemistry*, *191*, 36–44.

Gnanasambandam, R., Hettiarachchy, N. S., & Coleman, M., (1997). Mechanical and barrier properties of rice bran films. *Journal of Food Science*, *62*(2), 395–398.

Han, S. W., Chee, K. M., & Cho, S. J., (2015). Nutritional quality of rice bran protein in comparison to animal and vegetable protein. *Food Chemistry*, *172*, 766–769.

Hinton, J. J. C., & Shaw, B., (1954). The distribution of nicotinic acid in the rice grain. *British Journal of Nutrition*, *8*, 65–71.

Ina, S., Ninomiya, K., Mogi, T., Hase, A., Ando, T., Matsukaze, N., Ogihara, J., Akao, M., Kumagai, H., & Kumagai, H., (2016). Rice (*Oryza sativa japonica*) albumin suppresses the elevation of blood glucose and plasma insulin levels after oral glucose loading. *Journal of Agricultural and Food Chemistry*, *64*(24), 4882–4890.

Ju, Z. Y., Hettiarachchy, N. S., & Rath, N., (2001). Extraction, denaturation and hydrophobic properties of rice flour proteins. *Journal of Food Science*, *66*(2), 229–232.

Juliano, B., (1985). Polysaccharides, proteins and lipids of rice. *Rice: Chemistry and Technology*, 59–174.

Khoi, B. H., Dien, L. D., Lásztity, R., & Salgó, A., (1987). The protein and the amino acid composition of some rice and maize varieties grown in North Vietnam. *Journal of the Science of Food and Agriculture*, *39*(2), 137–143.

Morita, T., & Kiriyama, S., (1993). Mass production method for rice protein isolate and nutritional evaluation. *Journal of Food Science*, *58*(6), 1393–1396.

Pan, S. J., & Reeck, G. R., (1988). Isolation and characterization of rice alpha-globulin. *Cereal Chemistry (USA)*, *65*(4), 316–319.

Phongthai, S., Lim, S. T., & Rawdkuen, S., (2016). Optimization of microwave-assisted extraction of rice bran protein and its hydrolysates properties. *Journal of Cereal Science*, *70*, 146–154.

Prosky, L., Asp, N. G., Schweizer, T. F., DeVries, J. W., & Furda, I., (1988). Determination of insoluble, soluble, and total dietary fiber in foods and food products: Interlaboratory study. *Journal-Association of Official Analytical Chemists*, *71*(5), 1017–1023.

Saunders, R. M., (1985). Rice bran: Composition and potential food uses, *Food Reviews International*, *1*(3), 465–495.

Sawai, H., & Morita, Y., (1968). Studies on rice glutelin. 2. Gross-structure of glutelin from rice endosperm. *Agricultural and Biological Chemistry*, *32*(4), 496.

Sawai, H., & Morita, Y., (1968). Studies on rice glutelin. *Agricultural and Biological Chemistry*, *32*(1), 76–80.

Schmidt, C. G., Cerqueira, M. A., Vicente, A. A., Teixeira, J. A., & Furlong, E. B., (2015). Rice bran protein-based films enriched by phenolic extract of fermented rice bran and montmorillonite clay. *CyTA-Journal of Food*, *13*(2), 204–212.

Sharif, M. K., Butt, M. S., Anjum, F. M., & Khan, S. H., (2014). Rice bran: A novel functional ingredient. *Critical Reviews in Food Science and Nutrition*, *54*(6), 807–816.

Shih, F. F., (1996). Edible films from rice protein concentrate and pullulan. *Cereal Chemistry*, *73*(3), 406–409.

Shin, Y. J., Jang, S. A., & Song, K. B., (2011). Preparation and mechanical properties of rice bran protein composite films containing gelatin or red algae. *Food Science and Biotechnology*, *20*(3), 703–707.

Shyur, L. F., Wen, T. N., & Chen, C. S., (1994). Purification and characterization of rice prolamins. *Bot. Bull. Acad. Sin.*, *35*, 65–71.

Steenson, D. F., & Sathe, S. K., (1995). Characterization and digestibility of Basmati rice (*Oryza sativa* L. var. *Dehraduni*) storage proteins. *Cereal Chemistry*, *72*(3), 275–280.

Tanaka, K., Sugimoto, T., Ogawa, M., & Kasai, Z., (1980). Isolation and characterization of two types of protein bodies in the rice endosperm. *Agricultural and Biological Chemistry*, *44*(7), 1633–1639.

Vandenplas, Y., De Greef, E., & Hauser, B., (2014). Paradise study group safety and tolerance of a new extensively hydrolyzed rice protein-based formula in the management of infants with cow's milk protein allergy. *European Journal of Pediatrics*, *173*(9), 1209–1216.

Verma, D. K., & Srivastav, P. P., (2017). Proximate composition, mineral content and fatty acids analyses of aromatic and non-aromatic Indian rice. *Rice Science*, *24*(1), 21–31.

Verma, D. K., Mohan, M., Prabhakar, P. K., & Srivastav, P. P., (2015). Physico-chemical and cooking characteristics of Azad basmati. *International Food Research Journal*, *22*(4), 1380–1389.

Wen, T. N., & Luthe, D. S., (1985). Biochemical characterization of rice glutelin. *Plant Physiology*, *78*(1), 172–177.

Zhou, Z., Robards, K., Helliwell, S., & Blanchard, C., (2002). Composition and functional properties of rice. *International Journal of Food Science & Technology*, *37*(8), 849–868.

CHAPTER 13

Application of Natural Biopolymer Films as Edible Coatings on Cut Fruits and Vegetables

ARADHITA BARMANRAY and INDU BHARTI

Department of Food Technology, Guru Jambheshwar University of Science and Technology, Hisar, Haryana–125001, India,
E-mail: dhitaray@gmail.com

ABSTRACT

The production of fresh-cut fruits and vegetables (CFV) is increasingly becoming popular in today's world as consumers are becoming more aware about the healthy eating habits and giving emphasis on adding more varieties of fresh products to their diet. Modern hectic lifestyle demands quick and hygienic cooking practices of food; hence the new processed food market trend has enabled the industry to look out for more new and innovative ways to increase the storability and shelf life of these products. CFV acts as a convenient and more reliable replacement for fresh fruits and vegetables with an additional benefit of reduced wastage, and their processing results in loss of color, texture, and moisture, but if not dealt properly; can also lead to rapid quality deterioration. As a result, packaging and storage of CFV are becoming more and more challenging. These reasons have prompted the development of "Edible Biopolymer Films" which not only improve the general appearance of the products but also creates suitable techniques for application of a layer of any edible substances on the surface of CFV by providing modified atmosphere, less moisture loss and retarding gas transfer. These films are prepared from biopolymer sources including polysaccharides, proteins, and lipids which may also act as carriers for ingredients like antimicrobial and antioxidant agents, thus extending their shelf life while retaining the freshness of the products as well. The materials most commonly used to form edible films include starch, cellulose, alginate, carrageenan, zein, gluten, whey, carnauba, beeswax, chitosan, and fatty acids from natural sources whereas, additives such as antimicrobials, antioxidants, and nutrients are also added to the coating formulation to help preservation and quality retention of fresh-cut products. The applications of these films are affected by their mechanical, barrier, and thermal properties, which have been discussed in this chapter.

13.1 INTRODUCTION

The production of minimally processed (MP) ready-to-eat fruits represents a bigger challenge, as these products deteriorate faster than their intact counterpart, mainly due to the damage caused to cells and tissues by cutting, slicing, and removal of their natural protective skin. Despite respiration, fruits, and vegetables demonstrate wound-induced ethylene production and increased surface area per unit volume exacerbate water loss resulting in undesirable changes in product flavor, texture, and nutritional quality as reported by Gonzalez-Aguilar et al., (2010). Olivas and Canovas (2005) stated that biopolymer films may be good substitute to help in the preservation of MP fruits, which are able to provide barriers to moisture, oxygen, and carbon dioxide, improving mechanical as well as handling properties while carrying additives, avoiding volatiles loss and even contributing to the production of aroma volatiles. However, Siddiqui et al., (2011) mentioned that physiology of MP fruits and vegetables is generally typical because plant tissues are wounded when exposed to stress conditions, leading to faster deterioration, that are affected by temperature, ethylene, oxygen (O_2), carbon dioxide (CO_2) concentrations and increased respiration rate. These reasons have prompted the development of "Edible Biopolymer Films" which not only improve the general appearance of the products but also help in applying a layer or coating of any edible material on the exposed surface of cut fruits and vegetables (CFV) by providing modified atmosphere, less moisture loss and retarding gas transfer. These films may also be prepared from various biopolymer sources including polysaccharides, proteins, and lipids which may also act as carriers for ingredients like antimicrobial and antioxidant agents, thus extending their shelf life while retaining the freshness of the products as well.

13.2 NATURAL BIOPOLYMER FILMS

Biopolymer films are thin films prepared from edible materials that act as barriers to control the unfavorable exchange of moisture, oxygen, carbon dioxide, flavor, aroma between the food components and the atmosphere around them which are used to enhance the shelf life, quality, and reduce the microbial load of the fresh produce (Janjarasskul and Krochta, 2010). Generally, an edible film is often known as a preformed thin coat layer or solid sheet of edible material placed on, or in between food components (Lin and Zhao, 2007) whereas, Vargas et al., (2008) defined edible biopolymer film as a thin layer of natural material that covers the surface of the food and can be eaten as such being part of the whole product. Further, Ayala-Zavala et al., (2008) added that compounds most commonly used to form edible films include gluten, starch, cellulose, alginate, carrageenan, zein protein, whey, chitosan, carnauba, beeswax, and other fatty acids. Whereas, additives such as antimicrobials, antioxidants, and nutrients may also be added to the coating formulation to extend the shelf life and quality enhancement of fresh-cut products.

13.2.1 USES AND BENEFITS OF BIOPOLYMER FILMS

Biopolymer films have an intrinsic ability to control mass transfer between food components and its environment which also accomplish a number of specific requirements, for example, function as a gas and water barrier properties, good mechanical strength and adhesion, reasonable physicochemical and microbial stability, effective carrier capability for antioxidant, antimicrobial, color, and nutritional additives, safety for human consumption, acceptable sensorial characteristics, low cost of raw materials and simple technology for production (Sonti, 2000). As reported by Toivonen and Brummell (2008), CFV has increased water transpiration rates, which may lead to a decrease in firmness, metabolic alterations, nutritional changes, and water loss in fresh-cut products. So biopolymer coatings may reduce the respiration rate of CFV by decreasing the oxygen transmission rate, thereby decreasing the internal oxygen concentration. Ramos et al., (2012) stated that the ability of these films to maintain the quality of fresh cut-products might vary depending on the composition, thickness of the film, type of the product, food surface coverage area, and storage conditions. Lin and Zhao (2007) also stated that films incorporated with antimicrobials could effectively protect fresh CFV against microbial contamination by retaining preservatives on the surface of these products where they are needed, by avoiding diffusion into the tissue. Raybaudi-Massilia et al., (2008) emphasized that changes in the color of fresh-cut products can be prevented by the use of biopolymer films, as these films have the ability to decrease oxygen transmission rate (Figure 13.1).

- Produce a modified atmosphere in the fruit and vegetable
- Reduce decay
- Delay ripening of climacteric fruits
- Reduce water loss
- Delay color changes
- Improve appearance
- Reduce aroma loss
- Reduce the exchange of humidity between fresh produce
- Carriers of volatile precursors
- Impart color and flavor
- Carriers of nutraceuticals

FIGURE 13.1 Potential uses of edible films on cut fruits and vegetables (Adapted from Olivas and Barbosa-Canovas, 2005).

13.2.2 CLASSIFICATION/TYPES OF BIOPOLYMER FILMS

Components used for the formulation of biopolymer films can be classified into three classes: hydrocolloids (as proteins, polysaccharides, and alginates), lipids (as fatty acids, acyl-glycerol, and waxes) and composite materials (Sonti, 2000; Vargas et al., 2008).

13.2.2.1 POLYSACCHARIDES FILMS

Polysaccharides films are made up of starch, starch derivatives, alginate, chitosan, pectin, microbial gums, and cellulose derivatives; as reviewed by Sonti (2000). Baldwin et al., (1995) examined that these films exhibit excellent gas permeability properties resulting in the desirable modified atmosphere which enhance the shelf life of the products due to compact arrangements of polymer chains. Lin and Zhao (2007) also commented on the hydrophilic nature of polysaccharides, advantages of using these materials are more apparent as a gas barrier rather than retarding water loss. Hydrophobic compounds such as surfactants could be added to hydrophilic components to improve the moisture barrier properties which could also reduce the surface tension of the solution to improve the wettability and adhesion of the film (Ribeiro et al., 2007).

Dutta et al., (2009) explained that chitosan is of high-interest polysaccharide, a natural polymer consisting of 1, 4-linked 2-amino-deoxy-b-D-glucan, which is deacetylated derivatives of chitin, studied as a major constituent of the exoskeleton of crustaceans and is one of the most abundant polysaccharides found in nature. It forms clean, flexible, and tough films with good oxygen barrier properties, which may be applied as packaging or coating, particularly as biopolymer films, to enhance the shelf life of a wide range of processed food items. Wu et al., (2005) also observed the advantage of chitosan over a variety of biomaterials because of its antimicrobial activity against a wide variety of microorganisms, including fungi, algae, and bacteria. Chitosan-based coatings have shown effectiveness in delay ripening and decreasing respiration rates of fruits and vegetables (Valencia-Chamorro et al., 2011). Chien et al., (2007) observed the effectiveness of chitosan coating in preserving the overall quality and extending the shelf life of sliced mango fruit.

Starch is a natural polysaccharide most commonly used in the formulation of biopolymer 8*films because it is inexpensive, abundant, biodegradable, and easy to use (Vargas et al., 2008). The most important sources of starch extraction are potato, corn, wheat, and rice. Starch mostly contains two types of glucose polymers: a linear chain molecule termed as amylase and a branched polymer of glucose termed as amylopectin. High amylose starch such as corn starch is good for film formation, whereas free-standing films can be produced from aqueous solutions of gelatinized amylose (Bourtoom, 2008).

13.2.2.2 PROTEINS FILMS

Proteins that can be used in the formulation of edible films included whey and casein protein derived from animal sources whereas wheat gluten, soy protein, peanut protein, corn-zein, and cottonseed protein obtained from plant sources (Gennadios, 2002). Protein-based films have more interesting mechanical and barrier properties than polysaccharides. Ramos et al., (2012) stated that proteins as natural polymers are able to form amorphous three-dimensional structures stabilized mainly by noncovalent interactions. The functional properties of these materials are highly dependent on structural heterogeneity, thermal sensitivity, and hydrophilic behavior of proteins. Lim et al., (2002) also observed that protein-based films are brittle and susceptible to cracking due to the strong, cohesive energy density of the polymers. The addition of compatible plasticizers can improve the extensibility and viscoelasticity of the films (Sothornvit and

Krochta, 2001). One of the potential uses of protein films is in reducing lipid oxidation due to the film's excellent barrier properties (Bourtoom et al., 2006). Caseinate and whey protein-based coatings have been applied to raisins/frozen peas/ peanuts to provide a barrier to oxygen and moisture transfer for extending the shelf life of the products (Chen, 1995). Xu et al., (2001) reported that the soy proteins coating are able to retard the senescence of kiwifruit whereas, Rakotonirainy et al., (2001) studied on zein based coatings, and found that these coatings maintained the firmness and color of the broccoli florets.

13.2.2.3 LIPID FILMS

Edible lipids are being used to develop edible coatings that include waxes, triglycerides, fatty acids, fatty alcohols, acetylated monoglycerides, and sucrose fatty acids esters (Vargas et al., 2008). Morillon et al., (2002) stated that lipid-based edible coatings have a low water vapor permeability (WVP), which is of great use in coating of fresh fruits and vegetables to prevent them for desiccation. In comparison to the waxes, the triglycerides or neutral lipids form a uniform and stable layer on food surface due to their high polarity. Perez-Gago and Krochta, (2001) reported that the lipid films are generally opaque, rigid, waxy in taste and may not adhere to hydrophilic cut surfaces which limit the utility of lipid edible coatings on fresh-cut fruits. Bourtoom (2008) evaluated that film-forming formulation, including a lipid dispersed on a hydrocolloid matrix, is then the best strategy to produce an edible film for MP fruits and vegetables. Olivas and Canovas (2005) described that beeswax; acetylated monoglycerides, fatty acids, etc. are some of the lipids that have been successfully employed to coat cut- fruits and vegetables and always applied in combination with a polysaccharide or protein.

13.2.2.4 COMPOSITE FILMS

Recent studies regarding the development of these films have been focused on composite or bilayer films such as polysaccharides, proteins with or without lipids together to improve the functionality of the films. The fact emphasized that each individual film material has some unique but limited functions, which when combined may increase their benefits of applications. Polysaccharides and proteins are polymeric and hydrophilic in nature; thus, good film- formers with excellent aroma, oxygen, and lipid barriers at a lower relative humidity (Lin and Zhao, 2007). In composite films, the polysaccharide or protein component provides the film integrity and entraps the lipid substance, whereas the added lipid layer entraps the moisture barrier property (Krochta, 1997).

The composite film can be categorized as a bilayer or a stable emulsion for which lipid generally forms an added layer on the polysaccharide or protein layer, whereas the lipid in the emulsion of the composite film is dispersed and entrapped in the protein or polysaccharide matrix (Figure 13.2).

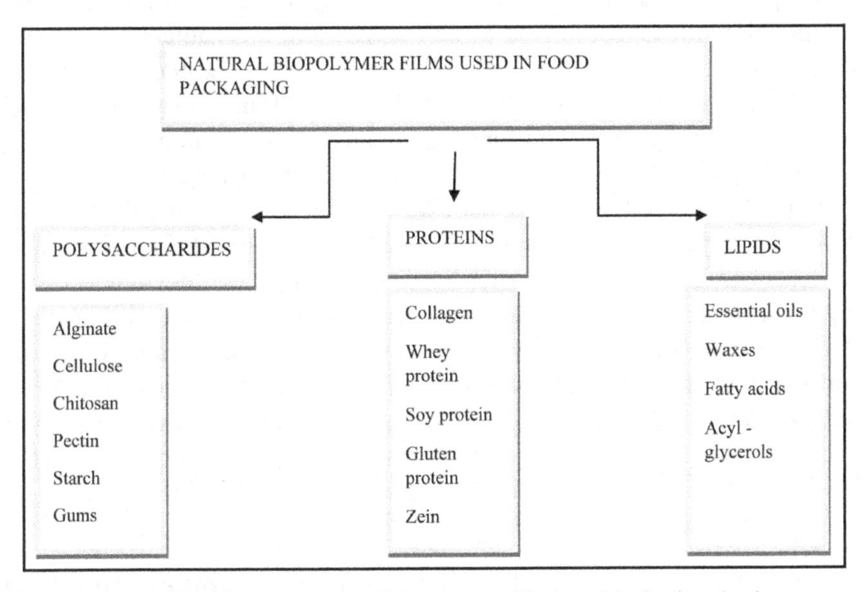

FIGURE 13.2 Classification of natural biopolymers films used in food packaging.

13.2.3 DEVELOPMENT OF BIOPOLYMER FILMS

Biopolymer films can be produced from natural materials from food sources with film-forming ability. Bourtoom (2008) reported that during manufacturing, film materials must be dissolved or dispersed in solvents such as water, alcohol or mixture of water and alcohol or a mixture of other solvents. Plasticizers, colors, flavors, antioxidants, and antimicrobial agents can be added in this process. To facilitate dispersion, adjusting the pH and/or heating the solutions may be done for a specific polymer. The film-forming solution is casted and then dried at the desired temperature and relative humidity to obtain free-standing films.

These films are produced by continuous film casting and mold casting method. The continuous film casting method is accomplished by coating a film forming solution onto a stable base material or a conveyor belt and then passing it through a drying chamber. These methods are inexpensive that can be used as laboratory scale edible film production techniques. According to Rossman (2009), these films are easy to produce as well as handle, since they require less heat and nontoxic solvents when compared to traditional petroleum-based food packaging material (e.g., polyethylene, polypropylene).

Biopolymer films also required to be organoleptically and functionally compatible with foods, having a high barrier and mechanical efficiencies, should be free from any toxics and safe for health. Depending on their composition, the functionality of these film materials may vary because each component confers different properties on the composite matrix.

Films made of hydrocolloids usually have strong mechanical and gas barrier properties, but also have poor water vapor barrier properties and high permeability to moisture (Lin and Zhao, 2007). In contrast, films composed of lipids exhibit good water vapor barrier properties, but they tend to show poor mechanical properties and high oxygen permeability. Combining these two components into one matrix allows them to physically and/ or chemically interact and may result in films with improved properties (Diab et al., 2001). These films are excellent vehicles for incorporating a wide variety of additives, such as antioxidants, antifungal agents, antimicrobials, bioactive compounds, drugs, colors, and other nutrients (Simi and Abraham, 2010).

13.2.3.1 EDIBLE FILM FORMULATION AND EXTENSION OF SHELF LIFE

One of the distinctive characteristics of biopolymer films is its capability to incorporate ingredients into the matrix to improve its functionality, as mentioned under:

- By incorporating plasticizers and emulsifiers, it improves mechanical properties as well as stabilize the composite film-forming solutions;
- Improves film adhesion;
- Enhanced quality, safety, and stability of coated foods by incorporating antioxidants, antimicrobial agents, natural plant extracts, nutraceuticals, flavoring, and coloring agents (Figure 13.3).

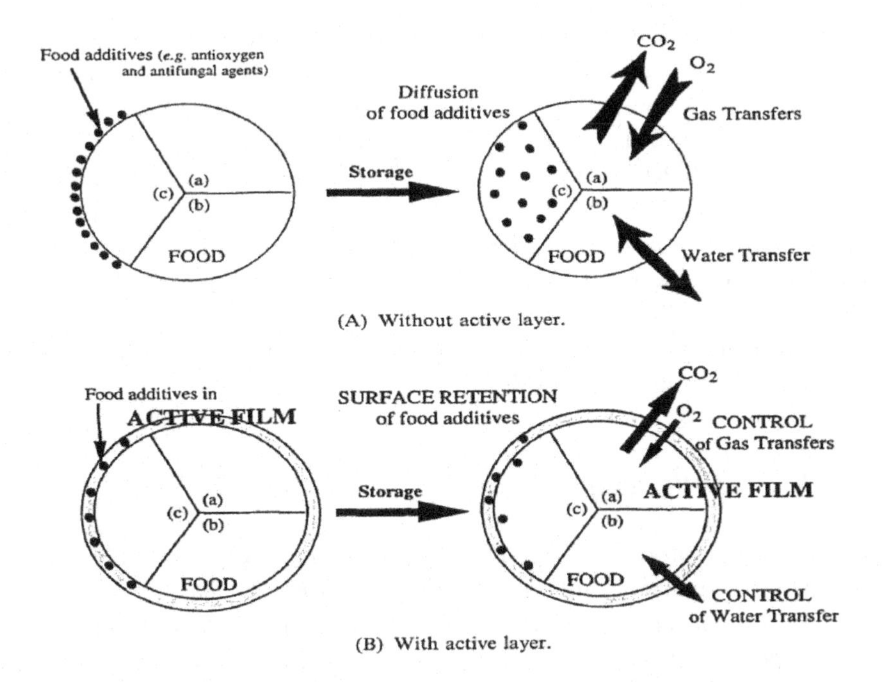

FIGURE 13.3 Schematic representation of food preservation with or without biopolymer films/ coatings.
(*Source:* Cuq et al., 1995)
Adapted and changed (no permission needed)

13.2.3.1.1 Plasticizers/Emulsifiers

Plasticizers and emulsifiers are added to the film to give it an elastic structure. The plasticizer was added between the molecular chains of polymers which resulted in the physical and chemical bonding due to the cohesiveness of the film, and it effectively soften/stretch the structure (Kokoszka and Lenart, 2007). The most commonly applied plasticizers are polyols (glycerol, sorbitol, glycol polyethylene), sugars (honey, glucose) and lipids (monoglycerides, phospholipids) as studied by Park et al., (2000). It was very important for the selection of right plasticizer for the edible film as its strong effect could significantly modify the physical and chemical properties of the polymers.

A plasticizer is necessary for making films based on polysaccharides and protein because the structure of such films is often brittle and stiff due to extensive interactions between polymer molecules (Krochta, 2002).

Sothornvit and Krochta (2000) reported that water might also function as a plasticizer, but easily lost due to the dehydration at low relative humidity environment and also improve tensile strength (TS) of the film.

According to Lin and Zhao, (2007), emulsifiers are surface-active agents that are able to reduce the surface tension of water-lipid or water-air interface). These additives are essential for the formation of protein or polysaccharide films containing lipid emulsion particles and can also modify surface energy to control the adhesion and wettability of the coating surfaces (Krochta, 2002). Addition of emulsifiers into the whey protein-coating layer increases the hydrophilicity and coatability of peanut surfaces, thus producing a better oxygen barrier of the coatings (Lin and Krochta, 2005).

13.2.3.1.2 Antioxidants

A wide series of compounds have been incorporated into edible film formulation to enhance antioxidant properties as mentioned by Siddiqui et al., (2011) as various acids (like acetic, ascorbic, sorbic, benzoic), potassium sorbate, potassium sorbate and sodium benzoate, essential oils from spices and herbs (Tovar et al., 2005). Rojas-Grau et al., (2009) observed that certain phenolic compounds like BHA (Butylated-hydroxy-anisole) or BHT (Butylated-hydroxy-quinone), tocopherols or acids such as propyl gallate have antioxidant properties which inhibit oxidation of fats and oils in foods. Zheng and Wang (2001) investigated that spice extracts from oregano, pimento, and rosemary were rich in phenolic compounds and volatiles, which are significant contributors to the antioxidant properties. Antioxidants can be added into the film forming solutions to protect against oxidative rancidity, degradation, and discoloration of certain foods. Lee et al., (2003) also reported that carrageenan and whey protein coatings contain anti-browning agents such as ascorbic acid (AA) and citric acid (CA), which effectively prolonged the storage life of apple slices. Prunenda et al., (2008) incorporated water-soluble extracts from Mexican oregano into soy protein isolate (SPI) films and concluded that the addition of extracts did not enhance the barrier properties of the film, but its addition provided important antioxidant properties to the film.

13.2.3.1.3 Antimicrobials

Antimicrobials are additives used to prevent biological deterioration and allowed for food applications that comprise of natural and synthetic compounds with minimum toxicological effects on human and environment. Antimicrobial compounds include organic acids and their salts (propionates, sorbates, benzoates, etc.), inorganic compounds (carbonates and bicarbonates, etc.) along with chitosan, enzymes, bacteriocins, parabens, polypeptides, essential oils and other natural extracts as mentioned by Valencia-Chamorro et al., (2011). A large number of antimicrobials have been added to biopolymer films to control microbiological growth, thereby enhancing the shelf life of food. These chemical compounds which are used for edible film formation must be classified as food grade additives or compounds generally recognized as safe (GRAS) by the relevant regulations. Chitosan-based coatings were shown to protect highly perishable fruits like strawberries and fresh-cut green pepper from fungal decay (Vargas et al., 2006; Devlieghere et al., 2004; Raymond et al., 2012). Park et al., (2004) demonstrated that when chitosan coatings were incorporated with lysozyme, increased the antimicrobial activity against *E. coli* and *Streptococcus faecalis* whereas, Park et al., (2005) commented that chitosan coating containing potassium sorbate enhance the antifungal activity against *Cladosporium* and *Rhizopus* on fresh strawberries. Rojas-Grau et al., (2006) observed that essential oil in the edible film was highly effective against *E. coli* O157:H7 and it also decreased WVP, but it did not alter the tensile properties of the films. Jin and Zhang (2008) studied the biodegradable polylactic acid (PLA) polymer and its application as a material for antimicrobial food packaging. They conducted experiments on the retention of nisin activity when incorporated into the PLA polymer and antimicrobial effectiveness against foodborne pathogens. Clove oil was found to be more effective against *S. enteritidis* in full-fat than in low-fat cheese slurry (Smith-Palmer et al., 2001). Gutierrez et al., (2009) investigated that the combination of a biopolymer with natural bacteriocin has the potential for use in antimicrobial food packaging.

13.2.3.1.4 Natural Plant Extracts

Natural compounds are becoming popular ingredients in active biopolymer film research owing to consumer's demand for more natural food ingredients. It is evident that many plant extracts contain antimicrobial properties, and several authors all over the world have reviewed the health benefits of natural plant extracts and essential oils. In recent days, plant extracts have drawn attention due to its ability to enhance the shelf life of fresh produce instead of synthetic agents. In general, plant extracts, which contain high concentrations of phenolic compounds, have strong antioxidant properties (Valdes et al., 2014).

Cowan (1999) classified useful antimicrobial phytochemicals into several categories; polyphenolic compounds, (simple phenols and phenolic acids, tannins, coumarins, quinones, flavonoids, flavonols, flavonones) terpenoids, essential oils, alkaloids, lectins, polypeptides, and other compounds. Shen et al., (2014) reported that the extracts of blueberry have shown an antimicrobial activity against the growth of *Salmonella Enteritidis* and *Listeria monocytogens.* Perumalla and Hettiarachchy (2011) investigated that green tea extract inhibits the growth of various strains of *Staphylococcus* and some gram-negative bacteria.

Essential oils are popular plant extracts being examined as antimicrobials in biopolymer films. Seydim and Sarikus (2007) studied the whey protein biopolymer films incorporated with oregano, rosemary, and garlic essential oils and by measuring zones of inhibition, they found that the oregano essential oil was the most effective against various bacteria at 2% inclusion vs. the garlic and rosemary. Olive leafs are an important source of active compounds, and the extract of olive leaf showed high antimicrobial activity, caused by their high concentration in polyphenols. Ozge Erdohan et al., (2013) studied that olive leaf extract added in polylactic acid matrix reflected the increase in the amount of the extract which may cause a considerable increase in inhibitory zones against test pathogens.

13.2.3.2 METHODS AND MECHANISM OF FILM FORMATION

The formulation of edible film packages must include at least one biopolymer to form a cohesive and continuous matrix. These films made with various ingredients have improved the complementary functional properties of each component and to minimize their disadvantages. Two general mechanisms for the formation of biopolymer films are as under:

- *Dry Process* such as thermoplastic extrusion is based on thermoplastic properties of biopolymers, under low water content conditions when plasticized and heated above their glass transition temperature used in edible film formulation. The use of thermoplastic biopolymers has been suggested either to form biodegradable packaging materials for various food/non-food sector application, or a component for composite or complex packaging materials with specific properties such as the selection of gas or control release of active compounds.

- *Wet Process* is based on a film-forming solution or dispersion when biopolymers are first dispersed or solubilized into a liquid phase. The wet process generally preferred to form biopolymer films or to apply coatings directly on the food product surface (Figure 13.4).

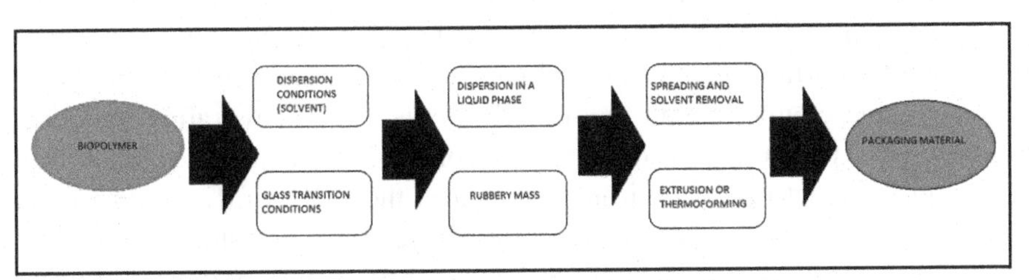

FIGURE 13.4 General mechanism of biopolymer edible film formation.
(*Source*: Guilbert and Gontard, 1995)
[Adapted and Changed (no permission needed)].

13.2.4 PROPERTIES OF BIOPOLYMER FILMS

The application of biopolymer films is affected by its mechanical, barrier, and thermal properties. Factors affecting the properties of the film have been outlined by Janjarasskul and Krochta (2010). These factors are:

1. The structure and chemistry of the biopolymer (molecular weight, branching, and polarity).
2. Film forming methods, e.g., solvent casting versus thermal processing.
3. Nature of the additives added during the film formation such (e.g., type, and concentration of plasticizers).
4. Film forming parameters such as temperature, pressure, solvent type, dilution, application, and drying techniques.

13.2.4.1 BARRIER PROPERTIES

One of the functions of biopolymer films is to act as a barrier that separates and protects the products from the exposure of the environment. These films have been commercially used to protect fruits and vegetables from pathogenic microbial contamination (Quintavalla, 2002). Krochta et al., (1994) investigated that the extent of the barrier is influenced by the chemical properties of the materials used. However, environmental conditions, such as temperature, relative humidity, and the stress of handling the product by consumers can influence the barrier performance of the package. For the application of food, the oxygen (O_2), carbon dioxide (CO_2) and WVP of biomaterials should be considered.

Lin and Zhao (2007) investigated that most biopolymers possess good properties as semi-permeable barriers to gases, but they are weak barriers to water vapor. The transference of water vapor is generally due to the hydrophilic portion of the film and depends on the ratio of hydrophilic to hydrophobic compounds in the film. To prevent the dehydration of foods, films used as packaging or coatings must control the transference of humidity from the product to the environment; hence, the WVP of these films must be as low as possible (Ma et al., 2008). In relation to maturity, fruit, and vegetable products require a coating that forms a semi-permeable barrier that permits the release of CO_2 and diminishes O_2 penetration. Pol et al., (2002) observed that the moisture sensitivity of biopolymer films could be measured by water vapor transmission rate (WVTR), contact angle, moisture sorption isotherms, and water solubility tests.

To improve barrier properties, incorporation of hydrophobic compounds into biopolymer films were studied by the researchers. Ghasemloua et al., (2011) incorporated 15–35% (w/w) of oleic acid into the hydrophilic film with an emulsifier (Tween 80) and the results showed that adding oleic acid decreased the moisture content of the film from 17.95% to 12.36%, along with reduction in film's solubility in water and WVTR. The increased hydrophobicity of the film happened due to the incorporation of the lipid. Siripatrawan and Harte (2010) reported that chitosan-based edible film incorporated with green tea extract results in low WVP. This was due to the interactions between the green tea polyphenols and chitosan. In addition, by reducing the available hydroxyl groups, the polyphenols also acted as crosslinking agents which tied up the hydroxyl groups of chitosan to prevent hydrogen bonding with water. Gonzalez et al., (2011) investigated that the effect of cross-linking on SPI film with genipin, a naturally-derived crosslinker, showed that the WVP value decreased by 29.5% when the film was incorporated with 2.5% (w/w of SPI) of genipin because cross-linking agents increases the film's density by lowering the void spaces.

13.2.4.2 MECHANICAL PROPERTIES

The mechanical property of an edible film is important because it determines how the film will behave during handling and storage. Ghanbarzadeh and Almasi (2011) stated that mechanical property is a critical parameter that determines the durability of the films and hence its ability to enhance the mechanical integrity of foods for packaging applications, films must be durable, stress-resistant, flexible, and elastic. Tensile or tension test is therefore commonly used to determine the mechanical properties of a material. This is a simple test where a sample of the material is pulled in the opposite direction, and the applied force measured.

TS and percentage elongation describes the mechanical resistance of the films. The TS (MPa) is the maximum strength measuring the resistance of the film, whereas the percentage of elongation at break (EB, %) is a measure of the stretching capacity or flexibility of the film prior to breaking (Krochta and De Muldcxzaer-Johnston, 1997).

The magnitude of TS and % EB depends closely on the relative humidity. As relative humidity increases, TS decreases while elongation increases with the increase in the amount of water in the coating (Olivas and Canovas, 2005).

The applications of biopolymer films are limited because of their poor mechanical properties compared to synthetic films. Several researchers have investigated the mechanical properties of biopolymer films, the factors that affect these properties, and the means to improve their tensile properties. Ghanbarzadeh and Almasi (2011) observed that the use of cross-linking agents blended with composite films have been used as a means to produce a high strength starch-based films with improved mechanical properties. In this study, corn starch films with different concentrations (0–20%, w/w) of CA as a cross-linking agent and carboxymethylcellulose (CMC) were produced by the solvent casting method. The TS was improved significantly as the percentage of CA increased from 0 to 10%. CA, with its carboxyl groups, acted as a cross-linking agent by forming strong hydrogen bonds with the hydroxyl groups of the starch. The study reported that the TS increased from 6.57 MPa for films without CMC to 16.11 MPa for films containing 20% CMC (Ghanbarzadeh and Almasi, 2011).

Drying of the film forming solution to obtain biopolymer films is a very important parameter that affects the mechanical properties of the films. The effect of drying temperatures on polysaccharide-based chitosan films results in higher TS and elongation at break at slow drying cycles in lower drying temperature (Fernandez-Pan et al., 2010).

Mayachiew and Devahastin, (2008b) also obtained chitosan biopolymer films of higher TS at lower temperatures of drying. Similarly, Srinivasa et al., (2007) reported that films obtained by a faster drying method, i.e., infrared (IR) drying had lower TS (i.e., 49.58 MPa) compared to that of ambient-dried films which had TS of 56.78 MPa. Faster drying at high temperature resulted in a less ordered structure with few hydrogen bonds due to a more intense structural collapse. On the contrary, Fernandez-Pan et al., (2010) stated that during slow drying, the polymer chains are able to rearrange and form a more ordered structure resulting in a higher TS.

13.2.4.3 FUNCTIONAL PROPERTIES

13.2.4.3.1 Thermal Properties

The thermal behavior of edible packaging is important since storage and end-use conditions of biopolymer films can vary from sub-zero to cooking temperatures. Packaged foods can be stored under these temperature conditions for a few days to several weeks.

Differential Scanning Calorimetry (DSC) is the method commonly employed to evaluate the thermal behavior of polymers. DSC is a technique that measures the thermal properties of a given material to establish a link between temperature and the specific physical properties of substances. It directly measures the enthalpy associated with a process of interest.

In the DSC experiment, heat energy is introduced simultaneously into a sample cell containing the material of interest and a reference cell. Temperatures of both cells are raised identically over a time, and the difference in the energy required to match the temperature of the sample to that of the reference would be the amount of excess heat absorbed (endothermic process) or released by the molecules in the sample (exothermic process). Different thermal transitions of polymers, including the glass transition temperature (T_g), crystallization, and melting temperatures can be detected by DSC. Below the T_g, the polymer will exist in the glassy state where the molecules are held tightly together by intermolecular forces. In this state, molecular mobility is restricted, but when the polymer is heated above T_g, the polymer gains more energy and moves from the glassy state to the rubbery state. In the rubbery state, segmental mobility begins but molecular mobility is forbidden.

In biopolymer film applications, DSC can be used to determine how different type of plasticizers or their concentrations have altered the properties of the material. Kim and Ustunol, (2001) examined DSC was used to investigate changes in the glass transition temperature (T_g) of glycerol-plasticized whey protein isolate films and that of sorbitol-plasticized films. The results showed that glycerol-plasticized films had a lower T_g than those of the sorbitol-plasticized films. This difference was due to the lower molecular weight of glycerol, as explained by the authors. Al-Hassan and Norziah (2012) observed that DSC was used to measure the reductions in T_g, T_m, and ΔH_m when a starch-gelatin composite film was plasticized with sorbitol. The researchers explained that sorbitol interacted with the hydroxyl groups of both starch and gelatine, resulting in a reduced interaction between the starch and gelatin. DSC can also be used to determine the miscibility or compatibility of polymer blends by determining the T_g of the blend. A single T_g gives information on compatibility of the blends. A single T_g indicates a homogenous blend, whereas two separate T_g values signify that there is heterogeneity (Al-Hanssan and Norziah, 2012).

13.2.4.3.2 Spectrometric Properties

IR spectroscopy is useful to study the hydrogen bonding and other interactions as well as the miscibility of polymer blends (Suyatma et al., 2004). The IR beam from an FTIR unit passes through the sample and causes vibration, stretch, bending, and contraction of the chemical bonds. When this occurs, the excited chemical bonds absorb the IR radiation at specific wave numbers (cm^{-1}) as studied by Goddard and Hotchkiss (2007). During the test, the sample is held onto a highly reflective crystal (e.g., diamond) and then exposed to the IR beam to determine the path of the reflected beam (Tanaka et al., 2001). FTIR (Fourier transform IR) can be used to determine the type of polar bonds and functional groups of hydrocolloids based films. It can also be used to determine the presence of plasticizers and emulsifiers in composite films. Turhan et al., (2001) studied the FTIR analysis on methylcellulose (MC) films. The spectra of the films were interpreted in terms of symmetry distortion of hydroxyl stretch at 3466 cm^{-1}, which is the measure of hydrogen bonding interaction in the polymer matrix. Results showed that (Poly ethyl glycol) PEG-400 reported as a good plasticizer for MC films because of the formation of high hydrogen bond capacity. Although good miscibility does not attain in high molecular weight PEG. Silva et al., (2008) investigated the biodegradable films based on blends of gelatine and polyvinyl alcohol. They found that the spectra of the films prepared with blends of gelatine and PVA were almost similar to that of pure gelatine films, indicating there were no major changes in the functional groups of the gelatine, induced by interactions between the PVA and the gelatine.

This behavior could be considered as there is no specific interaction between chitosan and PLA, and they are therefore immiscible and therefore confirms the findings of mechanical and thermal properties of these films (Suyatma et al., 2004).

13.2.4.3.3 Morphological Properties

Scanning Electron Microscope (SEM) may be used to evaluate film homogeneity, surface smoothness, layer structure, and thickness. It also used to detect pores and cracks. These properties are relevant to the film functionality because of their great impact on the appearance on the final product. SEM images of plasticized films containing lipid exhibited the smooth surface as compared to compact structure, indicative of homogenous dispersion of lipid within the film matrix (Garcia et al., 2009).

Salleh et al., (2009) investigate the SEM on the surface of the pure starch film, chitosan films, and their blends. They reported that the pure starch film exhibit characteristics patterns on the film surface. These patterns represent the withered ghost granules of starch. However, the pure chitosan exhibits the smooth homogenous surface with continuous matrix without cracks, which shows the good structural integrity. Zhong et al., (2012) studied the microstructural properties of β-chitosan and kudzu starch composite film and they observed that all the films presented continuous and compact surfaces without pores or cracks, and the kudzu starch generally enhanced the roughness of pure β-chitosan film, which might be caused by the formation of helical structures of starch during film forming process.

Berkun et al., (2008) also studied the two types of films consisting of the sodium salt of carboxymethyl cellulose (NaCMC) and hydroxypropyl cellulose (HPC) as film-forming materials and glycerine used as a plasticizer were prepared and characterized. SEM was used to depict the structure of the prepared biopolymer films. It can be observed that the both films have porous structures. However, HPC based biopolymer films have a relatively more homogenous porous structure in comparison with the NaCMC based films. The cross-sectional micrographs of the film samples with magnification at 5000× indicate that HPC films contain many small pores with dimensions of approx 0.5–1 μm.

13.2.5 CUT/ MINIMALLY PROCESSED (MP) FRUITS AND VEGETABLES

The increased concern about health, the availability of lesser time for food preparation and enhanced purchasing power have recently contributed towards the growth in the demand of cut/MP fruits and vegetables (Kader, 2002). MP fruits and vegetables are fresh fruits and vegetables processed to enhance their functionality without greatly changing their fresh like properties (Siddiqui et al., 2011). Consumers usually judge the quality of fresh CFV on the basis of freshness and appearance at the time of purchase. Gonzalez-Aguilar et al., (2010) emphasized that the fresh produce should have a shelf life sufficient for distribution, feasible within the region of consumption. Ahvenainenand Hurme (1994) stated that nowadays, fresh cut MP fruits attract consumers because they are freshly processed, nutritious, low priced, and ready-to-eat and hence, termed as 'quick' and convenient' products. Minimal processing operations like (peeling, trimming, sectioning, slicing, and coring) alter the integrity of fruits bringing about the changes on product quality such as browning, off-flavor development, texture breakdown and presence of microorganisms on the fruit surfaces may compromise the safety of fresh CFV (Rojas-Grau et al., 2009). Fresh cut produces deteriorate faster because of biochemical changes and microbial spoilage, which may result in degradation of color, texture, and flavor of the produce (Olivas et al., 2007). Other types of deterioration include chemical, enzymatic changes, microbial deterioration and physical damage (improper harvesting, handling/processing, and packaging) can also affect the shelf life of processed CFV (Thybo et al., 2006).

MP fruits and vegetables are preserved by various techniques like refrigeration, chemical preservatives, additives, bio-preservatives, mild heat treatment, reduction of water activity, disinfectants, and nonthermal techniques. Edible coatings (multilayer coating, osmotic membrane coating) are also being used in minimal processing of fruits and vegetables (Ayala-Zavala and Gonzalez-Aguilar, 2010). Biopolymer films offer a possible method of extending the post-harvest storage life of fresh produce. Fresh CFV can be coated in these edible materials, which provide a semi-permeable barrier to gases/water vapors; therefore, these are used as carrier additives.

13.2.6 APPLICATION OF BIOPOLYMER FILMS ON CUT FRUITS AND VEGETABLES (CFV)

Food products undergo numerous physical, chemical, and microbial changes during storage. The stability of food is a function of changes occurring in the food components, such as food proteins, lipids, carbohydrates, and water due to the environmental and processing factors (exposure to light, temperature, moisture, etc.). The protective coating or barrier provided during processing, storage, and handling not only retards detrimental deterioration of food, but may also enhance its quality (Su cha and Chinnan, 2004).

Olivas and Canovas (2005) observed that processing of fresh-CFV usually involves the stimulation of wounding stresses in the cut tissues. Consequently, these tissues have to face mechanical injury, browning, degradation, oxidation of pigments, water loss, translucency, whitening, and surface dehydration, which leads to an increase in their respiration rate. Respiration involves the oxidative breakdown of a complex substance, normally present in plant cells such as sugars, starch, and organic acids to simpler molecules, as the result of which energy, carbon dioxide, and water are given out. Rojas-Grau et al., (2009) suggested that atmosphere low in O_2 and high in CO_2 have been used to increase the shelf life of fresh CFV, thereby reducing respiration, transpiration, and ethylene production. Biopolymer films can be applied as either a compliment or an alternative to modified atmosphere packaging (MAP) to enhance the shelf life of CFV.

Biopolymer film technology is the most promising method to preserve the quality of fresh CFV. Along with an increased market demand for quality food; these films with unique functionality will certainly become more important in the future. Corbo et al., (2006) also emphasized that mentioned films may help to reduce the deleterious effects with minimal processing. These films not only retard food deterioration and enhance its quality, but also improve its safety because of their natural biocide activity. Films based on edible compounds from several sources have been evaluated to maintain the visual appearance of CFV. Research and development efforts are leading to an improvement of the functional characteristics of the biopolymer films, which depends on the properties of fruits and vegetables to be preserved or enhanced. These can be achieved by a precise optimum control of gas permeability, texture, and color changes by means of quantitative/qualitative changes in film forming formulation.

Edible coatings are continuous biopolymeric matrices formed as films and directly applied on the upper surface of fruits and vegetables. Avena-Bustillos et al., (1997) stated that biopolymer films are prepared as solutions/emulsions from protein, polysaccharides, lipids, and are applied on produce surface by different mechanical procedures such as dipping, spraying, brushing or electrostatic deposition, etc. (Amelia et al., 2006). The physical and chemical characteristics of biopolymer films formulation like the film thickness, homogeneity, adhesiveness depend on the surface structure and morphology of the fruits and vegetables (Miller and Krochta, 1997).

13.2.6.1 METHODS OF APPLICATION OF BIOPOLYMER FILM

Biopolymer films can provide either transparent or translucent (milky/opaque) appearance, but consumers generally prefer clear, invisible coatings. Coatings can be obtained in various ways:

1) By dipping the products into, or by brushing or spraying it with film-forming solutions, so as to develop the film directly on fresh produce.
2) By developing the stand-alone film from solutions or through thermo-formation for subsequent covering of fresh fruits and vegetables.

The simplest way to apply the film is directly from the solution. The fresh produce absorbs the appropriate amount of coating material necessary to form the desired layer, depending on the concentration of the film forming solution, which when dried forms a protective bilayer on the produce. In some of the cases, the plasticizers need to be added to film forming solution to keep the developing film from becoming brittle. Biopolymer should have good adhesion to rough surfaces (Hershko et al., 1996). Application of a uniform film layer to CFV surfaces is generally difficult. For better uniformity surfactants should be added to film forming solution to reduce the surface tension. This strategy will also reduce the water activity and in turn, reduce the water loss. A free-standing film was formed from a film-forming solution by evaporation. It should be pointed out that characteristics of stand-alone films might differ from those films created on food surface (i.e., those produced by dipping in or spraying). Films obtained by evaporation have a lower water permeability rate as compared to films prepared by spraying.

13.2.6.2 SHELF LIFE EXTENSION OF CUT FRUITS AND VEGETABLES (CFV)

Olivas and Canovas (2005) stated that edible film technology is a promising method to preserve the quality of fresh and MP fruits and vegetables. Research and development efforts are leading to an improvement of the functional characteristics of the films, which depends on the properties of the fresh produce to be preserved or enhanced. This can be achieved by a perfect optimum control of gas permeability, firmness, and color changes by means of quantitative and qualitative changes in film formulation. Biopolymer films can affect the quality of fresh-coated produce in several ways, since there are many mechanisms involved. Shaidi et al., (1999) reported these mechanisms as controlled moisture transfer between produce and the surrounding environment, controlled release of additives like antimicrobials substances, flavor compounds, and antioxidants, along with the reduction of internal oxygen partial pressure with a decrease in fresh produce metabolism. Therefore, some of the effects that can be observed in coated fruits during storage are the reduction in respiration rate (Wong et al., 1994), a decrease in weight loss (Baldwin et al., 1999), a delay in the occurrence of enzymatic browning (Baldwin et al., 1999; McHugh and Senesi, 2000) and in general a significant extension of fresh produce shelf life.

Baldwin et al., (1995) evaluated that fresh-cut mango coating used polysaccharide-based materials, which limit gas exchange, thus creating a modified internal atmosphere in the fruit slices, decreasing respiration rate and delaying senescence. Sothornvit and Rodsamran (2008) experimented with mango puree films and observed an increase in shelf life extension of 'Namdok Mai' mango slices by 2 to 3 days. Gonzales-Aguilar et al., (2008) found that anti-brownings (AA+ CA) maintained shelf life and nutritional quality of fresh-cut mangoes ('Keitt,' 'Kent,' and 'Ataulfo').

Diab et al., (2001) reported that by applying coating of pullulan/sorbitol/sucrose fatty acid ester (SE) on strawberries resulted in large changes in internal fruit atmosphere composition which were beneficial for enhancing shelf life of this fruit; the coated fruit showed much higher levels of CO_2, a large reduction in internal O_2, better firmness, color retention and reduced rate of weight loss. Ali et al., (2011) narrated the effect of chitosan coating on physicochemical characteristics of Eksotika II papaya fruit stored at 12±1°C and 85–90% relative humidity. The present study reported the effective control of chitosan in reducing weight loss, texture, and delayed changes in peel color of papaya fruit during storage of 5 weeks. The sensory evaluation also confirmed the efficacy of chitosan. It has been reported that when apple slices are coated with whey protein, the respiration rate decreased by 20% (Lee et al., 2003). Film formulations may be enhanced with the addition of volatile precursors such as fatty acids which can be incorporated by fruits into their metabolism, promoting the synthesis of aroma compounds (Olivas et al., 2003). Lee et al., (2003) studied on the coating of apple slices with films based carrageenan and whey protein carrying anti-browning agents like calcium chloride, which successfully maintained the color and texture of apple slices. Texture enhancers such as calcium chloride may be added to biopolymer films/coatings to enhance fruit quality during storage by inhibiting the loss of firmness of cut fresh produce. Eissa (2007) investigated that the application of a chitosan-based coating on fresh-cut mushrooms contributed to extend their shelf life, maintained quality, and controlled physiological decay. Rojas-Grau et al., (2007) observed the usefulness of alginate and gellan gum edible coatings along with antimicrobial effects of plant essential oil (lemongrass, oregano oil and vanillin) to extend the shelf life of fresh-cut apples (Table 13.1).

TABLE 13.1 Edible Coating on Minimally Processed Fruits

Coating	Composition	**Application**	**References**
Double layer: Polysaccharide/lipid	Carrageenan, pectin, cellulose, alginate, monoglycerides	Fresh–cut apple cylinders	Wong et al., 1994
Chitosan	Chitosan	Sliced mango fruit	Chien et al., 2007
Protein/polysaccharide	WPC, glycerol, CMC, $CaCl_2$	Fresh cut apple cubes	Lee et al., 2003
WPC/ WPI	Glycerol	Apple cubes	Sonti et al., 2003
Soy protein	Soy protein, glycerol, malic acid, lactic acid	Cantaloup melon cubes	Eswaranandam et al., 2006
Rosemary, oregano, olive, capsicum, garlic, onion, cranberry	Chitosan	Squash slices	Ponce et al., 2008
Clove-oil, cinnamon, lemongrass, eugenol	Alginate	Apple slices	Raybaudi-Massilia et al., 2008
Ascorbic acid	Alginate	Papaya	Tapia et al., 2008

The applications of edible films/coatings on CFV are shown in Table 13.2.

13.3 CONCLUSION REMARKS

The application of Biodegradable biopolymer films is one of the innovations of packaging technology aimed to improve food quality and to retain the freshness of food. In order to maximize the potential of this new technology, it is important to consider the consumer concerns and acceptance. The use of natural ingredients instead of synthetic chemicals is a worldwide inclination affecting most consumer products. Due to this consumer preference, more research and development should focus on the use of natural active agents.

TABLE 13.2 Application and Functions of Different Type of Edible Films/Coatings on Fruits and Vegetables

Commodity	Films applied	Composition	Functions	References
Melon (fresh cut)	Chitosan	Chitosan-lactic acid-calcium lactate	O_2/CO_2 barrier, stabilization of cell wall	Zsivanovits et al., (2012)
Papaya	Aloe vera gel	Aloe gel- papaya leaf extract	Eco-friendly, antifungal natural additive, extend shelf life	Marpudi et al., (2011)
Tomatoes	Soy protein	Soy protein-CMC-oleic acid	Antifungal, O_2/CO_2 /H_2Obarrier	Nandane et al., (2011)
Banana	Composite gum Arabic	Gum Arabic-chitosan	Antibrowning, O_2/CO_2 /H_2Obarrier	Maqbool et al., (2011)
Citrus (Murcott tangor) fruit	Low molecular weight Chitosan	Chitosan-glacial acetic acid-Tween 80	Antifungal, O_2/CO_2 barrier	Chien et al., (2007)
Eksotika II Papaya	Chitosan	Chitosan-glacial acetic acid-Tween 80	O_2/CO_2 /H_2Obarrier	Ali et al., (2011)
Cucumber	Gum Arabic	Gum Arabic-glycerol monostearate	Delayed softening, overall acceptability was good	Al-Juhaimi et al., (2012)

Their use as crude extracts or in partially purified forms may provide additional functionalities beyond the desired activity.

Fruits and vegetables are living organisms which continue to respire/transpire after harvesting, are being characterized by a respiration and transpiration rates. Respiration rate of fresh fruits and vegetables are often good indicators of their storage life; the higher the rate, the shorter the shelf-life. Moreover, the shelf-life of fresh-cut products is limited, as they are more prone to microbiological deterioration. So in order to reduce the effects of microbiological, chemical, and physical deterioration, more attention should be given on packaging rather than processing. The packaging operation must allow establishing an optimal atmosphere inside the packaging for the best preservation of the products. Biopolymer films may prove beneficial to extend the shelf-life of fresh-cut fruits by reducing moisture loss, solute migration, gas exchange, respiration, and oxidative reaction rates, as well as by suppressing physiological disorders. The film containing useful additives like antioxidants and antimicrobials was used to protect against oxidative rancidity, degradation, and discoloration. Naturally derived biological compounds and other natural products may have application in controlling pathogens in fresh produce. The use of biopolymer films as antimicrobial carriers represents an interesting approach for the external incorporation of plant essential oils and other phytochemicals onto food system surfaces. The agents can then diffuse into the food to control target microorganisms. Hence, the demand of natural plant extracts or their original compounds (phenolic acids, tannins, flavonoids, etc.) in film formulations has now become more pronounced in recent years.

KEYWORDS

- **biopolymer films**
- **cut fruits and vegetables**
- **edible coating**
- **storability**

REFERENCES

Ahvenainen, R., & Hurme, E., (1994). Minimal processing of vegetables. In: Ahvenainen, R., Mattila-Sandholm, T., & Ohlsson, T., (eds.), *Minimal Processing of Foods* (Vol. 142, pp. 17–35). Espoo, Finland, VTT Symposium.

Al-Hassan, A. A., & Norziah, M. H., (2012). Starch–gelatin edible films: Water vapor permeability and mechanical properties as affected by plasticizers. *Food Hydrocolloids*, *26*, 108–117.

Amelia, A. E., Abu-Ali, J. M., & Barringer, S. A., (2006). Improved functionality of food additives with electrostatic coating. *Innovative Food Science and Emerging Technologies*, *7*(3), 176–181.

Avena-Bustillos, R. I., Krochta, J. M., & Saltveit, M. E., (1997a). Water vapor resistance of red delicious apples and celery sticks coated with edible case innate-acetylated monoglyceride films. *Journal of Food Science*, 62(2), 351–354.

Ayala-Zavala, J. F., & Gonzalez-Aguilar, G. A., (2010). Optimizing the use of garlic oil as antimicrobial agent on fresh cut tomato through a controlled release system. *Journal of Food Sciences*, *75*, 398–405.

Baldwin, E. A., Burns, J. K., Kazokas, W., Brecht, J. K., Hagenmaier, R. D., Bender, R. J., & Pedid, E., (1999). Effect of two edible coatings with different permeability characteristics on mango *Mangifera indica* L. ripening during storage. *Postharvest Biology & Technology*, *17*, 215–226.

Baldwin, E. A., Nisperos-Carriedo, M. O., & Baker, R. A., (1995). Edible coatings for lightly processed fruits and vegetables. *Horticulture Sciences*, *30*(1), 35–38.

Berkun, D., Balkose, D., Tihminlioglu, F., & Altinkaya, S. A., (2008). Sorption and diffusion of water vapor on edible films. *Journal of Thermal Analysis and Calorimetry, 94*(3), 683–686.

Bourtoom, T., (2008). Edible films and coatings: Characteristics and properties. *International Food Research Journal, 15*(3), 237–248.

Bourtoom, T., Chinnan, M. S., Jantawat, P., & Sanguandeekul, R., (2006). Effect of select parameters on the properties of edible film from water-soluble proteins in surimi wash-water. *LWT- Food Science and Technology*, 39(4), 406–419.

Cha, D. S., & Chinnan, M. S., (2004). Biopolymer-based antimicrobial packaging: A review. *Critical Reviews in Food Science and Nutrition, 44*, 223.

Chen, H., (1995). Functional properties and applications of edible films made of milk proteins. *Journal of Dairy 1 Sciences, 78*, 2563–2583.

Chen, M., Yeh, G., & Chiang, B., (1995). Antimicrobial and physicochemical properties of methylcellulose and Chitosan films containing a preservative. *Journal of Food Processing and Preservation, 20*, 379–390.

Chien, P. J., Sheu, F., & Yang, F. H., (2007). Effects of edible chitosan coating on quality and shelf life of sliced mango fruit. *Journal of Food Engineering, 78*, 225–229.

Corbo, M. R., Del-Nobile, M. A., & Sinigaglia, M., (2006). A novel approach for calculating shelf-life of minimally processed vegetables. *International Journal of Food Microbiology, 106*, 69.

Cowan, M., (1999). Plant products as antimicrobial agents. *Clinical Microbiology Review, 12*(4), 564–582.

Cuq, B., Gontard, N., & Guilbert, S., (1995). Edible films and coatings as active layers. In: Rooney, M. L., (ed.), *Active Food Packagings* (pp. 111–142). Glasgow: Blackie Academic & Professional.

Devlieghere, F., Vermeulen, A., & Debevere, J., (2004). Chitosan: Antimicrobial activity, interactions with food components and applicability as a coating on fruit and vegetables. *Food Microbiology, 21*, 703.

Diab, T., Biliaderis, C. G., Gerasopoulos, D., & Sfakiotakis, (2001). Physicochemical properties and application of pullulan edible films and coatings in fruit preservation. *Journal of the Science of Food and Agriculture, 81*, 988–1000.

Dutta, P. K., Tripathi, S., Mehrotra, G. K., & Dutta, J., (2009). Perspectives for chitosan-based antimicrobial films in food applications. *Food Chemistry, 114*(4), 1173–1182.

Eissa, H. A. A., (2007). Effect of chitosan coating on shelf life and quality of fresh-cut mushroom. *Journal of Food Quality, 30*, 623–645.

Fernandez-Pan, I., Ziani, K., Pedroza-Islas, R., & Mate, J. I., (2010). Effect of drying conditions on the mechanical and barrier properties of films based on chitosan. *Drying Technology, 28*(12), 1350–1358.

Friedman, M., Henika, P. R., & Mandrel, R. E., (2002). Bactericidal activities of plant essential oils and some of their isolated constituents against *Campvlobacterjejuni, Escherichia coli 0157, H7, Listeria monocytogenes* and *Salmonella enterica*. *Journal of Food Protection, 65*, 1545–1560.

Friedman, M., Kim, S. Y., Lee, S., Han, P. G., Hanis-Lee, K. R., & Kozukue, N., (2005). Distribution of catechins, theaflavins, caffeine, and theobroittine in 77 teas consumed in the United States. *Journal of Food Science, 70*, 550–559.

Garcia, M., Pinotti, A., Martino, M., & Zaritzky, N., (2009). *Edible Films and Coatings for Food Applications*. New York: Springer (Chapter 6).

Gennadios, A., (2002). *Protein-Based Films and Coatings*. Boca Raton, FL: CRC Press LLC.

Ghanbarzadeh, B., & Almasi, H., (2011). Physical properties of edible emulsified films based on carboxymethyl cellulose and oleic acid. *International Journal of Biological Macromolecules, 48*(1), 44–49.

Ghasemlou, M., Khodayian, F., & Oromiehie, A., (2011). Physical, mechanical, barrier and thermal properties of polyol plasticized biodegradable edible film made from kefiran. *Carbohydrate Polymer Journal, 84*(1), 477–483.

Goddard, J. M., & Hotchkiss, J. H., (2007). Polymer surface modification for the attachment of bioactive compounds. *Progress in Polymer Science, 32*(7), 698–725.

González, A., Strumia, M. C., Ines, C., & Igarzabal, A., (2011). Cross-linked soy protein as material for biodegradable films: Synthesis, characterization and biodegradation. *Journal of Food Engineering, 106*, 331–338.

Gonzalez-Aguilar, G. J., Ayala-Zavala, G., Olivas, L., De-la-Rosa, & Alvarez-Parrilla, E., (2010). Preserving quality of fresh-cut products using safe technologies. *Journal of Consumer Protection and Food Safety, 5*, 65–72.

Guilbert, S., & Gontard, N., (1995). Edible and biodegradable food packaging. In: Ackermann, P., Jagerstad, M., & Ohlsson, T., (eds.), Foods and Packaging Materials—Chemical Interactions (pp. 159–168). Cambridge, U. K.: The Royal Society of Chemistry.

Gutierrez, J., Barry-Ryan, C., & Bourke, P., (2009). Antimicrobial activity of plant essential oils using food model media: Efficacy, synergistic potential and interactions with food components. *Food Microbiology, 26*, 142–150.

Han, J. H., & Gennadios, A., (2005). Edible films and coatings: A review. Ch. 15. In: Han, J. H., (ed.), *"Innovations in Food Packaging"* (p. 239). Elsevier Academic Press, Amsterdam.

Hershko, V., Klein, H., & Nuvissovitch, A., (1996). Relationships between edible coating and garlic skin *Journal of Food Science, 61*, 769–777.

Janjarasskul, T., & Krochta, J. M., (2010). Edible packaging materials. *Annual Review of Food Science and Technology, 1*, 415–448.

Jin, T., & Zhang, H., (2008). Biodegradable polylactic acid polymer with nisin for use in antimicrobial food packaging. *Journal of Food Science, 73*, 127–134.

Kader, A. A., (2002). Quality parameters of fresh-cut fruit and vegetable products. Ch. 2. In: Lamikanra, (ed.), *"Fresh-Cut Fruits and Vegetables- Science, Technology and Market"* (p. 11). Boca Raton: CRC Press.

Kim, S. J., & Ustunol, Z., (2001). Thermal properties, heat sealability and seal attributes of whey protein isolate/lipid emulsion edible films. *Journal of Food Science, 66*(7), 985–990.

Kokoszka, S., & Lenart, A., (2007). Edible coatings – formation, characteristics and use – a review. *Polish Journal of Food and Nutrition Sciences, 57*(4), 399–404.

Krochta, J. M., (2002). Proteins as raw materials for films and coatings: Definitions, current status, and opportunities. In: Gennadios, A., (ed.), *Protein-Based Films and Coatings* (pp. 1–41). Boca Raton, Fla.: CRC Press.

Krochta, J. M., & De-Mulder-Johnston, C., (1997). Edible and biodegradable polymer films: Challenges and opportunities. *Food Technology, 51*, 61–74.

Krochta, J. M., Baldwin, E. A., & Nisperos-Carriedo, M. O., (1994). *Edible Coatings and Films to Improve Food Quality*. Lancaster, Pa.: Technomic Publishing.

Lee, J. Y., Park, H. J., Lee, C. Y., & Choi, W. Y., (2003). Extending shelf-life of minimally processed apples with edible coatings and antibrowning agents. *Lebens. Wissen. Technol., 36*, 323–329.

Lim, L. T., Mine, Y., Britt, I. J., & Tung, M. A., (2002). Formation and properties of egg white protein films and coatings. In: Gennadios, A., (ed.), *Protein-Based Films and Coatings* (pp. 233–252). Boca Raton, Fla.: CRC Press.

Lin, D., & Zhao, Y., (2007). Innovations in the development and application of edible coatings for fresh and minimally processed fruits and vegetables. *Comprehensive Reviews in Food Science and Food Safety*, *6*, 60–75.

Lin, S. Y. D., & Krochta, J. M., (2005). Whey protein coating efficiency on surfactant-modified hydrophobic surfaces. *Journal of the Science of Food and Agriculture*, *53*, 5018–5023.

Ma, Y. Q., Chen, J. C., Liu, D. H., & Ye, X. Q., (2008). Effect of ultrasonic treatment on the total phenolic and antioxidant activity of extracts from citrus peel. *Journal of Food Science*, *73*(8), 115–120.

Mayachiew, P., & Devahastin, S., (2008b). Comparative evaluation of physical properties of edible chitosan films prepared by different drying methods. *Drying Technology*, *26*, 176–185.

McHugh, T. H., & Sensi, E., (2000). Apple wraps: A novel method to improve the quality and extend the shelf life of fresh-cut apples. *Journal of Food Science*, *65*, 480–485.

Miller, K. S., & Krochta, J. M., (1997). Oxygen and aroma barrier properties of edible films: A review. *Trends in Food Science and Technology*, *8*, 228–237.

Moore, M. E., Han, I. Y., Acton, J. C., Ogale, A. A., Barmore, C. R., & Dawson, P. L., (2003). Effects of antioxidants in polyethylene film on fresh beef color. *Journal of Food Science*, *68*(1), 99–104.

Morillon, V., Debeafort, F., Blond, G., Capelle, M., & Voilley, A., (2002). Factors affecting the moisture permeability of lipid-based edible films: A review. *Critical Review and Food Science and Nutrition*, *42*, 67–89.

Nam, S. H., Choi, S. P., Kang, M. Y., Koh, H., Kozukue, N., & Friedman, M., (2006). Antioxidative activities of bran extracts from 21 pigmented rice cultivars. *Food Chemistry*, *94*, 613–620.

Negi, P. S., (2012). Plant extracts for the control of bacterial growth: Efficacy, stability and safety issues for food applications. *International Journal of Food Microbiology*, *156*, 7–17.

Olivas, G. I., Mattinson, D. S., & Barbosa-Canovas, G. V., (2005). Alginate coatings for preservation of minimally processed 'Gala' apples. *Postharvest Biology and Technology*, *45*, 89–96.

Olivas, G. I., & Barbosa-Canovas, G. V., (2005). Edible coatings for fresh-cut fruits. *Critical Reviews in Food Science and Nutrition*, *45*, 657–670.

Olivas, G. I., Rodriguez, J. J., & Barbosa-Canovas, G. V., (2003). Edible coatings composed of methylcellulose, stearic acid, and additives to preserve quality of pear wedges. *Journal of Food Process Preservation*, *27*, 299–320.

Oussalah, M., Caillet, S., Salmieri, S., Saucier, L., & Lacroix, M., (2004). Antimicrobials and antioxidants effects of milk protein- based film containing essential oils for the preservation of whole beef muscle. *Journal of Agriculture and Food Chemistry*, *52*(18), 5598–5605.

Ozge-Erdohan, Z., Cam, B., & Turhan, K. N., (2013). Characterization of antimicrobial polylactic acid-based films. *Journal of Food Engineering*, *119*, 308–315.

Park, S. I., Daeschel, M. A., & Zhao, Y., (2004). Functional properties of antimicrobial lysozyme chitosan composite films. *Journal of Food Science*, *69*, 215–221.

Park, S. K., Hettiarachchy, N. S., Ju, Z. Y., & Gennadios, A., (2000). Formation and properties of soy protein film and coatings. In: Gennadios, A., (ed.), *Protein-Based Films and Coatings* (pp. 123–137). Corporate BLvol, Boca Raton, Florida.

Park, S., Stan, S. D., Daeschel, M. A., & Zhao, Y., (2005). Antifungal coatings on fresh strawberries (Fragaria × ananassa) to control mold growth during cold storage. *Journal of Food Science, 70*, 202–207.

Pascall, M. A., & Lin, S. J., (2013). The application of edible polymeric films and coatings in the food industry. *Journal Food Processing and Technology, 4*, 2.

Perez-Gago, M. B., & Krochta, J. M., (2001). Lipid particle size effect on water vapor permeability and mechanical properties of whey protein/ beeswax emulsion films. *Journal of Agriculture and Food Chemistry, 49*, 996–1002.

Perumalla, A. V. S., & Hettiarachchy, N. S., (2011). Green tea and grape seed extracts-potential applications in food safety and quality. *Food Research International, 44*, 827–839.

Pol, H., Dawson, P., Acton, J., & Ogale, A., (2002). Soy protein isolate/corn-zein laminated films: Poultry meat to inhibit development of thiobarbituric acid reactive substances. *Journal of Agriculture and Food Chemistry, 51*, 1602–1607.

Pruneda, E., Peralta-Hernandez, J. M., Esquivel, K., Lee, S. Y., Godinez, L, A., & Mendoza, S., (2008). Water vapor permeability, mechanical properties and antioxidant effect of Mexican oregano-soy based edible film. *Journal of Food Science, 73*, 488–493.

Quintavalla, S., & Vicini, L., (2002). Antimicrobial food packaging in meat industry. *Microbiology, 41*, 464–469.

Rakotonirainy, A. M., Wang, Q., & Padua, G. W., (2001). Evaluation of zein films as modified atmosphere packaging for fresh broccoli. *Journal of Food Science, 66,* 1108–1111.

Ramos, L. O., Fernandes, J. O., Silva, S. I., Pintado, M. E., & Malcata, F. X., (2012). Edible films and coatings from whey proteins: A review on formulation, and on mechanical and bioactive properties. *Critical Reviews in Food Science and Nutrition, 52*, 533–552.

Raybaudi-Massilia, R. M., Rojas-Grau, M. A., Mosqueda-Melgar, J., & Martin-Belloso, O., (2008). Comparative study on essential oils incorporated into an alginate-based edible coating to assure the safety and quality of fresh-cut Fuji apples. *Journal of Food Protection, 71*, 1150–1161.

Raymond, L. V., Zhang, M., & Roknul-Azam, S. M., (2012). Effect of chitosan coating on physical and microbial characteristics of fresh-cut green peppers (Capsicum annuum L.). *Pakistan Journal of Nutrition, 11*, 806.

Ribeiro, C., Vicente, A. A., Teixeira, J. A., & Miranda, C., (2007). Optimization of edible coating composition to retard strawberry fruit senescence. *Postharvest Biology and Technology, 44*(1), 63–70.

Rojas-Grau, M. A., Raybaudi-Massilia, R. M., Soliva-Furtuny, R. C., Avena-Bustillos, R. J., McHugh, T. H., & Martin-Belloso, O., (2007). Apple puree-alginate edible coating as carrier of antimicrobial agents to prolong shelf life of fresh-cut apples. *Postharvest Biology and Technology, 45*, 254–264.

Rojas-Grau, M. A., Soliva-Fortuny, R., & Martin-Belloso, O., (2009). Edible coatings to incorporate active ingredients to fresh-cut fruits: A review. *Trends in Food Science and Technology, 20*, 438–447.

Rojas-Grau, M., Sobrino-Lopez, A., Tapia, M., & Martin-Belloso, O., (2006). "Browning inhibition in fresh-cut "Fuji" apple slices by natural antibrowning agents." *Journal of Food Science, 71*(1), 59–65.

Rossman, J. M., (2009). Edible films and coatings for food applications. In: Embuscado, M. E., & Huber, K. C., (eds.), *Edible Films and Coatings for Food Applications*. Springer Science Business Media, USA.

Salleh, E., Muhamad, I. I., & Khairuddin, N., (2009). Structural Characterization and Physical Properties of Antibacterial (AM) Starch-Based Films. *World Academy of Science, Engineering and Technology*. 31, 428–436.

Sanchez-Gonzalez, L., Vargas, M., Gonzalez-Martinez, C., Chiralt, A., & Chafer, M., (2011). Use of essential oils in bioactive edible coatings. *Food Engineering Reviews, 3*, 1–16.

Seydim, A. C., & Sarikus, G., (2007). Antimicrobial activity of whey protein based edible films incorporated with oregano, rosemary and garlic essential oils. *Food Research International, 40*, 949.

Shahidi, F., & Naczk, M., (2004). *Phenolics in Food and Nutraceuticals*. CRC Press, Boca Raton, FL.

Shahidi, F., Arachchi, J. K. V., & Jeon, Y. J., (1999). Food applications of chitin and chitosan. *Trends in Food Science and Technology, 10*(2), 37–51.

Shen, X., Sun, X., Xie, Q., Liu, H., Zhao, Y., & Pan, Y., (2014). Antimicrobial effect of blueberry (*Vaccinumcorymbosum* L.) extracts against the growth of *Listeria monocytogenes* and *Salmonella enteritidis*. *Food Control, 35*, 159–165.

Siddiqui, W. M., Chakraborty, I., Ayala-Zavala, J. F., & Dhua, R. S., (2011). Advances in minimal processing of fruits and vegetables: A review. *Journal of Scientific and Industrial Research, 70*, 823–834.

Silva, G. G. D., Sobral, P. G. A., Carvalho, R. A., Bergo, P. V. A., Mendieta-Taboada, O., & Habitante, A. M. Q. B., (2008). Biodegradable films based on blends of gelatin and poly (vinyl alcohol): Effects of PVA type or concentrations on some physical properties of films. *Journal of Polymer Environment, 16*, 276–285.

Simi, C. K., & Abraham, T. E., (2010). Biodegradable biocompatible xyloglucan films for various applications. *Colloid and Polymer Science, 288*, 297–306.

Siripatrawan, U., & Harte, B. R., (2010). Physical properties and antioxidant activity of an active film from chitosan incorporated with green tea extract. *Food Hydrocolloids, 4*, 770–775.

Smith-Palmer, A., Stewart, J., & Fyfe, L., (2001). The potential application of plant essential oils as natural food preservatives in soft cheese. *Food Microbiology, 18*, 463–470.

Sonti, S., (2000). *Consumer Perception and Application of Edible Coatings on Fresh-Cut Fruits and Vegetables* (p. 143). A thesis, Osmania B.S, University College of Technology.

Sonti, S., Prinyawiwatkul, W., No, H. K., & Janes, M. E., (2003). Maintaining quality of fresh-cut apples with edible coating during 13 days refrigerated storage. *IFT Annual Meeting Technical Program Abstracts, 76–86.*

Sothornvit, R., & Krochta, J. M., (2001). Plasticizer effect on mechanical properties of beta-lactoglobulin films. *Journal of Food Engineering, 50*, 149–155.

Sothornvit, R., & Krochta, J. M., (2000). Plasticizer effect on oxygen permeability of b-lactoglobulin films. *Journal of Agricultural Food Chemistry, 48*, 6298–6302.

Sothornvit, R., & Rodsamran, P., (2008). Effect of a mango film on quality of whole and minimally processed mangoes. *Postharvest, Biology and Technology, 47*, 407–415.

Srinivasa, P. C., Ramesh, M. N., & Tharanathan, R. N., (2007). Effect of plasticizers and fatty acids on mechanical and permeability characteristics of chitosan films. *Food Hydrocolloid, 21*, 1113–1122.

Su-Cha & Chinnan, (2004). Biopolymer-based: Antimicrobial packaging a review. *Critical Reviews in Food Science and Nutrition, 44*(4), 223–237.

Suyatma, E. N., Copinet, A., Tighzert, L., & Coma, V., (2004). Mechanical and barrier properties of biodegradable films made from chitosan and poly (lactic acid) blends. *Journal of Polymer and the Environment, 12*, 1–6.

Tanaka, M., Iwata, K., & Sanguandeekul, R., (2001). Influence of plasticizers on the properties of edible films prepared from fish water-soluble proteins. *Fisheries Science, 67*(2), 346–351.

Thybo, A. K., Christinansen, J., Kaack, K., & Petersen, M. A., (2006). Effect of cultivars, wound healing and storage on sensory quality and chemical components in pre-peeled potatoes. *LWT Food Science and Technology, 39*, 166–176.

Toivonen, P. M. A., & Brummell, D. A., (2008). Biochemical bases of appearance and texture changes in fresh-cut fruit and vegetables. *Postharvest Biology Technology, 48*, 1–14.

Tovar, L., Salafranca, J., Sanchez, C., & Nerin, C., (2005). Migration studies to assess the safety in use of a new antioxidant active packaging. *Journal of Agricultural and Food Chemistry, 53*(13), 5270–5275.

Turhan, K. N., Sahbaz, E., & Guner, A., (2001). A spectrophotometric study of hydrogen bonding in methylcellulose-based edible films plasticized by polyethylene glycol. *Journal of Food Science, 66*, 59–62.

Valdes, A., Mellinas, A. C., Ramos, M., Garrigos, M. C., & Jimenez, A., (2014). Natural additive and agricultural wastes in biopolymer formulations for food packaging. *Frontiers Chemistry, 2*(6), 1–10.

Valencia-Chamorro, S. A., Palou, L., Delŕio, M. A., & Pérez-Gago, M. B., (2011). Antimicrobial edible films and coatings for fresh and minimally processed fruits and vegetables: A review. *Critical Reviews in Food Science and Nutrition, 51*(9), 872–900.

Vargas, M., Albors, A., Chiralt, A., & Gonzalez, M. C., (2006). Quality of cold-stored strawberries as affected by chitosan-oleic acid edible coatings. *Postharvest Biology and Technology, 41*, 164.

Vargas, M., Pastor, C., Chiralt, A., McClements, D. J., & Gonzalez-Marinez, C., (2008). Recent advances in edible coatings for fresh and minimally processed fruits. *Critical Reviews in Food Science and Nutrition, 48*(6), 496–511.

Vargas, M., Albors, A., Chiralt, A., & González-Martínez, C., (2009). Characterization of chitosan–oleic acid composite films. *Food Hydrocolloid, 23*, 536–547.

Wong, D. W. S., Tillin, S. J., Hudson, J. S., & Pavlath, A. E., (1994). Gas exchange in cut apples with bilayer coatings. *Journal of Agricultural and Food Chemistry, 42*, 2278–2285.

Wu, T., Zivanovic, S., Draughon, F. A., Conway, W. S., & Sams, C. E., (2005). Physicochemical properties and bioactivity of fungal chitin and chitosan. *Journal of Agricultural and Food Chemistry, 53*(10), 3888–3894.

Xu, S., Chen, X., & Sun, D. W., (2001). Preservation of kiwifruit coated with an edible film at ambient temperature. *Journal of Food Engineering, 50*, 211–216.

Zheng, W., & Wang, S. Y., (2001). Antioxidant activity and phenolic compounds in selected herbs. *Journal of Agricultural and Food Chemistry, 49*(11), 5165–5170.

Zhong, Y., Li, Y., & Zhao, Y., (2012). Physicochemical microstructural and antibacterial properties of β-chitosan and kudzu starch composite films. *Journal of Food Science, 77*(10), 280–286.

CHAPTER 14

Dry Fractionation of Legumes

SNEHASIS CHAKRABORTY and CHANDRIMA SHRIVASTAVA

Department of Food Engineering and Technology,
Institute of Chemical Technology, Mumbai 400019 India,
E-mails: sc.chakraborty@ictmumbai.edu.in, chandrima95@gmail.com

ABSTRACT

The rising demand for plant-based protein isolates triggers the food industry to find out a sustainable technique for fractionation of protein source such as legumes. Wet fractionation of legume protein is the common practice in which a huge amount of water is involved in this process. Moreover, the addition of chemicals and consumption of extra energy during dehydration have made it less sustainable. Dry fractionation involves only two basic operations: milling and separation. Milling fragments the legume into protein bodies and starch granules. Subsequently, a suitable dry separation technique fractionates the mixture into a protein-rich fraction. There are different dry separation techniques employed so far, including sieving, air classification, and electrostatic separation. There are literature using different pre- and post-treatment along with dry fractionation to improve the protein concentration in the fractionate. The morphology of different legumes leads to varying amount of protein-enriched fraction. This chapter explores the principle and potential of dry fractionation technique applied to legumes. Further, a summary of research works carried out on this area along with some pre- and post-treatments have been discussed. The attempt is to provide an overview of this technique to reestablish its potential towards a sustainable technology.

14.1 INTRODUCTION

Proteins are the building blocks of the human body. Consuming the required amount of protein through diet is indispensable for our health and growth.

Today both, the global food consumption pattern and the dietary practices of individuals, are largely governed by the per capita income and the preference towards nutrient-dense food. As a result the global demand for proteins is expected to be double in the next decade, considering the rate of increase in population along with a rise in the per capita income. To meet this increasing protein demand, the sole dependence on animal products, such as milk, eggs, cheese, and meat, seems inadequate. On the other hand, from economic as well as environmental fronts, consumer preferences are increasingly getting inclined towards plant-based protein sources as opposed to those from animals.

In the last decade, protein from legumes has been gaining an increasing interest to meet this rising protein demand. Amongst plant-based sources, legumes contain the highest amount of protein (typically 25 to 40%). It is also reported that 6 kg of the livestock of plant protein needs to be fed for getting an equivalent of 1 kg of animal protein (Pimentel and Pimentel, 2003). This entails the requirement of a far higher quantity of plant protein for obtaining an equivalent amount of animal protein. The existing methods of processing plant-based proteins are the conventional wet fractionation processes. These conventional wet fractionation processes to isolate legume proteins employ copious amounts of water; involve exposure to chemicals and an energy-intensive drying step. These harsh treatments result in the loss of the functionality of the isolated protein, thereby limiting its applications. These techniques are also placing a heavy strain on the environmental resources and generating high BOD waste streams, giving rise to concerns like sustainability and impact on the ecosystem. This has provided the impetus to explore more sustainable methods for the production of proteins.

Dry fractionation provides a more sustainable alternative for producing protein concentrates from legumes. The dry fractionation is the compartmentalization of the protein into the friable protein bodies (PB), and the starch into denser, less-brittle granules, enabling size reduction and the consequent fractionation of the legume flour into a protein-enriched fraction and a protein-depleted fraction. This technique has implications on the chemical composition and functional properties of the resultant protein concentrate. The enriched protein concentrate fractions thus produced could contain 30–80% protein. Though the protein enrichment is relatively modest in comparison to its wet counterpart, dry fractionation ensures the retention of the functional properties of the protein concentrate due to the absence of pH shifts, exposure to chemicals and denaturation due to the drying step. This allows the application of dry-fractionated protein concentrates in high-protein beverages and meat analogs.

Dry fractionation has emerged as a sustainable alternative to wet processes in the last three decades. This technique was extensively investigated in the late 1970s and early 1980s, after which it received lesser attention for two decades due to the relatively modest protein enrichment it provides. However, there has been a renewed interest in employing dry fractionation on legumes in recent years owing to its long-term sustainability, energy-, and resource-intensive nature, and the ability to yield functional ingredients. This has also driven the exploration of alternative separation techniques with newer driving forces. It is expected that further exploration of dry fractionation for the sustainable production of protein concentrates with high ingredient functionality will open several doors towards the development of novel protein foods.

This chapter highlights the different separation techniques used in dry fractionation of legumes, followed by a comparison between wet and dry fractionation processes in view of the fractionation efficiency and product quality. The criticality of several pre- and post-treatment of dry fractionation techniques applied to legumes have also been discussed. The importance of exergy calculation for designing a sustainable and more efficient fractionation process for legumes has also been highlighted. This will help in reestablishing the dry fractionation process in integration with other pre- or post-treatment for obtaining a high-quality protein fraction from a specific legume through a sustainable technology.

14.2 DRY FRACTIONATION

The principle of dry fractionation is that the milling of the legumes can detach the PB mechanically from the other cellular components like starch and fiber in the grounded mass or flour. In this sense, dry fractionation mainly consists of a primary milling followed by a physical separation process without any involvement of water (dry). The primary milling or size reduction is crucial for the physical disentanglement of the friable PB from the relatively less brittle starch granules. In addition, other cellular components are also detached from the PB during milling. Typically the particle size of the starch granules after milling is greater than the fragmented protein matrix. The flour thus obtained by size reduction is then subjected to subsequent separation or fractionation. The dry separation processes that could be used include air classification, sieving, and electrostatic separation, which differ with respect to the driving force employed. The legume morphology and composition also play a crucial role in the separation. The separation results in two fractions, one of which is rich in protein (protein-enriched fraction) and the other portion has lesser protein (protein-depleted fraction).

The most common separation technique is based on the fact that milling causes the physical detachment of the larger, less-brittle starch granules (typically about 20 μm) from the smaller and more friable PB (1–3 μm), and the particle size difference could be exploited for separation (Aguilera et al., 1982; Tyler and Panchuk, 1982). Taking a closer look, the dry fractionation simply needs milling, followed by dry separation techniques to produce two fractions with different protein content. In contrast, the wet fractionation of legumes consists of several steps, including milling, extraction, solubilization, precipitation, neutralization, and drying. For starch-rich legumes, the post-milling extraction step is omitted and suspending the pellet in water is followed by centrifugal separation using a hydroclone. The major steps involved in both wet and dry fractionation/separation processes for protein isolation have been presented in Figure 14.1. The separation of protein isolate from oil-rich legumes through wet processes requires a greater number of sequential operations than dry fractionation. It is also clear from Figure 14.1 that unlike the wet counterpart, dry fractionation has no waste streams emerging at any step, and it is devoid of energy or resource-intensive processes like spray drying.

The first stage is the optimal milling of the legume, essential for optimal disentanglement. The size reduction step is followed by the separation of flour constituents based on one or more driving forces such as the difference in particle size, density, or charging characteristics.

14.2.1 MILLING

Dry milling of legumes is performed after the steps like dehusking and splitting of cotyledons. Further, the cotyledons are milled into flour through any of the mechanism such as impact, abrasive, and others using a suitable milling machine. Milling leads to an increase in the surface to volume ratio of the substrate and the breakdown of cell wall structures, leading to increased accessibility to cellular contents (Maaroufi et al., 2000). Milling is a very critical step, as extensive milling of legumes may lead to severe damage to starch granules. Consequently, the PB and starch granule fragments will have a similar size, which may be detrimental for the separation processes using the difference in particle size as the driving force. For inadequate milling, the majority of the PB will stay entangled with the starch granules, leading to a poor separation.

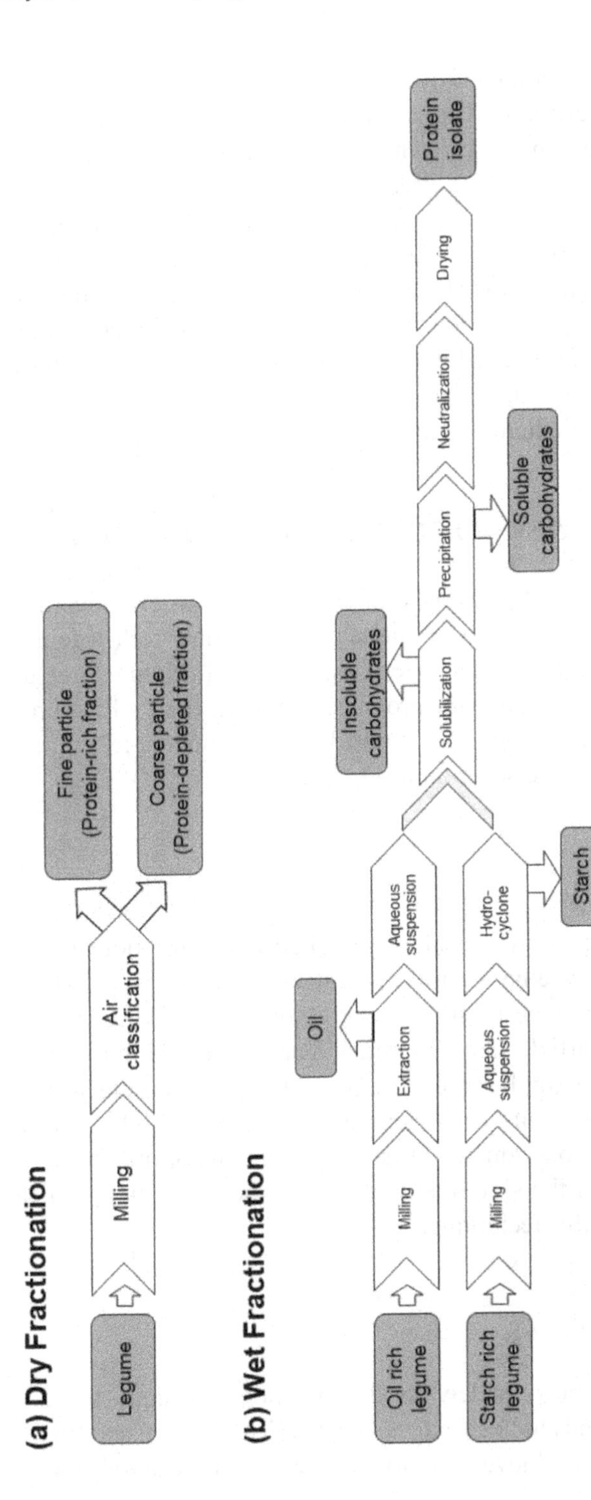

FIGURE 14.1 Schematic illustration of the steps involved in wet and dry fractionation processes (Adapted from Schutyser et al., 2015).

The extent of milling is crucial as it directly determines the particle size distribution of the flour. Several studies have used different milling techniques such as impact, jet, and pin milling to achieve the optimal disentanglement. Some studies have also compared the effect of two milling techniques on the final yield and purity of the enriched fraction. Wright et al. (1984) compared the relative milling efficiencies of a one-rotating stud-disc pin mill, and a two-disc counter-rotating pin mill. The counter-rotating pin mill resulted in a higher yield and protein content. Pelgrom et al., (2013) compared jet and impact milling for peas and inferred that damaged starch in the concentrate fraction was more after jet milling, hence impact milling was concluded to result in optimal disentanglement and minimum damaged starch.

14.2.2 SEPARATION TECHNIQUES USED IN DRY FRACTIONATION

The separation technique to be employed after milling is largely governed by the composition and morphology of the legume. For fiber-rich legumes like lupine, the difference in particle size is not the ideal driving force as fiber fragments are fractionated with proteins, reducing the yield. Three major separation techniques have been studied so far, namely sieving, air classification, and electrostatic separation.

14.2.2.1 SIEVING

Sieving is the simplest separation technique applied to legume flour based on the particle size difference between PB and starch granules. Typically, a stack of sieves with decreasing mesh size is employed for sieving. Starch granules usually have a particle size >20 μm, and are embedded in a friable matrix of PB typically ranging between 1–3 μm. The mean particle size of the fractions thus separated onto each sieve is measured on the basis of a mass fraction as opposed to a number of particles. One major drawback of this method is clogging of the sieves, also known as sieve blinding, which limits the applications of this technique.

14.2.2.2 AIR CLASSIFICATION

Air classification is the most widely used separation technique for dry fractionation of legumes and has been extensively applied on many starch-rich legumes. The separation is achieved due to a difference in the aerodynamic properties of starch and protein in the flour, particularly density and particle size difference. Air classification involves the dispersion of flour into a stream of air with a regulated flow rate. Most of the classifiers contain a rotating wheel or classifier, the speed of rotation of which governs the cut-point of separation. The working principle of an air classifier has been illustrated schematically in Figure 14.2.

Inside an air classifier, a particle is acted upon by two opposing forces, the centrifugal force due to the rotating classifier wheel, and the frictional force due to air drag (Figure 14.2). For finer particles, the drag force exceeds the centrifugal force; hence these are carried in the air stream while the larger particles are thrown out of the edge of the wheel and constitute the coarse fraction (Cloutt et al., 1987).

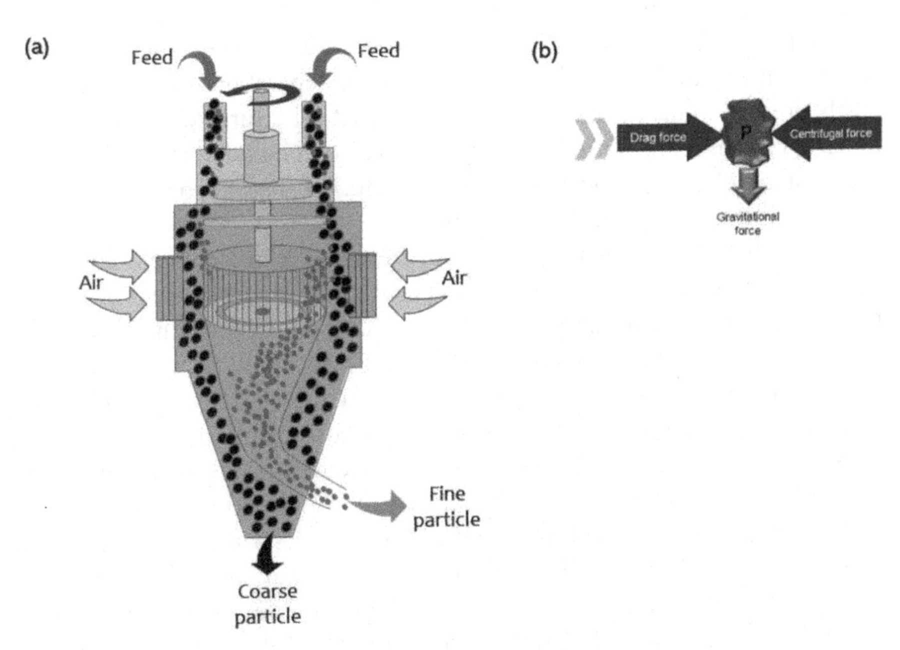

FIGURE 14.2 (a) Schematic diagram of the principle of a centrifugal air classifier; (b) free body diagram showing the forces acting on a particle in the air classifier.

The protein concentration in the enriched fraction obtained after air classification largely depends on the initial protein content of the flour, the dispersibility of the flour and the cut-point (Reichert, 1982; Pelgrom et al., 2013a). Pelgrom et al., (2013a) defined cut-point as the size at which a particle has an equal chance of moving into the fine fraction and the coarse fraction. It could also be defined as that particle size at which the drag force is equal to the centrifugal force. For protein-starch separation, a cut-point of 10–15 μm gives maximal separation for starch-rich legumes.

Air classification is effective for obtaining protein-enriched fractions from starch-rich legumes, but is unable to efficiently separate lipid-rich legumes like lupine. This is mainly because oil impairs the flour dispersibility and the presence of fine fiber fragments lowers the yield. Hence techniques that use another driving force must be employed.

14.2.2.3 ELECTROSTATIC SEPARATION

Electrostatic separation is a novel separation technique, which has been applied relatively recently in the separation of food powders. It works on the principle of imparting a surface charge to the particles of interest, and their consequent separation on the basis of this charge difference under the influence of an external electric field. Electrostatic separation has been extensively applied in the mining industry in the past. The surface charge can be imparted to particles in different ways, such as corona bombardment, tribo-electrostatic charging, and induction charging. Of these, tribo-electrostatic charging is the most widely used technique.

Tribo-electrostatic charging involves the contact of two materials with each other. Physical contact and friction cause the transfer of electrons from the surface of one material to the other due to the difference in work functions of the two surfaces in contact (Figure 14.3). Work function measures the minimum amount of energy required to remove an electron from the surface of a given solid. Thus, on physical contact, the particles with a higher work function will tend to accept an electron and become negatively charged. The charge attained by the particles is the negative of that attained by the tube, and is defined by specific charge and surface charge density. The charged particles are then carried to a separation chamber where they migrate towards the respective electrodes in the presence of an external electric field (Figure 14.3). The difference in the tribo-electrostatic properties of the protein, fiber, and starch due to the presence of different functional groups in their structures can be exploited to get an effective separation.

14.3 COMPARISON BETWEEN WET AND DRY FRACTIONATIONS

In case of wet fractionation, the flour containing the protein, fiber, and starch fractions is dispersed in water and/or other solvents in which the individual fractions have dissimilar solubilities. Afterwards, the slurry is treated through hydroclone to separate the protein-rich fraction from the starchy part. It is evident that wet fractionation produces protein isolates that have a high protein content (>80%). It is based on the principle of the differential solubility of various components of the legume matrix in different solvents. However, some major drawbacks of wet fractionation are the usage of copious amounts of water, the exposure to harsh chemical treatments to cause precipitation, and an energy-intensive drying step. This puts a heavy strain on resources and generates whey-like waste streams with a BOD almost as high as 30,000 mg·L^{-1}. On the other hand, the shifts in pH and the heat denaturation of protein during drying result in partial loss of the native functionality of the resultant protein isolates.

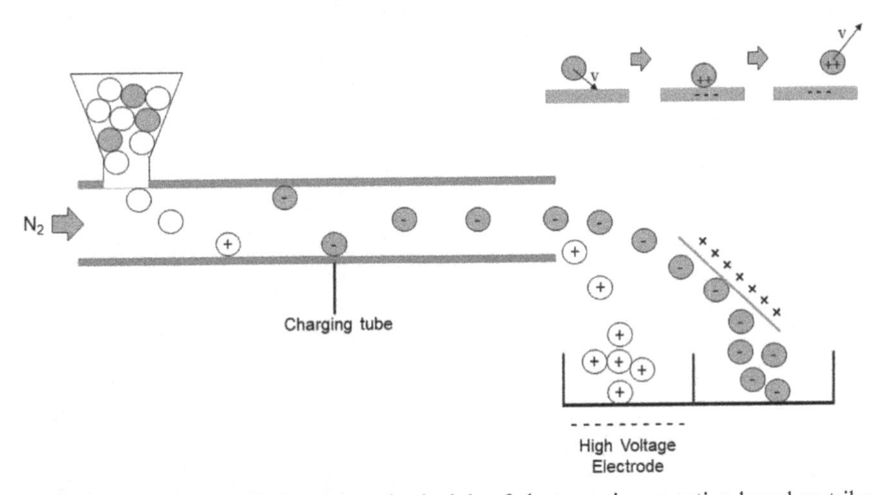

FIGURE 14.3 Schematic diagram and principle of electrostatic separation based on tribo-electrostatic charging (Adapted from Wang et al., 2014).

Dry fractionation provides a sustainable alternative to the conventional wet fractionation processes. The mass and energy flow for wet and dry fractionation processes are compared in Figure 14.4 through Sankey's diagrams. In this diagram, the width of the arrows is proportional to the quantity of flow. Berghout et al., (2015) used exergy as a parameter to evaluate the sustainability of the two processes and assess their environmental impact, as exergy is a measure of the maximum useful work for a system to come in equilibrium with its environment. Dry fractionation gave a physical exergy loss of 1 MJ·(kg protein-enriched flour)$^{-1}$, whereas the spray drying step of wet fractionation itself resulted in exergy losses of more than 12.8 MJ·(kg protein isolate)$^{-1}$. The spray drying step could be replaced by some other

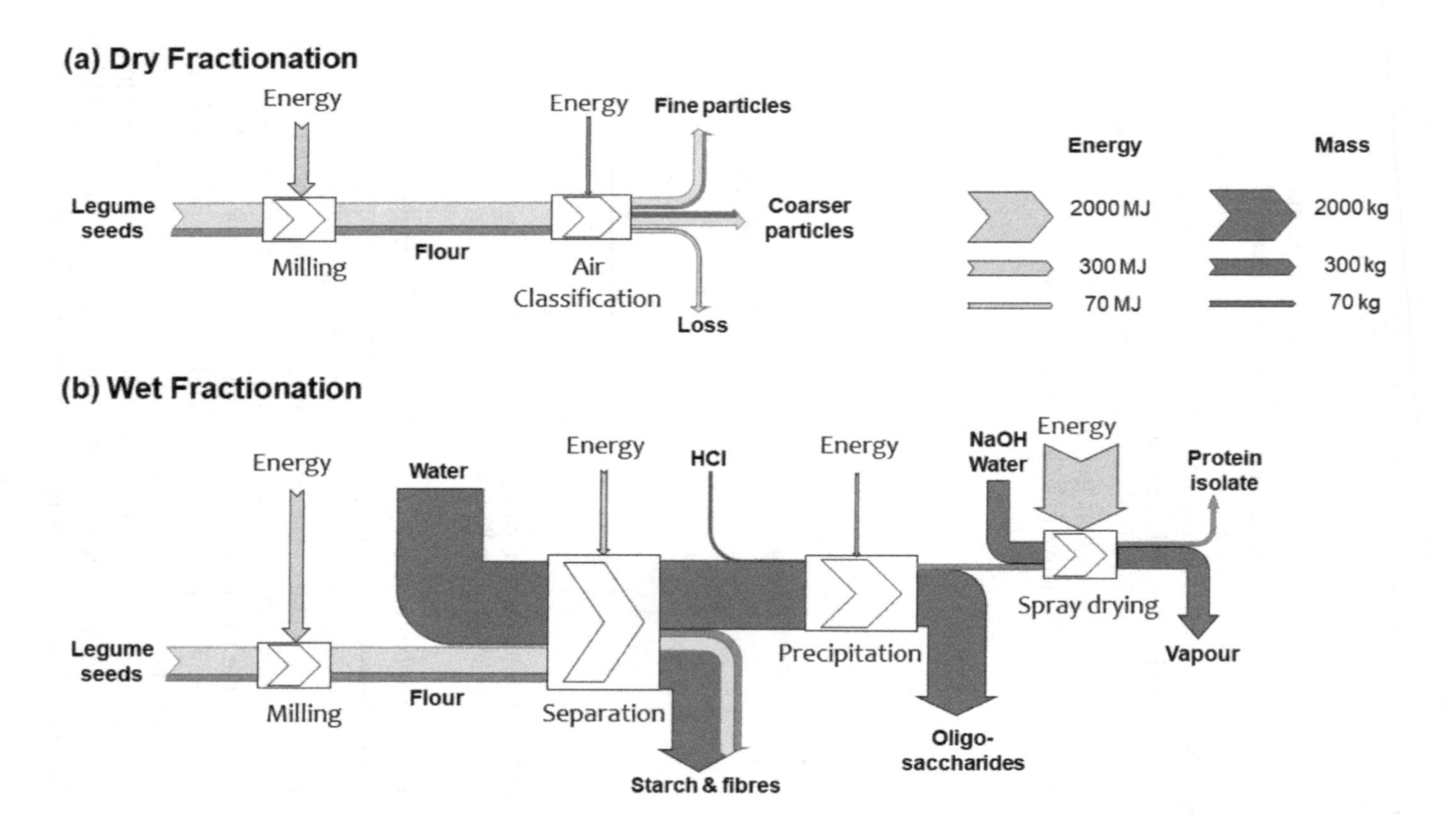

FIGURE 14.4 Sankey diagrams of different fractionation processes to separate protein from crude peas; (a) dry fractionation (b) wet fractionation (Adapted from Schutyser et al., 2015).

concentration step like ultrafiltration, but the resultant wet isolate needs to be stabilized after fractionation. Schutyser et al., (2015) calculated the energy required to produce one kg of recovered protein and reported a 93% reduction in the specific energy requirement for dry fractionation in comparison to its wet counterpart.

14.4 PRODUCT SUITABILITY

The separation yield, efficiency, and purity of the enriched fraction are largely governed by the composition and morphology of the legume of interest. The separation method employed also depends on the suitability of the legume.

14.4.1 COMPOSITION

In fiber-rich legumes such as lupine, milling results in fine fiber fractions which are co-fractionated with the proteins, thereby reducing the protein yield. The chemical composition of various legumes has been summarized in Table 14.1. It can be seen that lupine and soybean have high lipid content. This leads to cohesiveness in the flour after milling, resulting in the agglomeration and inefficient separation. These legumes thus need to be defatted attempting dry fractionation.

TABLE 14.1 Average Chemical Composition of Various Legumes (Schutyser et al., 2011)

Crop, DM%	Protein (%)	Carbohydrates (%)	Lipids (%)	Ash (%)
Lupine	40.4	45.1	10.9	3.7
Soybean	39.9	33.0	21.8	5.3
Fababean	29.3	65.5	1.7	3.5
Mung bean	28.3	66.1	1.8	3.8
Cowpea	26.7	68.2	1.4	3.7
Pea	25.6	68.4	1.9	4.1

The inherent protein content of the legume also affects the separation yield, purity, and efficiency. Reichert (1982) varied the protein content of peas at four levels between 14.5 to 28.5%. The protein separation efficiency increased on increasing the initial pea protein content. However, Pelgrom et al., (2015b) reported that in air classification of pea and lupine followed by electrostatic separation as a post-treatment, increasing the initial protein content decreased the separation efficiency as the Van der Waals interactions lowered the flour dispersibility.

14.4.2 MORPHOLOGY

Since the main driving force for air classification and sieving is the difference in size between the protein and starch bodies, the differential between the particle sizes of PB and starch granules is critical. Figure 14.5 shows a schematic diagram of the morphology of pea and lupine. In the case of pea, a clear distinction is seen in the starch granule size (> 20 μm) and the size of the PB (1–3 μm). Tyler et al., (1981) compared the separation efficiency of different grain legumes and concluded that mung beans and lentils were best suited for dry fractionation, whereas cowpea and lima beans had the lowest separation efficiency and yielded lower protein content. Tyler (1984) studied the effect of properties of eight legumes (mung beans, lentils, great northern beans, navy beans, faba beans, field peas, lima beans, cowpeas) on the impact milling quality. Despite keeping the seed moisture content, feed rate, and classifier cut-size constant, while also maintaining a sufficiently low-fat content, the differences in separation efficiency suggested that differences in legume composition affected the separation. This is attributed to differences in cell wall thickness and rigidity, degree of adhesion between cell contents, the extent to which proteinaceous material is broken down and degree of cohesion among individual cells (Tyler, 1984). Seeds containing more cell wall material were more resistant to disruption by pin milling.

Cloutt and Walker (1987) compared the separation for cowpea, faba bean, and pigeon pea. Faba bean showed the highest protein concentration in the enriched fraction, attributed to their very large starch granules, which resulted in a relatively uncontaminated protein-enriched fraction. As opposed to this, the protein-enriched fraction obtained after the fractionation of cowpea had greater starch content owing to a smaller starch granule size.

14.5 MEASUREMENT OF DRY FRACTIONATION EFFICIENCY

The efficiency of the dry fractionation process can be assessed in terms of four parameters. These are protein concentration or percentage purity (P_p), the yield of the enriched fraction (Φ_p), protein separation efficiency (η_p), and protein shift (δ). Protein concentration (P_p) is defined as the mass

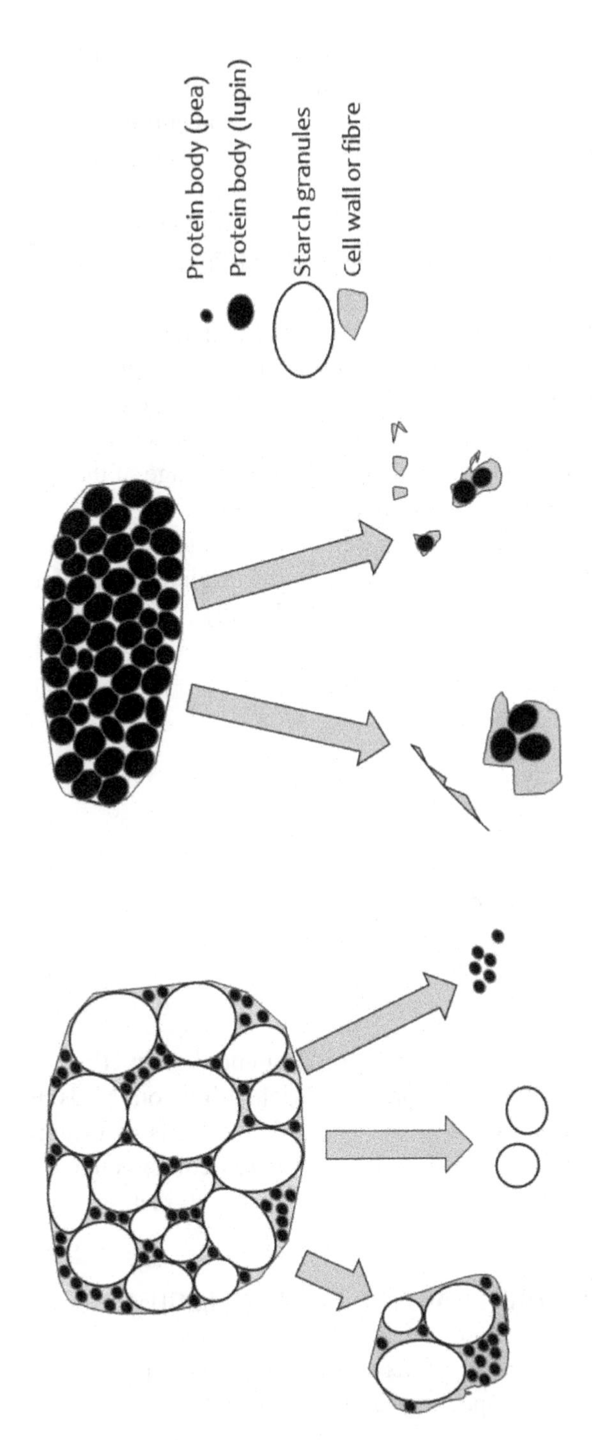

FIGURE 14.5 Schematic drawing of the cells of pea and lupine and the fragments obtained after milling (Adapted from Pelgrom et al., 2014).

percentage of protein in the protein-enriched fraction, as defined in Eq. (1). It is thus an indicator of the purity of the enriched fraction. Dry fractionation typically results in concentrates having a protein concentration in the range of 30–80%, as opposed to protein isolates, which have a higher purity of >80%. The yield of the protein-enriched fraction (Φ_p) is defined as the ratio of the entire dry matter mass of the protein-enriched fraction obtained after separation, to the dry matter mass of the original flour after milling, as defined in Eq. (2).

While both P_p and Φ_p can characterize the dry fractionation process, they cannot be used to represent the overall efficiency of the process, as they may have opposing effects on each other. Protein recovery or protein separation efficiency (η_p) has been used by several researchers, as it takes both yield and purity into account. It is defined as the percentage of the total flour protein recovered in the protein-enriched fraction, as shown in Eq. (3). Schutyser et al., (2011) also defined the parameter of protein shift (δ) to quantify the percentage of protein that shifted from the original milled flour to the protein-enriched fraction, which is defined in Eq. (4).

$$P_p = \frac{x_{p,fraction}}{x_{dm,fraction}} \cdot 100\% \tag{1}$$

$$\phi_p = \frac{M_p}{M_{original.flour}} \tag{2}$$

$$\eta_p = \frac{\phi_p \cdot P_p}{\phi_{flour} \cdot P_{flour}} \tag{3}$$

$$\delta = \frac{1}{P_{flour}} \left[\left(P_p - P_{flour} \right) \cdot \phi_p \right] \tag{4}$$

where x_p is the mass fraction of protein in the protein-enriched fraction, x_{dm} is the mass fraction of dry matter in the protein-enriched fraction, M_p is the mass of protein-enriched fraction (kg), $M_{original\ flour}$ is the mass of original flour (kg), Φ_p is the protein fraction yield, P_{flour} is the protein percentage in flour and P_p is the protein percentage in the protein-enriched fraction.

14.6 APPLICATIONS OF DRY FRACTIONATION ON LEGUMES

The studies on the application of dry fractionation on legume proteins are summarized in Table 14.2. Pea species was the most extensively studied

TABLE 14.2 Studies on Application of Dry Fractionation on Legume Proteins

Sample	Milling technique	Separation technique	Parameters varied	Responses analyzed	Specific observations	References
Navy bean	Pin milling	Triboelectrification	i. Tribo-charger material (PTFE, PVC, nylon & Cu) ii. Charger geometry	P_p, Φ_p	• PTFE showed the highest triboelectrification • A straight charger prevented overcharging, giving a protein concentration of 40.3%	Tabtabaei et al., (2016a)
Navy bean	Pin milling	Triboelectrification	i. Inlet flow rate (7–9 L/min) ii. Charger tube length (60–250 cm) iii. Plate voltage (–1 to –7.5 kV) iv. Plate angle (0–20°)	P_p, Φ_p	• A 3-way interaction existed between airflow rate, tube length & plate voltage; P_p unaffected by plate angle; • Optimum values of process variables are 7 L/min, –6.5 kV, 20° plate angle & 240 cm giving a protein content of 38.1%	Tabtabaei et al., (2016b)
Lupine	Pin milling	Air classification	i. Mill classifier wheel speed ii. Air classifier wheel speed	P_p, Φ_p, η_p, dispersibility, foam stability, digestibility	• Increasing air classifier speed increased P_p but decreased Φ_p • Finer milled flour dispersed better in air • Foam stability of protein-enriched fraction was higher than original lupine flour, but viscosity was lower • Comparable digestibility of heated & unheated concentrates	Pelgrom et al., (2014)

TABLE 14.2 *(Continued)*

Sample	Milling technique	Separation technique	Parameters varied	Responses analyzed	Specific observations	References
Fababean, field pea, lima bean, lentil, mung bean, lentil, navy bean, northern bean	Pin milling	Air classification (5x)	Legume type	P_p, Φ_p, η_p, proximate analysis, cell wall material	• Out of starch, protein, ash, fat, CF, and NDF, only CF significantly negatively correlated with η_p • Seeds containing more cell wall material were more resistant to disruption by milling • Particle size index values negatively correlated with η_p	Tyler (1984)
Yellow pea	Pin milling	Air classification	i. Number of millings ii. Type of pin mill iii. Feed rate iv. Air classifier wheel speed v. Additional classification cycles	P_p, Φ_p, starch, lipid, fiber	• Eight successive millings increased Φ_p from 24.6% to 36% • Counter-rotating pin mill had higher P_p & Φ_p than one stud rotating disc, and feed rate had no significant effect • High classifier speeds gave higher P_p (55.1%) but lower Φ_p; • Lipids and starch were also fractionated with proteins • Additional classification cycles doubled Φ_p, but P_p remained constant	Wright et al., (1984)

TABLE 14.2 *(Continued)*

Sample	Milling technique	Separation technique	Parameters varied	Responses analyzed	Specific observations	References
Chickpea, cowpea, faba bean, field pea, lentil, lima bean, mung bean, navy bean, northern bean	Pin milling	Air classification	Legume type	Anti-nutritional factors	• ANFs were concentrated in the fine fraction • Phytic acid content tripled in the protein fraction • Dietary fiber doubled in the protein fraction • Vicine and convicine found in high levels in the protein fraction of faba bean	Elkowicz and Sosulski (1982)
Field pea	Pin milling	Air classification (double pass)	Protein content, vane setting	Proximate analysis	• P_p was positively correlated with flour protein content • Lower vane settings increased P_p • Ash & sugar contents higher in protein concentrates	Reichert (1982)
Fababean, field pea	Pin milling	Air classification	Moisture content (3.8–14.3%)	Protein, starch, neutral detergent fiber (NDF), seed hardness	• As moisture content decreased, NDF of protein fraction increased, P_p decreased, Φ_p and η_p increased • Grinding time increased with an increase in seed hardness at lower moisture contents	Tyler and Panchuk (1982)

TABLE 14.2 *(Continued)*

Sample	Milling technique	Separation technique	Parameters varied	Responses analyzed	Specific observations	References
Fababean, field pea, lima bean, lentil, mung bean, lentil, navy bean, northern bean	Pin milling	Air classification	Legume type	Proximate analysis, η_p	• Based on η_p, mung bean & lentil were best suited for dry fractionation; cowpea & lima beans had the lowest η_p • Fat, ash & to a lesser extent crude fiber, showed a higher concentration in a protein-enriched fraction	Tyler et al., (1981)
Chickpea, field pea, lentil, northern, fava, lima, mung, & white pea bean	Pin milling	Air classification	Legume type	Water & oil absorption capacities, emulsification, whippability, foam stability	• Lipid & ash co-fractionated with proteins, crude fiber with starch particles • Aqueous dispersions had a pH of 6.2–6.4 • Protein-enriched fraction showed higher oil absorption, emulsification, whippability & foam stability	Sosulski and Youngs (1979)
Field pea, horse bean	Pin milling	Air classification	Legume type	Chemical analysis	• On two-pass air classification, Φ_p improved from 25–28% to 42–44% & P_p of 60–70% protein • Lipids, soluble sugars, P, Zn & Ca classified in the lighter fraction while Fe in the starch fraction, phytic acid doubled in the protein fraction	Vose et al., (1976)

TABLE 14.2 *(Continued)*

Sample	Milling technique	Separation technique	Parameters varied	Responses analyzed	Specific observations	References
Lupine	Impact milling	Electrostatic separation	i. Mill classifier wheel speed ii. Carrier gas flow rate (10–30 L/min) iii. Recycling, re-milling	P_p, Φ_p	• Φ_p decreased with increasing classifier wheel speed due to fouling & exposed lipid (2500 rpm optimum) • Increasing gas flow rate up to 20 L/min increased P_p • Recycling increased P_p from 51% to 65%, decreased Φ_p • Re-milling of fractions in filter bags doubled the yield	Wang et al., (2016)
Bean, chickpea, lentil, pea	Impact milling + Air jet sieving	Air classification	i. Milling classifier wheel speed (2500–8000 rpm) ii. Type of legume	Particle density, seed hardness, starch granule size, fat content, dispersibility	• Optimal detachment for maximal overlap of particle size distribution curve of flour & starch granules (4000 rpm) • Low seed hardness of lentils gave higher P_p; the high-fat content of chickpea flour reduced dispersibility & lowered P_p	Pelgrom et al., (2015a)
Lupine, yellow pea	Impact milling	Air classification	i. Pre-treatments (soaking, defatting, freezing, varying moisture content)	P_p, Φ_p, η_p,	• Lower moisture content increased η_p but reduced P_p • Defatting had no significant effect on the pea, but increased P_p for lupine	Pelgrom et al., (2015b)

TABLE 14.2 *(Continued)*

Sample	Milling technique	Separation technique	Parameters varied	Responses analyzed	Specific observations	References
			ii. Post-treatment (electrostatic separation)		• Soaking & freezing had a negligible effect • Protein enriched fraction was obtained by electrostatic separation even from a coarse air-classified fraction	
Yellow field pea	Impact milling, Jet milling	Air classification	i. Mill classifier wheel speed ii. Air classifier wheel speed	Particle size distribution, P_p, damaged starch, moisture, water holding capacity	• Damaged starch higher for the jet mill; optimal disentanglement for impact milling at 4000 rpm • Moisture content and particle size of fine fraction decreased with increasing classifier speed • Dry fractionated concentrate suspension at 25.7% (w/w) with P_p>30%, gave a gel with high WHC	Pelgrom et al., (2013b)
Cowpea, fababean, pigeonpea	Impact milling	Air classification	Cut-point	P_p, Φ_p	• For cowpea, P_p decreased from 50.6% at 10400 rpm to 29.6% at 6200 rpm • Fababean gave highest P_p (56.5–62.7%) at all cut-points • As P_p decreased, Φ_p increased in a linearly for all legumes	Cloutt et al., (1987)

TABLE 14.2 *(Continued)*

Sample	Milling technique	Separation technique	Parameters varied	Responses analyzed	Specific observations	References
Navy bean	Impact milling	Air classification (double-pass)	Roasting temperature (92–125°C)	Proximate analysis, anti-nutritional factors, particle size	• Maximum P_p (47.6%) for the harshest roasting treatment • Residual activity of trypsin inhibitor & hemagglutinin decreased in a protein-enriched fraction • Finer grinding & lower cut-point resulted in higher δ	Aguilera et al., (1982)

legume, as it is starch-rich and has a higher size differential between the starch granules and PB, providing the ideal driving force for air classification. Legumes like common beans, chickpea, cowpea, faba bean, lima bean, lentil, lupine, mung bean, navy bean, and northern bean have also been considerably studied.

Many studies focused on optimizing the milling parameters. Pelgrom et al., (2013) compared jet milling and impact milling for peas, varying the classifier wheel speeds between 2500 and 8000 rpm. Impact milling at 4000 rpm led to optimum disentanglement and the minimum damaged starch. The moisture content decreased with increasing the classifier wheel speed, which was established using the Guggenheim-Anderson-de Boer (GAB) model for describing the moisture isotherms. The study also reported that damaged starch was higher after jet milling; hence impact milling at 4000 rpm was used for their further experiments. Wright et al., (1984) studied the effect of feed rate to the classifier and reported that varying the feed rate had no significant effect on the outcome of classification.

Wang et al., (2016) varied the classifier speed of impact milling of lupine between 2500 and 8000 rpm, followed by electrostatic separation. Higher mill classifier speeds are expected to result in a higher yield, as finer milled flours are dispersed better in air, due to lower fluidization velocity at smaller particle sizes. However, the authors reported that the yield decreased with increasing classifier wheel speed due to more fouling and more exposed lipid. No protein enrichment was found at 8000 rpm, as the high lipid content of lupine resulted in an excessive agglomeration. Classifier wheel speed of 2500 rpm had the highest yield and was determined to be the optimal speed. Pelgrom et al., (2014) also studied the dry fractionation of lupine by air classification and reported that higher classifier speeds during milling increased the losses, as the smaller particle sizes implied a higher fouling and liquid bridging due to the liberation of oil. Pelgrom et al., (2015a) noted that optimal detachment in starch-rich legumes is reached when the particle size distribution curve of flour overlaps maximally with the curve of starch granules, which occurs at 4000 rpm for pea, thus resulting in maximum protein enrichment. For pea, milling at 2500 rpm resulted in insufficient detachment, while that at 8000 rpm showed poor separation. Hence intermediate speeds of the classifier wheel result in optimal disentanglement without agglomeration.

Pelgrom et al., (2015a) also studied the protein enrichment in pea, bean, chickpea, and lentil and observed that seed properties such as seed hardness, particle density, starch granule size, fat content, and flour dispersibility also affected the protein content of the enriched fraction. Lower particle diameter after milling due to lower seed hardness relates to higher protein content in protein-enriched fraction, as seen in lentils. This is because hardness is related to the adhesion between starch granules and protein matrix. The particle density of chickpea was much lower than others; consequently, it resulted in lower protein content in the protein-enriched fraction. Chickpea flour was also the least dispersible due to the cohesiveness resulting from its high-fat content. Lentil flour was the most dispersible, yielding the highest protein content.

Wu and Nichols (2005) reported that for the same intensity of the grinding, dehulled peas gave a higher protein fraction and a useful protein shift than whole peas. However, Kon et al., (1977) reported that dehulling did not improve the protein purity since the hulls got concentrated in the coarse fraction during air classification.

Tyler and Panchuk (1982) studied the effect of seed moisture content on the air classified field pea and faba beans, varying the moisture content between 3.8 and 14.3%. It was reported that a decrease in the initial moisture content led to an increase in the neutral detergent fiber (NDF) of the protein-enriched fraction and an increase in protein fraction yield and protein separation efficiency, whereas the protein content of protein-enriched fraction decreased. The moisture content was found to have an inverse relation with seed hardness. As seeds become harder, milling is more efficient as seeds are more brittle. Consequently, a higher yield of the enriched fraction is obtained. However, an increase in seed hardness also increases the grinding time. Tyler and Panchuk (1982) concluded that moistures below 10% are most suitable for milling. Pelgrom et al., (2013) studied the thermo-mechanical morphology of peas and reported that the increase in glass transition temperature is faster for proteins than for the starch. The temperature and the initial moisture content of the legume should thus be adjusted before milling, so that proteins are in the rubbery state, whereas the starch is in the glassy state. This maximizes the disentanglement of protein and starch. Furthermore, the presence of water has a plasticizing effect as it increases the mobility of polymer chains, thus reducing the glass transition temperature. If both, protein and starch, is in the glassy state, the fracture lines are not between the two entities, but through them, which would result in an inefficient separation. Conversely, if both protein and starch are in the rubbery state, the milling efficiency is drastically reduced.

The effect of different air classification cut-points by varying the classifier wheel speed and the vane setting has also been extensively studied. The cut-point can be increased by increasing the vane-setting or lowering the classifier speed. Wright et al., (1984) varied the classifier speed about the cut-point of separation for the air classification of peas. A high classifying speed resulted in a higher protein content in the enriched fraction (55.1% protein content at 11,000 rpm); however, the yield was low, only about 29%. Wu and Nichols (2005) studied the effect of varying the intensity of grinding on the air classification of field peas. The cut-point of air classification was varied between 15 and 30 μm. On increasing the grinding intensity, there was an increase in yield and protein content of the protein-enriched fraction and in the useful protein shift value. The effect of changing the air classifier cut-point by changing the classifier speeds was also studied on cowpea, faba bean, and pigeonpea by Cloutt and Walker (1987). For cowpea, the protein content of the enriched fraction was found to decrease from 50.6% at classifier speed of 10400 rpm, to 29.6% at 6200 rpm. Tyler (1984) reported that increasing the vane setting, and consequently, the cut-point, increased the yield but the protein purity of the enriched fraction was compromised. Pelgrom et al., (2013) varied the classifier wheel speed for air classification of yellow field pea between 5000 and 12000 rpm. The experimental cut-off was estimated using Tromp's curve, and the theoretical cut-off was determined using an equation with a fouling factor incorporating the drag and centrifugal force. Although the protein content of the enriched fraction was greater at higher speeds, there was significant fouling; hence, yields were low. Thus, lower speeds are preferred in air classification. Pelgrom et al., (2014) also studied the dry fractionation of lupine by air classification and reported that increasing the classifier speed increased the protein content of the enriched fraction, but decreased yield.

Several studies analyzed the effect of re-milling and successive air classification on the yield and protein purity. Re-milling of the starch-enriched (protein-depleted) fraction can be effective in reducing the number of agglomerates and in separating the PB attached to the starch. Wright et al., (1984) attempted two additional milling and air classification cycles for pea and succeeded in doubling the yield of the final protein-enriched fraction while simultaneously reducing the protein in coarser fractions; however, the protein content combining all runs was not very different from that of a single pass. Tyler (1984) also reported that despite five successive air classifications, the adherent protein, which is derived from membranes and stroma of chloroplasts in which starch granules developed, cannot be milled free of the starch granules. Wright et al., (1984) also subjected pass to two, four, and eight successive millings. No significant change in particle size distribution was observed after four passes but milling for eight times increased the yield from 24% to 36%. Wang et al., (2016) conveyed that re-milling and re-separation of refined lupine protein fractions in filter bags after electrostatic separation doubled the yield while maintaining the protein content around 57.3%. Consequently, they proposed a scheme in which the seeds were optimally milled; the starch-enriched fraction was used to carry out the subsequent electrostatic separation steps, optimally combined with re-milling and recycling. Vose et al., (1976) reported that on a two-pass air classification, the yield values obtained for air classified field peas and horse beans improved from 25% to 44% and from 28% to 42%, respectively.

In the recent past, the studies have also focused on the use of electrostatic separation. Only tribo-electrostatic charging has been explored so far to impart a charge to the dispersed flour particles. Tribo-electric charging is influenced by chemical, physical, and electrical properties of the particle and contact surface, the number of impacts, the velocity, and angle of impact, and environmental conditions such as temperature and relative humidity (Wang et al., 2014). Wang et al., (2016) varied the flow rate of the carrier nitrogen gas at 10, 20, and 30 $L{\cdot}min^{-1}$ for the electrostatic-separation of lupine. Increasing the gas flow rate from 10 to 20 $L{\cdot}min^{-1}$ enhanced the protein content of the enriched fraction, but a similar yield was obtained. For the single-component particles, the corresponding elevated gas velocities led to a higher specific charge by increasing the normal component of impact velocity (Wang et al., 2014). However, in the electrostatic separation of lupine, the higher flow rates did not improve the enrichment, as a balance between formation of agglomerates due to Coulombic interaction and the deagglomeration due to inertial forces needs to be maintained, which was achieved at 20 $L{\cdot}min^{-1}$.

Tabtabaei et al., (2016a) compared the triboelectrification effect of polytetrafluoroethylene (PTFE), polyvinyl chloride (PVC), nylon, and copper. The authors reported that PTFE showed the maximum tribo-electrification as it has the highest work function. The diameter, shape, and length of the tribo-charger were also found to be critical. Spiral-type tribo-chargers resulted in over-charging as opposed to straight chargers. To avoid overcharging, a straight charger, 60 cm long with a high diameter was used, and to compensate for low charge-to-mass ratio, the applied voltage was increased from –1 kV to –3 kV.

Tabtabaei et al., (2016b) used these findings to design a model system consisting of a fluidized bed flour reservoir, PTFE charging tube, and plate-type separation chamber to optimize the triboelectrification of navy beans, using a mixed-level full factorial design, varying the airflow rate (7–9 $L{\cdot}min^{-1}$), tribo-charger length (60–250 cm), plate voltage (–1 to –7.5 kV) and plate angle (0–20°). The authors reported that higher protein content of the enriched fraction was observed at lower airflow rates and plate voltages, while higher separation efficiency was obtained at higher airflow rates and plate voltage. Airflow in tribo-charger changed from laminar to turbulent when the inlet airflow rate was increased from 7 to 9 $L \cdot min^{-1}$, which significantly increased charge-to-mass ratio. However, there was a reduction in the protein content at a higher flow rate of 9 $L \cdot min^{-1}$, possibly due to agglomeration arising from severe interactions. The protein content of the enriched fraction was affected minimally by plate angle. The charge acquired by particles significantly increased when the tube length was on the higher side. A three-way interaction existed between airflow rate, tube length, and plate voltage, which have a significant effect on protein enrichment. For longer tubes (250 cm) and laminar flow, PSE, and protein enrichment was enhanced at higher plate voltages, while for shorter tubes, these were lowered. Among individual factors, the airflow rate was the most significant, followed by plate voltage and tube length. The authors reported the optimum values of process variables to be 7 $L \cdot min^{-1}$, –6.5 kV, 20 plate angle and 240 cm, resulting in a protein content of 38.1% in the enriched fraction.

Wang et al., (2014) designed an experiment to evaluate the influence of carrier gas velocity, particle size, charging tube length, carrier gas relative and material water activity on the electrostatic fractionation of single-component mixtures. Higher gas velocities led to a higher specific charge by increasing the normal component of the impact velocity. Smaller particles acquired a more specific charge due to the high surface-to-volume ratio. Wang et al., (2015) also studied the influence of composition of mixture, powder dosing rate, gas flow rate, applied voltage, and distance between electrodes on the charging and separation behavior, taking gluten-starch mixtures as the model. The authors noted that higher gas flow rates result in a higher impact velocity and turbulence (at $Re > 2100$), and therefore, a higher specific charge. It was concluded that optimal charging of particles was crucial to avoid agglomeration due to electrostatic force and to reduce the space charge effect (shielding effect) near the electrode surface.

Wang et al., (2016) studied the electrostatic separation of lupine and reported that lupine protein takes the positive charge. However, Tabtabaei et al., (2016a) reported that on tribo-charging of navy bean flours, the proteins acquired a negative charge, due to the presence of ionizable functional groups and polar molecules. Proteins contain several functional groups, including the amino group, carboxyl group, and groups on the side-chains of amino acids. PB from different sources with varying sizes, shapes, and surface properties possess different charge ability characteristics and exhibit different tribo-electric charging behavior. The negative charge acquired by navy beans could be attributed to the presence of more carboxyl groups than amino groups.

Most of the studies on dry fractionation observed a trade-off between the protein purity and the protein yield. Hence, either of these parameters cannot be used individually to assess the efficiency of the dry fractionation process. For wet fractionation processes, a hyperbolic curve is obtained for a plot of protein purity versus protein yield (Berghout et al., 2015). However, the curve obtained for dry fractionation is different. Cloutt and Walker (1987) reported that as the protein content of the fines decreased, the dry weight yield increased linearly for all legumes. It may thus be possible to establish a relationship between these two parameters and predict contents of the fines from the respective yield values, independently for specific varieties of each legume. A trade-off between the fine fraction yield and protein content was also reported by Wang et al., (2016), who performed a three-step electrostatic separation to increase protein enrichment from 51.4% to 65.0%, but this was accompanied by a decrease in yield.

14.7 EFFECT OF DRY FRACTIONATION ON COMPOSITION

14.7.1 CHEMICAL COMPOSITION

Besides protein and starch, other components of the legume, such as fiber, ash, fat, and sugars, are also fractionated in the separation. Sosulski and Youngs (1975) air-classified chickpea, white pea bean, northern bean, faba bean, field pea, lima bean, mung bean and lentil flour and reported that most lipid and ash appeared in the protein-enriched fraction, while crude fiber appeared mostly with the starch particles. Pelgrom et al., (2015c), Reichert (1982) and Tyler et al., (1981) also reported similar observations for mung beans, green lentils, northern beans, faba beans, field peas, navy beans, lima beans, and cowpeas. Aguilera et al., (1982) studied air classification of roasted navy beans and concluded that the hulls, due to their particle size, were present in the starch fraction, whereas fat was more concentrated in the protein fraction. Vose et al., (1976) reported that the oil moieties, as well as the soluble sugars, particularly verbascose and stachyose, were found to be classified in the lighter protein-enriched fraction. The partition of phosphorous, zinc, and calcium was also majorly into the protein-enriched fraction, while the iron was largely found in the starch fraction. Kon et al., (1977) also reported that the protein-enriched fraction obtained on air classifying white bean flour was 1.5 times enriched in bean sugars like sucrose, raffinose, and stachyose. The lipids fractionate with the proteins due to the presence of a lipid-rich coating on PB, and the co-fractionation of lipid-coated cell wall material with proteins. The co-fractionation of sugars in the protein-enriched fraction is attributed to the presence of these sugars in the cell wall material, which gets fractionated into the lighter fraction.

Vose et al., (1976) reported that the protein-enriched fraction of air-classified field peas and horse bean flours contained 60–70% protein. The starch in the coarse fraction was found to have retained a layer of proteinaceous material on their surface; this layer of adherent proteins was derived from the granular membrane within which the starch granule had developed. Pelgrom et al., (2014) studied the dry fractionation of lupine by air classification and inferred that characterization of proteins of the enriched and depleted fractions showed that the fine protein-enriched fraction had larger sub-units (50–500 kDa) originating from β-conglutin, whereas the coarse protein-depleted fraction majorly comprised of albumins (5–50 kDa).

The amino acid composition of the proteins largely determines its quality. Animal proteins are considered to be a complete source of proteins as they contain all the essential amino acids. On the other hand, most legumes are deficient in methionine. Han and Khan (1990) studied the amino acid composition of the protein concentrates and concluded that the concentrations of the essential amino acids in these fractions were similar to the requirements elaborated for humans. They also reported that in fractionated pea and bean flour, lysine, and leucine were abundant in the protein-enriched fraction, but methionine was the first limiting amino acid. The slight difference in the amino acid pattern of the proteins before and after fractionation is attributed to the difference in composition between the interstitial proteins concentrated in the protein-enriched fraction, and the adherent proteins are predominantly remaining in the starch fraction. Vose et al., (1976) also reported that the protein-enriched fraction of air classified field peas and horse bean flours contained adequate lysine, but showed a low methionine content.

Sosulski and Youngs (1979) also studied the fractionation pattern followed by pigments by dry fractionating green pea flour, resulting in a white protein fraction and a green starch fraction, concluding that chlorophyll majorly fractionated into the coarser fraction. However, re-milling and classification of the starch fraction increased the chlorophyll content in the fine protein fraction.

14.7.2 ANTI-NUTRITIONAL FACTORS (ANF)

Legumes also contain several anti-nutritional factors (ANF) such as trypsin inhibitors, hemagglutinin, phytic acid and saponins which interfere with digestion, absorption, and some other aspects of the metabolism of the nutrients contained in other foods. These ANF can typically be inactivated or eliminated during the heating and aqueous extraction stages which take place in wet fractionation. However, owing to the absence of these steps in dry fractionation, ANF persists in this separation method. Most of the ANF are proteinaceous or nitrogen-containing. Thus, they tend to follow the distribution of proteins in their relative distribution between fine and coarse fractions.

Elkowicz and Sosulski (1982) studied the air classification of eleven starch-rich legumes (lima bean, navy bean, great northern bean, chickpea, lentil, cowpea, field pea, faba bean, mung bean) and analyzed them for ANF. Trypsin inhibitor was highest in lima beans and segregated into the fine fraction with the majority of the protein. Akin to trypsin inhibitor activity, hemagglutinating activity was also concentrated in the protein fraction, highest in lima bean. Navy bean and cowpea flours had the highest levels of hemolytic activity (saponins), concentrated into protein fraction. Saponins are heat stable, and their removal involves aqueous extractions with hot water or dilute alkali. Vose et al., (1976) reported that the phytic acid content in the protein-enriched fraction nearly doubled from 0.7% to 1.9% in field peas, and from 1.8% to 2.7% in horse beans. Elkowicz and Sosulski (1982) also reported that phytic acid is present in protein fractions in a three times higher proportion than in the original flour and the dietary fiber content doubled in protein fraction, which affects the absorption of trace minerals in the gastrointestinal tract. Vicine and convicine were found in high levels in protein fractions of faba beans. The aqueous extraction is effective in the removal of these glycosides. Aguilera et al., (1982) also studied air classification of roasted navy beans and reported that the residual activity of trypsin inhibitor and hemagglutinin was found to decrease with increase in the roasting temperature, and was found to be lesser in the protein-enriched fraction than in the original flour. Before utilization of these protein-enriched fractions in food products, it is thus essential to evaluate the anti-nutritional effects of these factors and process them accordingly.

14.8 EFFECT OF DRY FRACTIONATION ON FUNCTIONAL PROPERTIES

The functional properties of the protein concentrate largely affect the behavior of proteins in a food system. The concentrates obtained by dry fractionation can be used in several food applications, owing to their solubility, foam stability, emulsification, water- and oil-binding properties. The two major reasons are the retention of functional properties due to the absence of heat denaturation and pH shifts in the dry process, and the presence of some amounts of residual starch in the protein concentrate.

Sosulski and Youngs (1975) analyzed the functional properties of eight legume flours and concluded that soybean and lupine flour gave the highest values of water and oil absorption, and oil emulsification. Northern bean showed a higher water absorption and emulsification capacity, while chickpea showed a good oil absorption. The starch fraction showed higher water absorption, and higher peak and cold viscosities, while the protein fraction showed a higher oil absorption, emulsification capacity, whippability, and foam stability. The authors also reported that pea bean, northern bean, chickpea, and lima bean gave more desirable values in functional property tests, while faba bean, field pea, mung bean, and lentil gave a higher protein fractionation by air classification.

Pelgrom et al., (2014) studied the functional properties of air classified lupine and compared it with heated protein concentrates on drawing a comparison to the harsh treatment applied to a protein during wet fractionation. Foam stability using the air-classified protein-enriched fraction was significantly higher than that of the original lupine flour. A decrease in foam stability was observed with the heated protein concentrate (corresponding to wet fractionation). Lupine protein concentrate suspensions exhibited a lower viscosity than the corresponding lupine flour. Thus, potential applications in high-protein beverages are possible. The authors also compared the digestibility of the concentrates. In the original lupine flour, a lesser amount of protein was digested in comparison to the concentrates. Heat treatment equivalent to that of wet fractionation impaired ANF enhancing digestibility, but the simultaneous denaturation of proteins led to aggregation, making them less accessible. Thus, the digestibility of heated and unheated concentrates was comparable.

Pelgrom et al., (2015c) analyzed the gelatinization and phase behavior of air classified yellow peas. The dry-fractionated samples phase-separated when suspended in water, resulting in four distinct layers. The first two layers were rich in protein. These layers of the fine fraction were extracted and concentrated by ultrafiltration, increasing the protein content from 30% to 67%. The viscosity of the suspended pea fractions increased with increasing the protein and fiber content. Gel strength increased exponentially on increasing starch, and the presence of protein and fiber weakened this network. The gel structure of the protein-rich fraction consisted of continuous protein phase with embedded cell wall particles and starch granules. Gel strength was very poor, possibly due to interruptions in the network from the high fiber, oil, and ash content, as these are also enriched in the lighter fraction. Gel prepared from ultra-filtered protein consisted of a continuous protein network with a higher gel strength than a fine fraction as fewer dispersed components were present that could disturb the network. The authors also investigated the influence of heating and cooling rates on gelatinization. Slow cooling allowed proteins more time to arrange in a network, and fast heating slowed down phase separation. The gel strength of the protein-enriched fraction was not influenced by the heating rate, but lowering the cooling rate increased its gel strength. The authors also compared enzymatic and thermal gelatinization and reported that gels of the enriched fraction made using transglutaminase were stronger than heat-set gels, as enzymatic cross-linking lead to a stronger network. Presence of other water-absorbing components like starch and water in the dispersed phase enhanced gel strength due to the concentration effect. Summing up these studies, it can be concluded that the solubility, oil absorption, emulsification, whippability, foam stability and digestibility of the protein-enriched fraction increase with respect to the flour, whereas the water absorption and viscosity are lowered after dry fractionation.

14.9 DRY FRACTIONATION IN COMBINATION WITH OTHER TECHNIQUES

14.9.1 PRE-TREATMENTS ADOPTED WITH DRY FRACTIONATION

Several pre-treatments have been studied to overcome the limitations and improve the yield and protein enrichment of the dry fractionation technique. These have been summarized in Table 14.3. These were primarily targeted in two major directions: the removal of certain specific components and the cellular structure weakening.

One major drawback of dry separation is the presence of undesirable flavor and odor which persist even after cooking the raw ground legumes. These off-odors are attributed to the presence of the lipoxidase enzyme, which catalyzes the oxidation of unsaturated fatty acids through the formation of hydroperoxides. Thermal treatment at 104–105°C for 6–8 min inactivates lipoxidase. Aguilera et al., (1982) roasted navy beans between 92 and 125°C in a particle-to-particle heat exchanger and applied a double-pass air classification. It was reported that the protein-enriched fraction showed a protein concentration of 48%, with the maxima corresponding the harshest roasting treatment. It was concluded that roasting tends to improve the separation of starch and protein, which could be attributed to lowering of the

TABLE 14.3 Pre-Treatments Adopted with Dry Fractionation of Legumes

Pre-treatment	Sample	Milling technique	Separation technique	Parameter	Effect of dry fractionation in yield/ quality	References
Roasting	Navy beans	Impact mill	Air classification	Temperature (92–125°C)	• Maximum protein concentration (47.6%) obtained for harshest roasting treatment • Roasting decreased residual trypsin inhibitor and hemagglutinin activity	Aguilera et al., (1982)
Soaking	Yellow peas, lupine	Impact mill	Air classification	-	• Dissolution of the protein bodies, decrease in density due to structural collapse • Had little effect on air classification	Pelgrom et al., (2015b)
Freezing	Yellow peas, lupine	Impact mill	Air classification	-	• Decrease in density due to structural collapse • Had little effect on air classification	Pelgrom et al., (2015b)
Defatting	Yellow peas, lupine	Impact mill	Air classification	Fat content	• Defatting had no significant effect on pea, but increased the protein content of the fine fraction in lupine	Pelgrom et al., (2015b)
Varying moisture content	Yellow peas, lupine	Impact mill	Air classification	Moisture content (5–15%)	• Lower moisture content increased the PSE but reduced the protein content in fines	Pelgrom et al., (2015b)
Flowability aids	Lupine	Pin mill	Air classification	i. 12 nm Aerosil ii. 43.7 μm potato starch	• Enhanced flour dispersibility, improving classification efficiency • Aerosil improved the PSE, but potato starch decreased the yield of a fine fraction as lupine particles adhered to starch granules forming aggregates	Pelgrom et al., (2014)

moisture content of the seeds, which led to optimal disentanglement during milling. Another major limitation of dry fractionation is the co-fractionation of ANF into the protein-enriched fraction. Aguilera et al., (1982) reported that roasting resulted in a decrease in the residual activity of ANF. It was found that the residual trypsin inhibitor activity and residual hemaglutinin activity decreased with increase in roasting temperature, and was found to be significantly lower in the protein-enriched fraction than in the original flour.

Pelgrom et al., (2015b) studied the effect of four pre-treatments, i.e., soaking, defatting, freezing, and varying the moisture content on pea and lupine, to study the effect of weakening the cellular structures. Pea and lupine were representatives of starch-rich and oil-rich legumes, respectively. Soaking resulted in the dissolution of the PB, which affected the original cellular structure of the grits. The firmness of pea and lupine seeds decreased after soaking and freezing followed by drying, as the absorbed water was replaced by air without collapsing the structure, thus decreasing the density. The authors reported that these pre-treatments reduced the energy needed for milling and did not lead to protein denaturation. For pea, defatting, soaking, and freezing did not affect particle size distribution after milling, but for lupine, these pre-treatments led to coarser particle, attributed to the change in particle density. Defatting decreased the cohesiveness of the powder, but had no significant effect on pea owing to a very low-fat content. Defatting of lupine significantly decreased the yield, but increased the protein content of the enriched fraction. The authors also observed that lower moisture content resulted in smaller particle sizes during milling due to higher brittleness, and increased the protein separation efficiency. However, there was a reduction in the protein purity of the fine protein-enriched fraction as the fracture surfaces were not between protein and starch, but through them. Pelgrom et al., (2013) also reported that increasing the moisture content during milling such that proteins and starch were in the rubbery and glassy state, respectively, would maximize the disentanglement of protein and starch due to the plasticizing effect of the presence of water. Thus, soaking, and freezing have little effect on air classification, but increasing moisture content and defatting of lupine increases the protein purity.

Pelgrom et al., (2014) also studied the effect of flowability aids on the flour dispersibility, and reported that for lipid-rich legumes like lupine, flowability aids increase the inter-particle distance, decreasing the attractive forces and hence decreasing clump formation. The flowability aids used were 12 nm Aerosil and 43.7 μm potato starch, both of which enhanced the flour dispersibility, thus improving the air classification efficiency. However, potato starch decreased the yield of a fine fraction as lupine particles adhered to starch granules forming aggregates.

14.9.2 DRY FRACTIONATION FOLLOWED BY POST-TREATMENT

Despite having tremendous potential as a sustainable process for protein enrichment, dry fractionation has certain limitations, such as the trade-off between yield and purity, and the incomplete disentanglement of the PB from the starch granules. To overcome these limitations, a few studies have been attempted that apply dry fractionation as a pre-treatment, followed by another technique.

Pelgrom et al., (2015b) studied electrostatic separation as a post-treatment on the air-classified fractions of pea and lupine. Interestingly, a high protein content fraction could also be obtained by the electrostatic separation of the coarse fraction of lupine. The authors thus concluded that electrostatic separation alone has the potential to be used as a separation process.

Another study used ultrafiltration as a concentration step after dry fractionation, to match the purity of protein isolates. Pelgrom et al., (2015c) suspended the dry-fractionated fine fraction of yellow pea samples in water and observed four distinct layers. The first two were rich in protein. These layers of the fine fraction were extracted and concentrated by ultrafiltration. The protein content increased from 30% to 67%, explained by the permeation of low molecular weight carbohydrates and the retention of proteins. This combination of dry fractionation plus aqueous phase separation uses 11 $MJ \cdot kg^{-1}$ protein isolate as opposed to conventional wet fractionation using 63 $MJ \cdot kg^{-1}$ isolate. Schutyser et al., (2015) also described a hybrid process which differed from wet fractionation as it involved the suspension of flour at a lower dilution, no exposure to chemicals and no dilution before spray drying. This considerably reduced the energy requirement per kg of recovered protein from 54 MJ for wet fractionation to 20 MJ, and improved the protein purity to 81%.

14.10 CONCLUDING REMARKS AND FUTURE RESEARCH NEEDS

The production of protein concentrates by the dry fractionation of plant proteins offers a sustainable, efficient, and economical alternative. The topic has gained renewed interest in the last decade. However, the studies so far present several research gaps and future research prospects. Mechanistic insights on the fracture behavior, physical interactions between constituents, and the microstructural architecture of legumes could be further explored to obtain the desired optimal milling behavior of the legume. Discrete element method modeling can be used to predict the particle flows precisely during milling and separation processes. The breakage pattern could be studied from the perspective of the bonding energies of specific components. Plant breeding could be looked at to obtain seeds with optimal hardness and composition, maximizing disentanglement and the driving force for separation. With multiple separation techniques being studied, attempts could be made to combine two or more such techniques, so that the drawbacks in the driving force of one technique could be compensated by the other, allowing the simultaneous optimization of the process concerning multiple parameters such as yield, protein separation efficiency, and purity.

The major benefit of dry-fractionated concentrates is the retention of functional properties. Since food is a complex matrix of several ingredients, there has been a recent shift from ingredient purity towards ingredient functionality. The applications of these functional ingredients in meat analogs, imitation milks, infant foods, bakery, and extruded products could be further explored to provide the impetus for further exploration of dry fractionation. Furthermore, the applications of starch concentrates can also be explored.

India is the largest producer of pulses. While the potentials of a large number of legumes have been extensively studied across the globe over the last four decades, for Indian varieties of legumes, the field of dry fractionation is relatively barren and open for exploration. The potential of Indian legumes such as kidney beans, black gram, pigeon pea, moth bean, lablab bean, and broad bean as dry fractionation substrates could further be explored and exploited in the future to sustainably meet the increasing global protein demand.

KEYWORDS

- **air classification**
- **electrostatic separation**
- **energy loss**
- **protein bodies**
- **sieving**
- **sustainability**

REFERENCES

Aguilera, J. M., Lusas, E. W., Uebersax, M. A., & Zabik, M. E., (1982). Development of food ingredients from navy beans (*Phaseolus vulgaris*) by roasting, pin milling, and air classification. *Journal of Food Science*, *47*(4), 1151–1154.

Berghout, J. A. M., Pelgrom, P. J. M., Schutyser, M. A. I., Boom, R. M., & Van der Goot, A. J., (2015). Sustainability assessment of oilseed fractionation processes: A case study on lupin seeds. *Journal of Food Engineering*, *150*, 117–124.

Boye, J., Zare, F., & Pletch, A., (2010). Pulse proteins: Processing, characterization, functional properties and applications in food and feed. *Food Research International*, *43*(2), 414–431.

Cloutt, P., Walker, A. F., & Pike, D. J., (1987). Air classification of flours of three legume species: Fractionation of protein. *Journal of the Science of Food and Agriculture*, *38*(2), 177–186.

Elkowicz, K., & Sosulski, F. W., (1982). Antinutritive factors in eleven legumes and their air-classified protein and starch fractions. *Journal of Food Science*, *47*(4), 1301–1304.

Gueguen, J., (1983). Legume seed protein extraction, processing, and end product characteristics. *Plant Foods for Human Nutrition*, *32*, 267–303.

Han, J. Y., & Khan, K., (1990). Physicochemical studies of pin-milled and air-classified dry edible bean fractions. *Cereal Chemistry*, *67*(4), 384–390.

Kon, S., Sanshuck, D. W., Jackson, R., & Huxsoll, C. C., (1977). Air classification of bean flour. *Journal of Food Processing and Preservation*, *1*(1), 69–77.

Maaroufi, C., Melcion, J. P., De Monredon, F., Giboulot, B., Guibert, D., & Le Guen, M. P., (2000). Fractionation of pea flour with pilot scale sieving. I. Physical and chemical characteristics of pea seed fractions. *Animal Feed Science and Technology*, *85*(1), 61–78.

Pelgrom, P. J., Berghout, J. A., Van der Goot, A. J., Boom, R. M., & Schutyser, M. A., (2014). Preparation of functional lupine protein fractions by dry separation. *LWT-Food Science and Technology*, *59*(2), 680–688.

Pelgrom, P. J., Boom, R. M., & Schutyser, M. A., (2015a). Method development to increase protein enrichment during dry fractionation of starch-rich legumes. *Food and Bioprocess Technology*, *8*(7), 1495–1502.

Pelgrom, P. J., Boom, R. M., & Schutyser, M. A., (2015b). Functional analysis of mildly refined fractions from yellow pea. *Food Hydrocolloids*, *44*, 12–22.

Pelgrom, P. J., Schutyser, M. A., & Boom, R. M., (2013). Thermomechanical morphology of peas and its relation to fracture behavior. *Food and Bioprocess Technology*, *6*(12), 3317–3325.

Pelgrom, P. J., Vissers, A. M., Boom, R. M., & Schutyser, M. A., (2013). Dry fractionation for production of functional pea protein concentrates. *Food Research International*, *53*(1), 232–239.

Pelgrom, P. J., Wang, J., Boom, R. M., & Schutyser, M. A., (2015b). Pre-and post-treatment enhance the protein enrichment from milling and air classification of legumes. *Journal of Food Engineering*, *155*, 53–61.

Pimentel, D., & Pimentel, M., (2003). Sustainability of meat-based and plant-based diets and the environment. *The American Journal of Clinical Nutrition*, *78*(3), 660–663.

Reichert, R. D., (1982). Air classification of peas (*Pisum sativum*) varying widely in protein content. *Journal of Food Science*, *47*(4), 1263–1267.

Schutyser, M. A. I., Pelgrom, P. J. M., Van der Goot, A. J., & Boom, R. M., (2015). Dry fractionation for sustainable production of functional legume protein concentrates. *Trends in Food Science & Technology*, *45*(2), 327–335.

Schutyser, M. A. I., & Van der Goot, A. J., (2011). The potential of dry fractionation processes for sustainable plant protein production. *Trends in Food Science & Technology, 22*(4), 154–164.

Sosulski, F., & Youngs, C. G., (1979). Yield and functional properties of air-classified protein and starch fractions from eight legume flours. *Journal of the American Oil Chemists' Society, 56*(3), 292–295.

Tabtabaei, S., Jafari, M., Rajabzadeh, A. R., & Legge, R. L., (2016a). Solvent-free production of protein-enriched fractions from navy bean flour using a triboelectrification-based approach. *Journal of Food Engineering, 174*, 21–28.

Tabtabaei, S., Jafari, M., Rajabzadeh, A. R., & Legge, R. L., (2016b). Development and optimization of a triboelectrification bioseparation process for dry fractionation of legume flours. *Separation and Purification Technology, 163*, 48–58.

Tyler, R. T., (1984). Impact milling quality of grain legumes. *Journal of Food Science, 49*(3), 925–930.

Tyler, R. T., & Panchuk, B. D., (1982). Effect of seed moisture content on the air classification of field peas and faba beans. *Cereal Chemistry, 59*(1), 31–33.

Tyler, R. T., Youngs, C. G., & Sosulski, F. W., (1981). Air classification of legumes [beans, lentils, peas]. I. Separation efficiency, yield, and composition of the starch and protein fractions. *Cereal Chemistry, 58*(2), 144–148.

Vose, J. R., Basterrechea, M. J., Gorin, P. A. J., Finlayson, A. J., & Youngs, C. G., (1976). Air classification of field peas and horsebean flours: Chemical studies of starch and protein fractions. *Cereal Chemistry, 53*(6), 928–936.

Wang, J., De Wit, M., Schutyser, M. A., & Boom, R. M., (2014). Analysis of electrostatic powder charging for fractionation of foods. *Innovative Food Science & Emerging Technologies, 26*, 360–365.

Wang, J., De Wit, M., Boom, R. M., & Schutyser, M. A., (2015). Charging and separation behavior of gluten–starch mixtures assessed with a custom-built electrostatic separator. *Separation and Purification Technology, 152*, 164–171.

Wang, J., Zhao, J., De Wit, M., Boom, R. M., & Schutyser, M. A., (2016). Lupine protein enrichment by milling and electrostatic separation. *Innovative Food Science & Emerging Technologies, 33*, 596–602.

Wright, D. J., Bumstead, M. R., Coxon, D. T., Ellis, H. S., DuPont, M. S., & Chan, H. W. S., (1984). Air classification of pea flour–analytical studies. *Journal of the Science of Food and Agriculture, 35*(5), 531–542.

Wu, Y. V., & Nichols, N. N., (2005). Fine grinding and air classification of field pea. *Cereal Chemistry, 82*(3), 341–344.

Index

B

C

D

E

F

G

H

I

K

L

M

R

S

X

Z